Geological Heritage and Biodiversity in Natural and Cultural Landscapes

Geological Heritage and Biodiversity in Natural and Cultural Landscapes

Editors

Antonio Miguel Martínez-Graña
José Ángel Sánchez Agudo

MDPI • Basel • Beijing • Wuhan • Barcelona • Belgrade • Manchester • Tokyo • Cluj • Tianjin

Editors
Antonio Miguel Martínez-Graña   José Ángel Sánchez Agudo
Geology   Botany
University of Salamanca   University of Salamanca
Salamanca   Salamanca
Spain   Spain

Editorial Office
MDPI
St. Alban-Anlage 66
4052 Basel, Switzerland

This is a reprint of articles from the Special Issue published online in the open access journal *Sustainability* (ISSN 2071-1050) (available at: www.mdpi.com/journal/sustainability/special_issues/ Geological_Heritage_Biodiversity).

For citation purposes, cite each article independently as indicated on the article page online and as indicated below:

LastName, A.A.; LastName, B.B.; LastName, C.C. Article Title. *Journal Name* **Year**, *Volume Number*, Page Range.

ISBN 978-3-0365-1167-2 (Hbk)
ISBN 978-3-0365-1166-5 (PDF)

Contents

About the Editors

Antonio Miguel Martínez-Graña

Antonio Martínez Graña is full professor of Geology and Vice Dean of the Faculty of Sciences in University of Salamanca. He teaches geology, geological engineering, environment, and agricultural engineering. His research interests cover a broad topics including external geodynamics in geomorphology, applied to the analysis of geological risks using GIS and remote sensing techniques, in environmental geology integrated in planning and land use planning as well as in geoenvironmental thematic mapping and geological heritage. He is the author of more than 20 educational publications and original teaching material published in prestigious publishers: Springer and McGrawHill. He has participated as a main researcher and collaborator in 39 research projects and contracts. He is also the author and coauthor of more than 184 publications in national and international books and scientific journals (6 books, 44 book chapters, 64 indexed in WoS, 31 articles in journals, and 39 congresses).

José Ángel Sánchez Agudo

José Angel Sánchez Agudo is a professor of botany and an academic secretary at the Faculty of Biology, where he teaches evolutionary biology, botany, and ecoinformatics applied to the conservation of biodiversity. His main lines of research are linked to the management of tools in the field of GIS and ecological niche models in various facets of knowledge of the processes that link biological events with the biotic and abiotic characteristics of the territory. As a result of this interest, he has participated in about 50 national and international projects and contracts, some as relevant in the field of conservation as the LIFE projects. His scientific production materializes in the form of more than 40 publications, indexed scientific journals, and book chapters and 45 scientific-technical reports. Currently, his interests have turned towards agroecology.

Preface to "Geological Heritage and Biodiversity in Natural and Cultural Landscapes"

This book highlights the importance of studies on geological and biological heritage and their involvement in geodiversity and biodiversity in an integrated way in the landscape, describing varied methodologies applied to specific territories (cultural and traditional heritage), taking into account the singularity and its geoconservation based on active processes (geological risks, anthropic environmental impacts, etc.) and dynamics of its ecosystems (biological invasions, bioindicators, ethnobotanical studies, interactions between species, agrobiodiversity, etc.). This book contributes to the methodology and application of both design approaches, and evaluation and decision-making frameworks in the context of the definition of strategies for natural heritage (geological and biological) and landscape.

Antonio Miguel Martínez-Graña, José Ángel Sánchez Agudo
Editors

 *sustainability*

Article

Geomorphological Analysis Applied to the Evolution of the Quaternary Landscape of the Tormes River (Salamanca, Spain)

J.L. Goy, G. Rodríguez López, A.M. Martínez-Graña **, R. Cruz and V. Valdés ***

Department of Geology, Faculty of Sciences, University of Salamanca, Square Merced s/n., 37 008 Salamanca, Spain; joselgoy@usal.es (J.L.G.); gloriardr@usal.es (G.R.L.); amgranna@usal.es (A.M.M.-G.); rqcruz@usal.es (R.C.)
* Correspondence: vvaldes@usal.es; Tel.: +34-923294400; Fax: +34-923294514

Received: 20 November 2019; Accepted: 13 December 2019; Published: 17 December 2019

Abstract: This paper presents a geomorphological analysis of the Tormes River during the Quaternary. The Tormes River formed in the center-west of the Iberian Peninsula in the province of Salamanca. It runs along a Cenozoic basin with basement materials and through Varisco, and consists of mainly granitic and metamorphic materials, leaving a wide stream of river terraces, both erosional and depositional, that confirm its evolution throughout the Quaternary. Geomorphological analyses using Geographic Information Systems tools, Digital terrain model high resolution (MDT05, LIDAR), Orthophotos (scale 1:5000), and geological maps (1:50,000 Series Magna) have allowed different morphologies and depositional terraces to be distinguished, namely, 19 levels of erosional terraces and 3 levels of erosion surfaces. Based on these correlations, the levels of terraces in the Tormes River between T1 (+140 m) and T7 (+75–80 m) are located in the Pleistocene, those between T8 (+58–64 m) and T14 (+18–23 m) in the Middle Pleistocene, those between T15 (+12–13 m) and T17 (+6–7 m) in the Upper Pleistocene, and those between T18 (+3 m) and T19 (+1.5 m) in the Holocene. The erosion surfaces are divided into six levels: S6 (+145 m), S5 (+150 m), S4 (+160 m), S3 (+170 m), S2 (+180 m) and S1 (+190 m) located in the Lower Pleistocene, This work performs a geomorphological mapping procedure applied to the evolutionary analysis of the landscape, so that it determines different geomorphological units allowing the relief and morphology of the terrain in past times, establishing a dynamic analysis of the landscapes.

Keywords: quaternary landscape; geomorphological analysis; depositional-erosional terraces; incision-displacement rates; Tormes River

---

1. Introduction

The system of stepped terraces of the Tormes River is the most frequent in the rivers of the Iberian Peninsula, with numerous examples noted: Pisuerga and Arlanzón-Duero Rivers, Arlanzón-Duero River, Tagus River, Duero River, Ebro River, Lozoya-Tajo Rivers, and Tagus River, among many others.

A synthetic scheme (maintaining the heights to scale) of the terrace levels of the Tormes River, in the sub-basin of Salamanca, where the heights relative to the thalweg have been indicated and as a result, the heights of the different escarpments between terraces have been drawn up successive. From the observation of these profiles, it is appreciated that there are a series of more significant escarpments, and that they are maintained in the three selected rivers. The reason for the selection of Arlanzón River (Duero Basin) and Tajo River (Tajo Basin), is that the first one has been mapped with several field scampies since the year 1983, and subsequent chronological studies of the terraces with sampling were carried out paleomagnetic and dating by ESR TL and OSL, being able to attribute to these escarpments a more precise chronology when determining the boundaries between the lower, middle, and upper Pleistocene and Holocene.

The second profile selected, Tajo River (Tajo Basin), downstream of Toledo, has been due to the fact that this sequence has numerous deposits of fauna (Macro and microfauna.), and lithic industry, which also allows the dating of terraces and their correlation; as an example, the level of +75 m is clearly located in the lower Pleistocene (presence of Equus stenonis) and that of +60m in the middle Pleistocene (southern Mammutus), so this type of correlation between escarpments can be used for related. The Tormes River belongs to the Duero River network and is located in the central-western basin of the Iberian Peninsula. This is the largest basin in the entire Iberian Peninsula, and it is called the Duero Basin. It is located in the province of Salamanca. The Tormes River runs for 80 km along a series of Cenozoic materials and through a Paleozoic basement, constituted primarily of granitic and metamorphic materials. The area is characterized by little sharp reliefs with flat and elevated surfaces and staggered plains in its central part. It is surrounded by a peripheral mountainous border with heights between 600 and 800 m. The study area occupies an area of 3200 km^2 in the Northeast part of the Province of Salamanca (Figure 1)

Figure 1. Location of the study area in the province of Salamanca (Spain).

The objective of this work was the realization of a detailed geomorphological cartography that allows different morphologies to be identified, and establishes the succession of depositional terraces of the Tormes River to determine the relationships among the different terraces and the time periods in which they were deposited [1–4]. From longitudinal and cross-sectional profiles, the different terraces of the same basin (Duero Basin) and similarly behaving basins (Tagus Basin) were altimetrically correlated. With these data, the incision and displacement rates of the Tormes River in the Salamanca sub-account were calculated by correlating them with the terraced sequences of the Arlanza River (Duero Basin) and the Tagus River (south of Toledo). Different three-dimensional geological models of the different stages of the Quaternary (Lower Pleistocene, Middle Pleistocene, Upper Pleistocene, and Holocene) were generated by analyzing the paleogeographic evolution of this sub-basin.

1.1. Geological Context

Geologically, this river belongs to the Central Iberian Zone [5], where sediments of the Cambrian-Ordovician age are deposited on the Varisc basement, and materials of the final Mesozoic and Cenozoic age are recognized as filling the Duero basin (Figure 2). The oldest known materials are from the Precambrian-Cambrian era and were formed by gneisses associated with slate and sandstone belonging to the so-called Greywacke Shale Complex. Particularly, in the study area, they are grouped into two formations: The Monterrubio Formation and the Aldeatejada Formation [6]. The Ordovician rocks include sandstone, black and gray slate, and the quartzite "Armoricana Quartzite", which are located within the Iberian Massif that is located discordantly in the nuclei of the synclines. For the Carboniferous and Devonian eras, there is hardly any sedimentary record, but igneous rocks, granites, aplites, and pegmatites emerge [7].

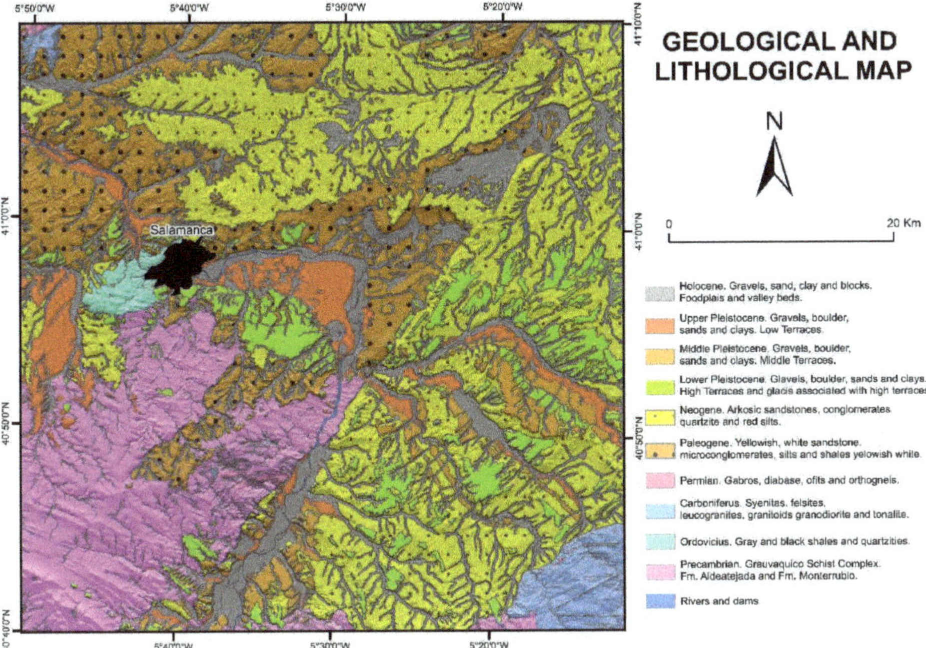

Figure 2. Geological and Lithological map of the study area. Classification according to the materials and their corresponding chronostratigraphic periods. It can be observed that the Tormes River changes from harder materials belonging to older ages (Varisc substrate) to softer materials belonging to the Neogene and Paleogene Duero basin, and then returns to the former.

At the end of the Mesozoic, a distensive stage occurred, generating a dismantling in the raised reliefs and producing wide alluvial fans in the NE direction that gave rise to siliceous sandstones (Salamanca Sandstone Formation) composed of thick sands and gravels at the base and formed sand roofs, silt, and clays.

Cenozoic materials belong to the South-West part of the Duero Basin. During the filling, horst and grabens reliefs were produced, representing two stages: One exorheic during the Paleogene and one endorheic during the Neogen. The Paleogene stage is characterized by yellowish white sandstones, microconglomerates, silt and white-yellow shales, and siliceous sandstones. They correspond to river sediments formed by canal and avenue deposits, and above them are those of the flood plain. In the Neogen, the arc, sandstones, conglomerates, and red silts stand out, as does the Red Unit of the Neogen, which formed during the endorheic stage of the basin. This is of conglomeric character with sand and clay intercalations. In the Quaternary, the sedimentary record is linked to the activity of the Tormes River and the development of its valley. Its initiation is linked to an exorheic stage due to the clogging of the basin in the previous phase.

1.2. Geomorphological Context

At the geomorphological level, a series of deposits located on the rooves of ancient alluvial fans that correspond to the onset of the river inset stand out (Figure 3 and Table 1). Embedded in these, the most important forms in the area are the quaternary erosive and depositional terraces, formed by accumulations of gravel, conglomerates, sand, and sometimes clays (sometimes crusted) with flat, stepped morphologies that are associated with the main rivers of the area, highlighting those of the Tormes river [8,9]. The valley funds are formed by deposits of sand, gravel, and scarce silty matrix. Alluvial fans and alluvial cones are formed in opening areas of secondary valleys that pour into the main river, causing the sedimentation of sands, silts, and gravels. With a conical shape, the most important ones are located on the low terraces. Pediments (glacis) are observed, which are siliciclastic sediments of streams formed by sands, silts, and gravels that give rise to relatively flat surfaces with less extension than terraces. Within the gravitational domain are colluvions (hillside deposits) that appear to be associated with the escarpments. Both rivers and structural components stand out, being abundant in the NE of the study area. Finally, within the endorheic domain, semiendorheic areas that form in carbonate rock dissolution environments stand out and are structures that generate temporary lagoons due to the waterproofing of their materials or the proximity of the water table [10].

The most significant tectonic effects are neotectonics, which are associated with large reactivated varisc fractures that affect the geomorphology of the study area. The main direction of major faults such as Alba-Villoria is NNE–SSW, which causes a relevant morphological scarp as a result of its recent activity, affecting Neogenic deposits by tilting them towards the NE, resulting in an asymmetry of the river valleys of the Tormes and its tributaries (Almar, Gamo, and Margañán). This causes sequences of terraces on the left banks of the rivers and escarpments on the right banks (Tormes River, Almar, Regamon, Trabancos, Zapardiel, etc.) due to the general balancing of the Sierra de Béjar towards the NE. In addition to this mega-structure, there are numerous modifications of river courses in the area, following the NEN–WSW (Torremencías, south of Mozárbez, La Vellés, etc.), NW–SE (Almar, Gamo and Margañán), EW (Tormes River in Cabrerizos, east of Salamanca), and WNW–ESE (Tormes River northwest of Salamanca) directions.

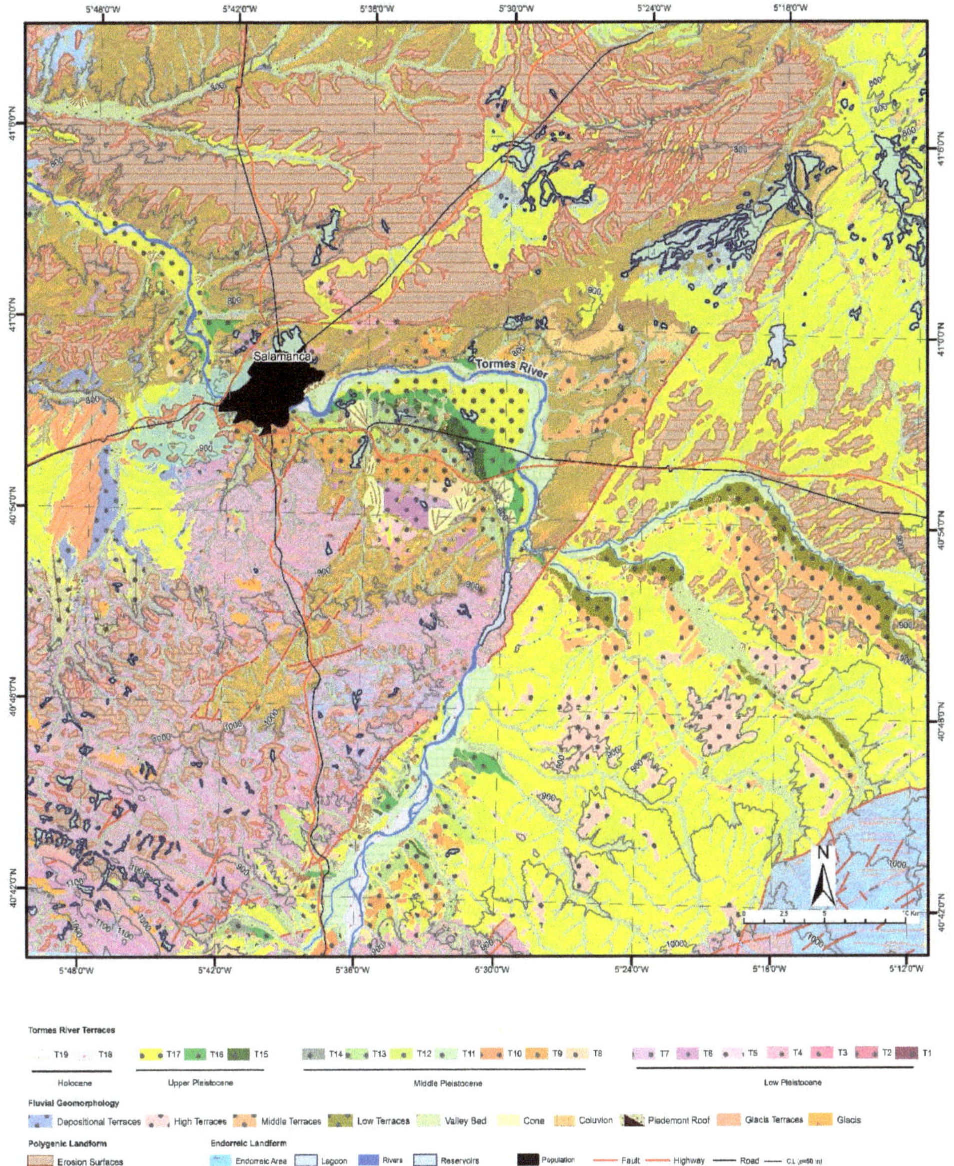

Figure 3. Geomorphological map of study area.

Table 1. Summary table of the Morphogenetic System (M.S.) present in the area.

	Domain	Landforms	Description	Age
Morphogenetic Fluvial System	Fluvial Domain	Fluvial terraces	There are flat surfaces in the area with small slopes that mark the position of the flatness of ancient floods that developed in the river located on its current course.	Holocene Upper Pleistocene Middle Pleistocene Lower Pleistocene
		Flood plain	Surface of the flat land adjacent to a main river formed by alluvium deposited by the river.	Holocene
		Valley wallpapers	They are the result of the silting up of valleys by fluvial processes although they may also occur beside colluvions.	Holocene
		Alluvial fans	Deposits of sediment—gravel, sand, and finer sediments—that accumulate in the flattest part where the relief is wide and reduces the slope of a river or stream.	Holocene Pleistocene
		Pediment (glacis)	Surfaces with a flat slope rooted to a mountain slope to link to a valley bottom or depression.	Lower Pleistocene
M.S. Depositional	Hillside Deposits	Colluvion	Slope deposits associated with the combined action of running water and gravity in sheds.	Holocene Upper Pleistocene
M.S. Polygenic	Polygenic Surface Forms	Erosion surfaces	Eroded surfaces that have been excavated well on alluvial deposits or on bedrock.	Holocene Pleistocene
M.S. Marshy	Endorheic Forms	Endorheic areas	Areas with poor drainage and therefore with temporary waterlogging.	Holocene

2. Materials and Methods

To elaborate the geomorphological cartography, the following stages were carried out:

Firstly, an analysis of the geological cartography was carried out. The geomorphological cartography was elaborated on the same scale, and once the base maps were elaborated, the interpretation of different sectors was carried out by means of photointerpretation and verification in the field, assigning the new chronology of terraces with deposits and erosives for this central sector of the Tormes River sub-basin.

The geomorphological cartography was completed with the differentiation of the different morphogenetic systems (Table 1), highlighting, above all, the river system, and within this, the sequences of erosive and depositional terraces. These cartographies were subsequently geo-referenced and digitized with GIS (ArcGis), which integrated them with the topographic cartography of the National Topographic Map at a scale of 1:50,000, and with the orthophotos (years 1956 and 2018) of the national plan of aerial orthophotography at a scale of 1:10,000.

In a second stage, a digital terrain model with LiDAR data (spatial resolution: 1 m) was generated to recognize the different morphologies present in 3D, as well as to calculate the different levels of interest. From the digital terrain model, auxiliary maps were made to show the location, drainage, elevation, orientation, and slope (intervals selected according to the geomorphological characteristics of the area), in addition to an auxiliary geological–lithological map showing where the materials (conglomerates, gravels, sands, etc.) that constitute the deposits of the terraces came from, the pre-quaternary materials found, and a map of erosive surfaces (terraces) in the southern sector of the study area (Figure 4).

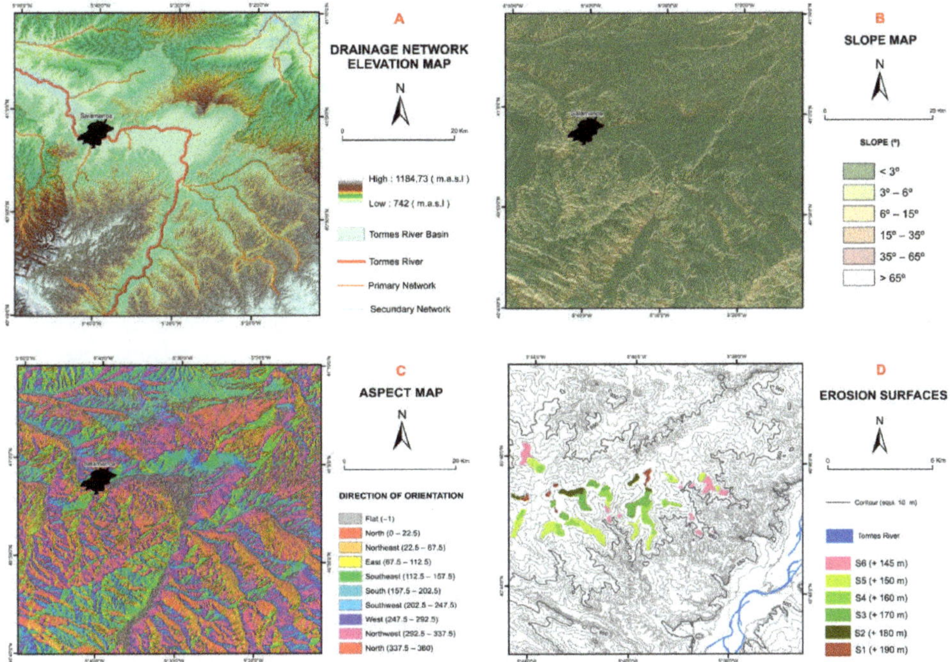

Figure 4. Auxiliary maps: (**A**) drainage (**B**) slopes (**C**) aspect and (**D**) erosive surfaces.

In a third stage, the enlistment (incision) and river displacement rates were calculated, taking into account the relative height of each group of erosive and depositional terraces with respect to the river thalweg and the age at which they were deposited, grouping them into four eras: Lower Pleistocene, Middle Pleistocene, Upper Pleistocene, and Holocene; so the fitting rate is the ratio between the meters of incision or displacement and the elapsed time. The displacement rates of the main channel of the Tormes River were also calculated in relation to the times indicated for the incision, in meters of displacement per 1000 years. Two transversal profiles of the entire sequence of terraces and nineteen longitudinal profiles were generated crossing each terrace.

Finally, paleogeographic reconstruction of the area was carried out using a technique that is currently being developed and is used to model the relief and the processes that occurred in the past [11,12]. To do this, a small area with the most complete sequence of depositional terraces was located. To obtain the different models, the altimetry of the area was determined in the form of dimensions (points), the height of each terrace in meters, and the incision suffered by them at each time. The width of the channel was estimated by an approximate average (Equation (1)):

$$Riverbed = \frac{\sum x1 + xn}{no.\ points} \tag{1}$$

where "x_1" and "x_n" are the distances between each of the terrace levels of each era, divided by the number of points, that is, the total number of terrace levels there are. Based on this estimate and once the data was obtained, the digital terrain model was constructed, and the necessary knowledge was applied in terms of color techniques, shading, exposure to light, etc., to give a better view, and 3D views were generated with the ArcScene module.

3. Results

Based on the cartography, the different morphologies present were grouped into domains associated with different active processes. The main morphologies that stand out in the area are the sequence of terraces deposited by the river itself arranged in a staggered way dating back to the Quaternary. It is important to highlight another type of terrace at higher levels, in this case, erosive, which is associated with a set of river processes constituting coatings of low thickness [13]. This type of terrace is presented in the same way and forms flat, inclined, and variable inclined lengths. A total of six levels were classified (S1 to S6), within which there were other sub-levels, highlighting that these surfaces are above the highest depositional terrace level (T1).

3.1. Sequence of River Terraces

A group of 19 river terraces with deposit (T1 to T19) in the northern sector and six previous erosive levels in the southern sector (S1 to S6) that formed during the Quaternary (Table 2) were differentiated. They were classified according to their relative heights with respect to the elevation of the thalweg of the Tormes River, and they were assigned a relative chronology corresponding to the time in which they were deposited. Most of them were found to be arranged in elongated bands in the WNW–ESE direction in the north and NE–SW in the south. It is important to note that the continuity of the terraces (from T6) is affected by the narrowing of the Tormes River valley in the Tejares area, downstream of Salamanca, where the Tormes River fits into the Varisc basement, causing a coining of them. Sedimentologically, the levels belonging to the Lower Pleistocene are characterized by siliceous ridges, gravels, and arcosic sands, presenting a white-greenish clay matrix. The immediately higher levels belonging to the Middle Pleistocene are constituted of siliceous conglomerates, sands, and red silts, presenting tabular bodies of erosive bases and flat roofs. Those belonging to the Upper Pleistocene have siliceous conglomerates, sand, and ocher limos. Lastly, the ones closest to the current riverbed, attributed to the Holocene, contain deposits of sand, silt, and ridges. However, it should be noted that level T18 presents a morphology in the form of bars, and T19 corresponds to nearest terrace and next to the flood plain of the current river. The number of levels of river terraces present in the valleys will depend, in general, on the age and importance of the drainage basin, and in some cases, on the neotectonic activity.

Table 2. Sequence of depositional terraces belonging to the Tormes River.

	Terraces	Relative Heigth (m)	Absolut Heigth (m)
Holocene	T19	+ 1.5	787.5
	T18	+ 3	789
Upper Pleistocene	T17	+ 6–7	792–793
	T16	+ 8–10	794–796
	T15	+ 12–13	798–799
Middle Pleistocene	T14	+ 18–23	804–809
	T13	+ 28–34	814–820
	T12	+ 38–42	824–828
	T11	+ 43	829
	T10	+ 50	836
	T9	+ 53	839
	T8	+ 58–64	844–850
Lower Pleistocene	T7	+ 75–80	861–866
	T6	+ 84	870
	T5	+ 94	880
	T4	+ 102	888
	T3	+ 109	895
	T2	+ 122	908
	T1	+ 140	926
	South Sector–Erosive Terraces		
	S6	+ 145	931
	S5	150	936
	S4	+ 160	946
	S3	+ 170	956
	S2	+ 180	966
	S1	+ 190	976

3.2. Longitudinal and Transversal Profiles

These two types of profile were made to study their arrangement according to their altitude. Of the 25 longitudinal profiles corresponding to each terrace, the most significant nine were chosen (Figure 5). As general characteristics, we can highlight the low slope, the little separation between the terraces, and the divergence when leaving the Paleozoic materials, which converges again when leaving the basin when fitting in the same materials.

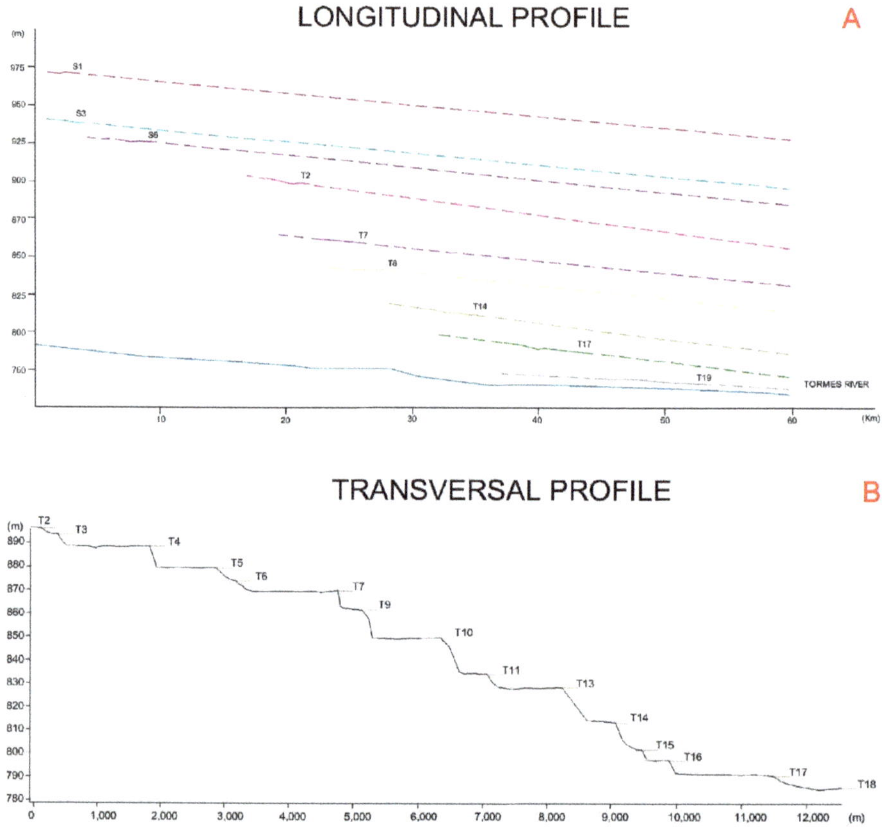

Figure 5. (**A**) Upper-longitudinal profile of the most significant terraces in the central area of the study. The trend that they would follow can be observed. It can be seen that for the most modern terraces, those that are closer to the riverbed tend to come together. (**B**) Lower-cross-sectional profile in the central area of the study. The staggering of the terraces and their extension can be observed.

A cross-sectional profile was also selected (Figure 5), cutting off most of the sequence of possible terraces where a difference in the heights of the steps shown in each of the terrace levels with respect to the previous one is observed due to climatic and/or neotectonic variations and the disparity in surface morphology as a result of erosion and depositional processes. The staggering is greater in the central zone than at the edges due to the hardness of the materials that the river has to erode, in one case, the soft (arc) materials of the Neogen, and in another case, the hard materials of the Paleozoic.

As for the extension they present, the most extensive terrace levels are T7, T10, T13, and T17, and the least extensive are T2, T6, T11, and T16. The staggering is also variable and can be correlated with terraces dated to other river valleys.

A synthetic scheme (maintaining the heights to scale) of the terrace levels of the Tormes River in the sub-basin of Salamanca (Figure 6) indicates the heights relative to the thalweg and, as a result, the heights of the escarpments between successive terraces. We start with the hypothesis that the formation of the river terraces of the Spanish rivers originated, in general, in cold times (glacial and/or seasonal times), and the endangered species and edaphic levels originated in warm moments (interglacial and/or interstate). This was deduced on the basis of numerous existing dates (OSL, paleomagnetism, ESR...) and the presence of cold geomorphological features on the different deposits (cryoclasty, cryoturbations, ice wedges, etc.) and cold fauna associations of vertebrates and microvertebrates of terraced deposits.

Figure 6. (**A**) Neogen Substrate: River sediments blushed with intense process of gleyzation processes and above river terraces. (**B**) River terrace of deposit + 140 m on the river thalweg with thick edges of slate fragments. (**C**) River terrace of + 10 m in the town of Francos, with graves of river channels on an alluvial fan of the Lower Pleistocene. (**D**) River terrace at +5 m interspersed with red soil and gleysols.

If we evaluate the escarpments between successive terraces, different values can be seen that indicate variations in the climatic conditions of each moment because the interglacial periods did not have the same intensity or duration. Observing the escarpments of the Tormes River in the sequence shown, there are significant differences of more than 15 m between them (18 m T1–T2 and 2.5 m T15–T16). This allowed us to select the most important sequences to use as correlation elements in the river sequences of the same basin and basins with similar characteristics (Duero and Tagus Basins).

In this case, the escarpments were selected between terraces T7 and T8 (16.5 m), T13 and T14 (10.5 m), and T16 and T17 (3 m).

Taking the sequence of the Arlanzón River as a reference for the correlation, within the same Duero basin when presenting paleomagnetic sampling and dating by ESR and TL [2,14–16], we can attribute these escarpments to a more precise chronology by being able to determine the boundaries between the lower, middle, and upper Pleistocene and Holocene. This type of correlation between escarpments can be used for related basins (Tagus Basin) (Figure 7).

3.3. Paleogeographic Reconstruction: Incision and Displacement Rates

The concept of paleogeographic reconstruction or reconstruction of past reliefs includes a series of techniques based on the study of the physical environment, from the manual extension of the tendency of the morphology in a profile to more complex models, such as the reconstruction of paleo concrete surfaces [17–23]. Since the surface to be reconstructed was a basin in this study, the TIN triangulation algorithm was chosen [24].

From the Lower Pleistocene to the Holocene, the relief gradually evolved so that, during the Lower Pleistocene, the Tormes River was located at a distance of about 8600 m with respect to the position currently occupied in the NE direction. During the Middle Pleistocene, it traveled 3700 m, and during the Upper Pleistocene, 500 m. At the same time that the displacement of the river in the NE direction was generated, the different levels of terraces caused by the river incision of the river were deposited, which corresponded to a displacement of the course of the river caused by neotectonic changes in an area south of the sierra (Sierra de Béjar), which tilted towards the NE and caused the river to move in the same direction (Figure 8).

Two types of rates were used in this work: The incision rate related to the river fit and the general survey of the area, and that of lateral displacement related to the evolution of the river valley and its main channel due to the neotectonic activity.

We used a height of +190 m as the approximate age for the S1 level 2.35 million years (m.y.) ago (Gelasian age), considering this level is more recent than the deposits of the alluvial fan type "raña" that we considered with the age of onset of the Quaternary (2.58 m). The beginning of the Calabriense (1.8 m.y.) corresponds to the T1 height (+140 m), constituting the first terrace with a basin deposit, coinciding with the data indicated by [2]. The beginning of the Middle Pleistocene (0.78 m) corresponds to the T8 height (+58–64 m), constituting the first positive value of the paleomagnetic scale (Brunhes) according to [11]. The beginning of the Upper Pleistocene (0.130 m) was place on T15 with a height of +12–13 m from the date of the T11 terrace of the Almanzor River with a value of 0.140 m.y. at a river height of +12–14 m [14].

In general, during the Lower Pleistocene, the incision rate was 0.0821 m/ka (meter/thousand year), while in the Middle Pleistocene, it was 0.0623 m/ka. Table 3 shows that in the study area, the incision decreased from the Lower Pleistocene to the Middle, increasing again in the Upper Pleistocene, and then increasing much more during the Holocene.

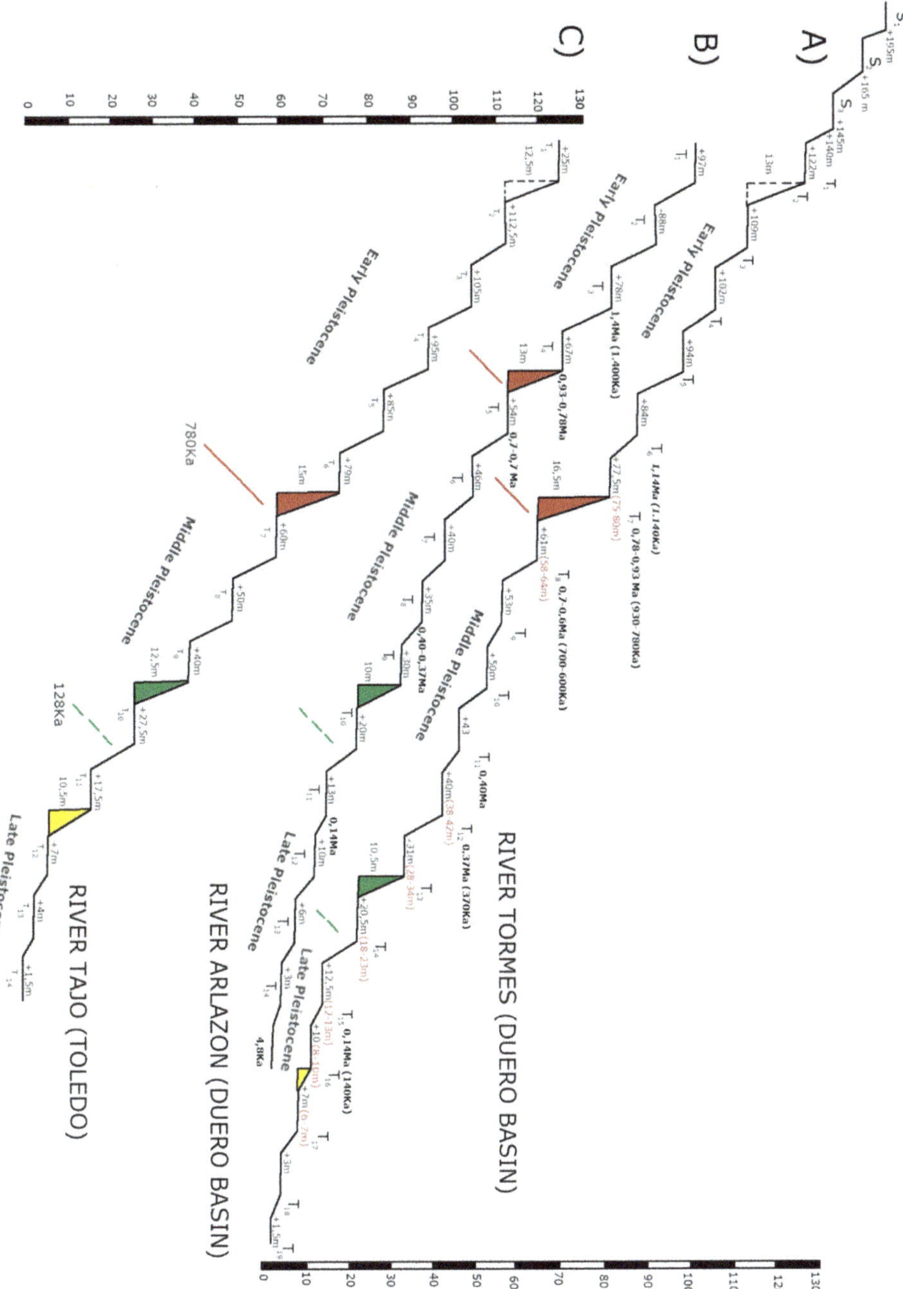

Figure 7. General chronological sequence of terraces of the Tormes River and comparison with other adjacent rivers in the Iberian Peninsula. (**A**) River Tormes (duero basin) (**B**) River Arlazon (duero basin) (**C**) River Tajo (Toledo)

The greatest fittings of the Tormes River occurred in the Holocene and the Upper Pleistocene, reducing towards the Lower Pleistocene and the Middle. The average incision rates for the entire Quaternary amounted to 0.0808 m/ka, considering not only the depositional terraces but also the erosive terraces belonging to the Gelasian (Lower Pleistocene), in which the height difference between the oldest and first T1 (Calabrian) was found to be 50 m with an age difference of 500 ka, so the fitting rate was 0.0909 m/ka. The rates for the Upper Pleistocene were found to be similar to those of the same time period in the Arlazón River (Duero Basin) [14].

The displacement rates were relatively high and variable as we took steps perpendicular to the riverbed from T1 to T19. Very high displacement was found in the Upper Pleistocene and Holocene (12.89 m/ka and 31.11 m/ka), and when we analyzed Diagonal distances, the displacement increased considerably (44.58 m/ka and 64.10 m/ka). These data indicate the existence of a general movement of neotectonic origin due to the readjustments of the Central System (Sierra de Béjar) that swung during the Quaternary, especially from the Upper Pleistocene.

Figure 9 shows the evolution of the Tormes River during the Quaternary. The thalweg at the end of the Lower Pleistocene was located in the central area of the valley with considerable engagement (0.0821 m/ka) 129 m from T1 that progressively decreased (0.0623 m/ka) to 48.5 m during the Middle Pleistocene, rectifying its direction (initial section SN, middle section SSE–NNE and final section EW) during the Upper Pleistocene (9.5 m with an incision rate of 0.859 m/ka) to reach higher values in the Holocene (0.25 m/ka).

Table 3. Incision rate and displacement during the Quaternary (in m/Ka).

	Incision Rate	Displacement	
Quaternary 0.0808	0.0808	Maximum diagonal 6.19	Perpendicular 6.06
Lower Pleistocene 0.0822	Gelasian 0.0909 / Calabrian 0.0774	5.37 m/ka	5.43
Middle Pleistocene 0.0746	0.0746	4.63 m/ka	6.85
Upper Pleistocene 0.0803	0.0803	44.58 m/ka	12.70
Holocene 0.2564	0.2564	64.10	29.57

A dismantling of the relief occurred as time passed, as did an increase in the incision in the Upper Pleistocene, while the river changed direction from ESE–WNW to E–W. The movement of the channel decreased from the Lower Pleistocene to the Middle Pleistocene, and significantly increased in the Upper Pleistocene. From the Middle Pleistocene, the tilting of S to N at the regional level accelerated the displacement of the river.

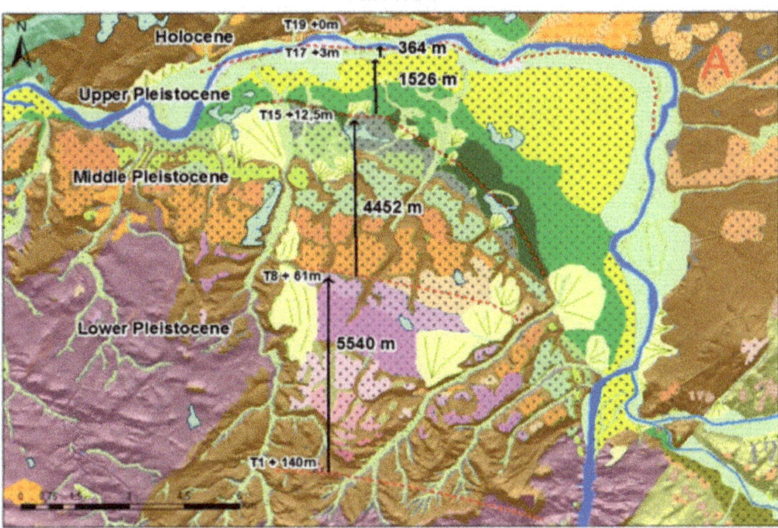

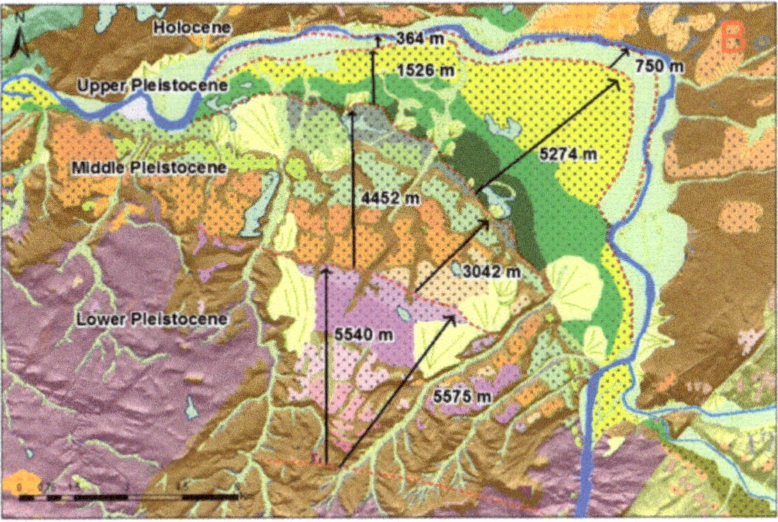

Figure 8. Sequence of depositional terraces belonging to the Tormes River. (**A**) incision and (**B**) displacement.

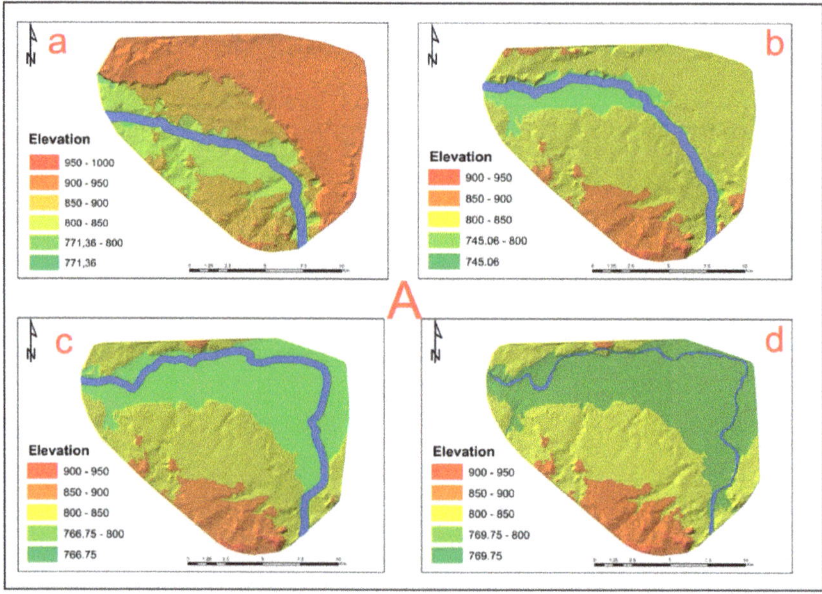

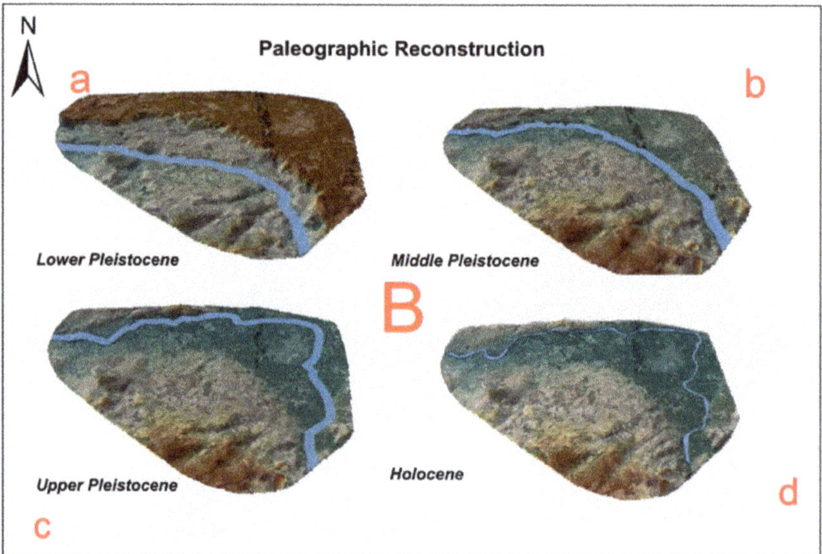

Figure 9. Reconstruction of the Quaternary landscape and paleogeographic evolution using the triangulation algorithm TIN (**A**), and ArcMap and ArcScene (**B**): (a) Situation of the Tormes River during the Lower Pleistocene; (b) situation of the Tormes River during the Middle Pleistocene; (c) situation of the Tormes River during the Upper Pleistocene; (d) situation of the Tormes River during the Holocene.

4. Discussion

The geomorphological mapping generated with GIS techniques from Lidar data with 1 m spatial resolution allowed differentiation of the different geomorphological systems and the spatial delimitation of erosive and depositional river terraces. Georeferenced auxiliary maps such as the geological and lithological map complement the general information of the geomorphological map.

From the general sequence of terraces and the support of the longitudinal and transverse profiles of the terrain, it can be seen that although the sequence is staggered, when the river enters the Neogene basin, the (oldest) terraces diverge, while at the exit, the terraces (more modern) converge. This is due to the difference in the competition or resistance of the Paleozoic materials (more resistant entrances and exits) and the Cenozoic formed by basin filling materials (less resistant materials).

The cartography presents more developed terraces on the ground floor (T4, T7, T10, T13, and T17) and fittings of greater importance (T1–T2 18 m, T2–T3 13 m, T7–T8 16.5 m, T13–T14 10.5 m). These developed escarpments allow correlations between sequences of river terraces between rivers of the same basin and similar basins where climatic and neotectonic conditions are similar (Duero and Tajo Basin).

The incision rates for the Tormes River in the surroundings of the city of Salamanca decrease from the Middle Pleistocene (0.0821 to 0.0746 m/ka) and increase in the Upper Pleistocene (0.0746 to 0.0859 m/ka), presenting values higher than those of the Tagus River (0.2 m/ka) [2], somewhat larger than those of the Arlanzon River (0.081 m/ka, Benito et al., 2018), and much lower than those of the Duero River in the Arribes del Duero, where the last 100 ka have values of 2–3 m/ka [25].

5. Conclusions

The morphogenetical and chronological evolution is then established based on the relative age of the lithological units and the geomorphologic domains, as well as their spatial distribution. The analysis of the generators of the relief and their actions allows knowledge of the morphodynamics of the different geoforms, being able to know the different landscapes that have been developed in each region and to establish their relative evolution in time from the relative or absolute morphochronology. The analysis of the landscape from precise morphogenetic systems has allowed reconstructions paleoclimatic and paleogeographical, which identify the evolutionary phases of the relief with the time.

As for the rates of engagement, they are very high for the same periods of time, with a "rest" occurring for the Middle Pleistocene, and a large increase occurring for the Upper Pleistocene. This is related to regional neotectonics and affects the large active faults (Fault of Alba-Villoria), which delimit the area by the East, and the tilting of the Sierra de Béjar towards the NE, which forces the rivers to travel in the same direction.

The detail and precision of this geomorphological cartography allows the paleogeographic characterization of the area by showing the paleoreliefs through 3D models during the Quaternary, visually evaluating both the incisions and the fitting of the Tormes River, and studying its paleolandscapes and the evolutionary territorial morphology of the paleo channels and paleovalleys.

Author Contributions: Conceptualization, J.L.G. and G.R.L.; methodology, G.R.L. and V.V.; software, G.L.R.; validation, J.L.G. and A.M.M.-G.; formal analysis, G.R.L.; investigation, J.L.G. and R.C.; resources, G.R.L.; data curation, G.R.L.; writing—original draft preparation, A.M.M.-G.; writing—review and editing, A.M.M.-G. and V.V.; visualization, G.R.L.; supervision, J.L.G.; project administration, A.M.M.-G.

Funding: This research was funded by projects Junta Castilla y León, grant number SA044G18.

References

1. Martín Serrano, A.; Santisteban Navarro, J.; Carral, P. Cartografía y Memoria del Cuaternario de la Hoja 12–19 (Salamanca). In *Mapa geológico de España E: 1/50.000 (2ª serie)*; 2000; p. 156.
2. Silva, P.G.; Roquero, E.; López-Recio, M.; Huerta, P.; Martínez-Graña, A.M. Chronology of fluvial terrace sequences for large Atlantic Rivers in the Iberian Peninsula (Upper Tagus and Duero drainage basins, Central Spain). *Quat. Sci. Rev.* **2016**, *166*, 188–203. [CrossRef]
3. Martínez-Graña, A.; Goy, J.L.; González-Delgado, J.A.; Cruz, R.; Sanz, J.; Bustamante, I. 3D Virtual itinerary in the Geological Heritage from Natural Parks in Salamanca-Ávila-Cáceres, Spain. *Sustainability* **2019**, *11*, 144. [CrossRef]
4. Martínez-Graña, A.M.; Goy, J.L.; Zazo, C.; Silva, P.G.; Santos-Francés, F. Configuration and evolution of the landscape from the geomorphological map in the Natural Parks Batuecas-Quilamas (Central System, SW Salamanca, Spain). *Sustainability* **2017**, *9*, 1458. [CrossRef]
5. Julivert, M.; Fonboté, J.M.; Ribeiro, A.; Conde, L. *Mapa Tectónico de la Península Ibérica y Baleares. Escala 1:1.000.000*; Instituto Geológico y Minero de España (IGME): Madrid, Spain, 1974; p. 113.
6. Diez Balda, M.A. *El Complejo Esquisto—Grauváquico, las Series Paleozoicas y la Estructura Hercínica al Sur de Salamanca*; Salamanca, Ed.; Universidad de Salamanca: Salamanca, Spain, 1986; p. 162.
7. López-Plaza, M.; López Moro, F.J.; Preto Gornes, E.; Sousa, L. *Patrimonio Geológico Tranfronteriço na Regiâo do Douro, Roteiro IV*; Preto Gornes, E., Alencoào, A.M., Eds.; UTAD: Vila Real, Portugal, 2005; pp. 79–94.
8. Martínez-Graña, A.M.; Goy, J.L.; Zazo, C. Geomorphological applications for susceptibility mapping of landslides in natural parks. *Environ. Eng. Manag. J.* **2016**, *15*, 1–12. [CrossRef]
9. Martínez-Graña, A.M.; Silva, P.G.; Goy, J.L.; Elez, J.; Valdés, V.; Zazo, C. Geomorphology applied to landscape analysis for planning and management of natural spaces. Case study: Las Batuecas-S. de Francia and Quilamas natural parks, (Salamanca, Spain). *Sci. Total Environ.* **2017**, *584–585*, 175–188. [CrossRef] [PubMed]
10. Martínez Graña, A.M. Estudio Geológico Ambiental Para la Ordenación de los Espacios Naturales de "Las Batuecas—Sierra de Francia" y "Quilamas". Aplicaciones Geomorfológicas al Paisaje, Riesgos e Impactos. Análisis Cartográfico Mediante SIG. Ph.D. Thesis, Universidad de Salamanca, Salamanca, Spain, 2010; p. 806.
11. Benito-Calvo, A. Análisis Geomorfológico y Reconstrucción de Paleopaisajes Neógenos y Cuaternarios en la Sierra de Atapuerta y el Valle Medio del río Arlanzón. Ph.D. Thesis, Universidad Complutense de Madrid, Madrid, Spain, 2004; p. 420.
12. Benito-Calvo, A.; Pérez-González, A.; Parés, J.M. Quantitative reconstruction of Late Cenozoic landscapes: A case study in the Sierra de Atapuerca (Burgos, Spain). *Earth Surf. Process. Landf.* **2008**, *33*, 196–208. [CrossRef]
13. Díez Herrero, A. Geomorfología e Hidrología Fluvial del río Alberche. Modelos y SIG Para la Gestión de Riberas. Ph.D. Thesis, Universidad Complutense de Madrid, Madrid, Spain, 2001; p. 587.
14. Benito-Calvo, A.; Ortega, A.I.; Navazo, M.; Moreno, D.; Pérez-González, A.; Parés, J.M.; Bermúdez de Castro, J.M.; Carbonell, E. Evolución geodinámica pleistocena del valle del rio Arlanzón: Implicaciones en la formación del sistema endokártico y los yacimientos al aire libre de la Sierra de Atapuerca (Burgos, España). *Bol. Geol. Min.* **2018**, *129*, 59–82. [CrossRef]
15. Roquero, E.; Silva, P.G.; Lopez Recio, M.; Tapias, F.; Cunha, P.P.; Morín, J.; Alcaraz-Castaño, M.; Carrobles, J.; Murray, A.S.; Buylaert, J.P. Geocronología de las Terrazas de Pleistoceno Medio y Superior del valle del Tajo en Toledo (España). In Proceedings of the XIV Reunión Nacional de Cuaternario Ibérico, Granada, Spain, 30 June–2 July 2015; pp. 8–12.
16. Sancho, C.; Calle, M.; Peña-Monné, J.L.; Duval, M.; Oliva-Urcia, B.; Pueyo, E.I.; Benito, G.; Moreno, A. Dating the Earliest Pleistocene alluvial terrace of the Alcanadre River (Ebro Basin, NE Spain): Insights into the landscape evolution and involved processes. *Quat. Int.* **2016**, *407*, 86–95. [CrossRef]
17. Abbot, L.D.; Silver, E.A.; Roberto, S.A.; Smith, R.; Infle, J.C.; Kling, S.S.; Haig, D.; Samll, E.; Gaqlewsky, J.; Sliter, W. Measurement of tectonic surface uplift rate in a young collisional mountain belt. *Nature* **1997**, *385*, 501–507. [CrossRef]
18. Small, E.E.; Anderson, R.S. Pleistocene relief production in Laramide mountain ranges, western United States. *Geology* **1998**, *26*, 123–126. [CrossRef]
19. Delmas, M.; Braucherb, R.; Gunnellc, Y.; Guilloub, V.; Calvet, C.; Bourlèsb, D.; Aster, T. Constraints on Pleistocene glaciofluvial terrace age and related soil chronosequence features from vertical 10Be profiles in the Ariège River catchment (Pyrenees, France). *Glob. Planet. Chang.* **2015**, *132*, 39–53. [CrossRef]

20. Hongshan, G.; Zongmeng, L.; Yapeng, J.; Baotian, P.; Xiaofeng, L. Climatic and tectonic controls on strath terraces along the upper Weihe River in central China. *Quat. Res.* **2016**, *86*, 326–334.

21. Karampaglidis, T.; Benito-Calvo, A.; Rodés, A.; Pérez-González, A.; Miguens-Rodríguez, L. Datación de dos terrazas rocosas del valle del Río Lozoya (Comunidad de Madrid, España) mediante los isótopos cosmogénicos ^{10}Be y ^{26}Al. *Cuatern. Geomorfol.* **2016**, *30*, 37–47. [CrossRef]

22. Karampaglidis, T.; Benito Calvo, A.; Pérez-González, A.; Baquedano, E.; Arsuaga, J.L. Secuencia geomorfológica y reconstrucción del paisaje durante el Cuaternario en el valle del río Lozoya (Sistema Central, España). *Bol. R. Soc. Esp. Hist. Nat. Sec. Geol.* **2011**, *105*, 149–162.

23. Roquero, E.; Silva, P.G.; Zazo, C.; Goy, J.L.; Masana, J. Soil evolution indices in fluvial terrace chronosequences of Central Spain (Tagus and Duero fluvial basins). *Quat. Int.* **2015**, *376*, 101–113. [CrossRef]

24. Leverington, D.W.; Teller, J.T.; Mann, J.D. A GIS method for the reconstruction of late Quaternary landscapes from isobase data and modern topography. *Comput. Geosci.* **2002**, *28*, 631–639. [CrossRef]

25. Antón, L.; Rodés, A.; De Vicente, G.; Pallas, R.; García-Castellanos, D.; Stuart, F.M.; Braucher, R.; Bourles, D. Quantification of fluvial incision in the Duero Basin (NW Iberia) from longitudinal profile analysis and terrestrial cosmogenic nuclide concentrations. *Geomorphology* **2012**, *165–166*, 50–61. [CrossRef]

 *sustainability*

Article

Effects of the Climate Change on Peripheral Populations of Hydrophytes: A Sensitivity Analysis for European Plant Species Based on Climate Preferences

Ricardo Enrique Hernández-Lambraño [1,2,*] , David Rodríguez de la Cruz [1,2] and José Ángel Sánchez Agudo [1,2]

[1] Grupo de Investigación en Biodiversidad, Diversidad humana y Biología de la Conservación, Campus Miguel de Unamuno, Universidad de Salamanca, s/n, E-37007 Salamanca, Spain; droc@usal.es (D.R.d.l.C.); jasagudo@usal.es (J.Á.S.A.)

[2] Departamento de Botánica y Fisiología Vegetal, Área de Botánica, Campus Miguel de Unamuno, Universidad de Salamanca, s/n, E-37007 Salamanca, Spain

* Correspondence: ricardohl123@usal.es

Abstract: Biogeographical theory suggests that widespread retractions of species' rear edges are expected due to anthropogenic climate change, affecting in a particularly intense way those linked to fragile habitats, such as species' rear edges closely dependent on specific water conditions. In this way, this paper studies the potential effects of anthropogenic climate change on distribution patterns of threatened rear edge populations of five European hydrophyte plants distributed in the Iberian Peninsula. We explored (i) whether these populations occur at the limit of the species' climatic tolerance, (ii) we quantified their geographic patterns of vulnerability to climate change, and in addition, (iii) we identified in a spatially explicit way whether these threatened populations occur in vulnerable environments to climate change. To do this, we simulated the climatic niche of five hydrophyte species using an ecological modelling approach based on occurrences and a set of readily available climatic data. Our results show that the Iberian populations studied tended to occur in less suitable environments relative to each of the species' optimal climates. This result suggests a plausible explanation for the current degree of stagnancy or regression experienced by these populations which showed high sensitivity and thus vulnerability to thermal extremes and high seasonality of wet and temperature. Climatic predictions for 2050 displayed that most of the examined populations will tend to occur in situations of environmental risk in the Iberian Peninsula. This result suggests that the actions aimed at the conservation of these populations should be prioritized in the geographic locations in which vulnerability is greatest.

Keywords: sensitivity; vulnerability; threatened species; hydrophyte plants; species' rear edges; climatic change; MaxEnt; CENFA

Citation: Hernández-Lambraño, R.E.; de la Cruz, D.R.; Agudo, J.Á.S. Effects of the Climate Change on Peripheral Populations of Hydrophytes: A Sensitivity Analysis for European Plant Species Based on Climate Preferences. *Sustainability* **2021**, *13*, 3147. https://doi.org/10.3390/su13063147

Academic Editors: Gioele Capillo and Ivo Machar

Received: 22 January 2021
Accepted: 10 March 2021
Published: 12 March 2021

1. Introduction

Biodiversity of freshwater habitats, especially in relation to wetland plants, is of conservation concern world-wide [1,2]. These species are threatened by multiple factors. Climate and land use changes are standing out and are the most important factors [3,4]. Risk of extinction due to anthropogenic climate change is a significant threat for these species [2], especially those populations inhabiting at the rear range edge [5].

Biogeographical theory suggests that widespread retractions of species' rear edges should be seen in response to anthropogenic climate change [5,6]. Indeed, climate is considered the most important driver of plant species distribution, with temperature mostly affecting the upper elevation/latitude edges and water availability the low elevation/latitude edges [7]. In this way, it is reasonable to expect that population loss and range retractions should be seen in the most drought-prone areas of a species' distribution [8]. This prediction is based on the fact that rear edge populations often occur at the limit of the species' ecological tolerance. Thus, they habitually occupy less favourable habitats [9,10] and are expected to decline in performance as climate warming pushes them to extirpation [5,11], although empirical evidence remains rare [12]. The study of these populations are critically important for the long-term conservation of genetic diversity, phylogenetic history, evolutionary potential, and species' response to ongoing climate change [6]. In this sense, identifying the vulnerability of threatened rear edge populations to climate change is vital for guiding effective conservation efforts [13].

Williams et al. [14] describe three fundamental aspects of the vulnerability of a species or individuals to climate change: (i) sensitivity, as the degree to which the species' persistence ability is determined by the climatic conditions of its habitat; (ii) exposure, as the degree to which the species will experience climate change across its distribution range; and finally (iii) adaptive capacity, as the ability to adapt to changes in climate, through dispersal, evolutionary responses, and phenotypic plasticity [15]. Under this scenario, climate-niche models based on ecological niche theory [16] may offer a spatially-explicit insight into geographic patterns of species vulnerability (e.g., sensitivity and exposure) to climate change. This approach uses known occurrence locations and spatially-explicit data on the environmental conditions believed to restrict the geographic distribution of the target species to predict habitat suitability across the landscape [17]. In recent years, a novel approach to model the species' climate niche based on Hutchinson's concepts, termed Climate-Niche Factor Analysis (CNFA), was proposed by Rinnan and Lawler [15]. CNFA quantifies species marginality and the specialization relative to the global distribution [18] and provides a spatially-explicit insight into geographic patterns of species vulnerability to climate change [15].

A particularly interesting situation for the study of the effects of the anthropogenic climate change on the persistence ability of threatened species populations living in the rear edge range concerns the hydrophytic flora of the Iberian Peninsula, where many Central European species occur at their northern range limits, often in peripheral isolates [19]. This pattern is closely associated with the climatic and geological history of the region with an important role of migratory processes caused by glacial and interglacial periods [20]. For this study we selected hydrophyte plants since it is known that aquatic and wetland habitats are among the most threatened worldwide mainly due to hydrological system alterations, especially those derived from global warming, pollution, and invasive species [21,22]. Moreover, many of these taxa are very scarce in the Iberian Peninsula and therefore have been included in the Red List of the Vascular Flora of Spain and Portugal [23–25]. Based on a comparative analysis of such species (n = 5), the aims of this paper are threefold: (i) we test whether threatened populations living on the rear range edge occur at the limit of the species' climatic tolerance; (ii) we quantify the geographic patterns of vulnerability to climate change; (iii) we attempt to identify in a spatially explicit way whether threatened rear edge populations occur in vulnerable habitats to climate change.

2. Materials and Methods

2.1. Study Species

Five hydrophyte plant species of the European Flora were selected to carry out this study (Table 1). Each of them has a different range of distribution, but they all have in common that they reach the Iberian Peninsula marginally with only a few populations which constitute the western limit of their European distribution range [19]. Moreover, due to the species' rarity within the Iberian territory, they are included in the Red List of the Vascular Flora [23–25]. We based our selection on listed species because their distribution is solidly documented [19] and also because of their conservation significance.

2.2. Study Area and Dataset

The study area covers most of Continental Europe including the British Islands. The limits used are according to those defined by Flora Europaea [27], excluding Iceland, Faroe Islands, Svalbard, and Novaya Zemlya (Figure 1). Partial territories in the Russian Federation west of the Urals were also included [9]. The Iberian Peninsula was established a priori as a geographical marginal territory due to its latitude, orographical and climatic characteristics. In addition, this territory has historically been one of the major Mediterranean refugia for Central European species [20].

Table 1. List of studied species, their biological features, and species' threat category.

Species	Flowering Month	Biological Type	Iberian Threat Category	IUCN Global Category
Callitriche platycarpa Kütz. in Rchb.	Early-summer	Hydrophyte	VU	LC
Eleocharis austriaca Hayek (=*Eleocharis mamillata* subsp. *austriaca* (Hayek) Strandh.)	Late-spring	Hydrophyte, Geophyte	CR	LC
Luronium natans (L.) Raf.	Mid-summer	Hydrophyte, Herb	EN	LC
Nymphoides peltata (S.G. Gmel.) O. Kuntze	Mid-summer	Hydrophyte, Herb	CR, EN	LC
Rumex hydrolapathum Huds.	Late-spring	Hydrophyte, Herb	CR	LC

Note: Iberian red list category according to the Spanish and Portuguese legislations (EN, endangered; CR, critically endangered; VU, vulnerable) [23–25] and to the IUCN global red list category (LC, least concern) [26].

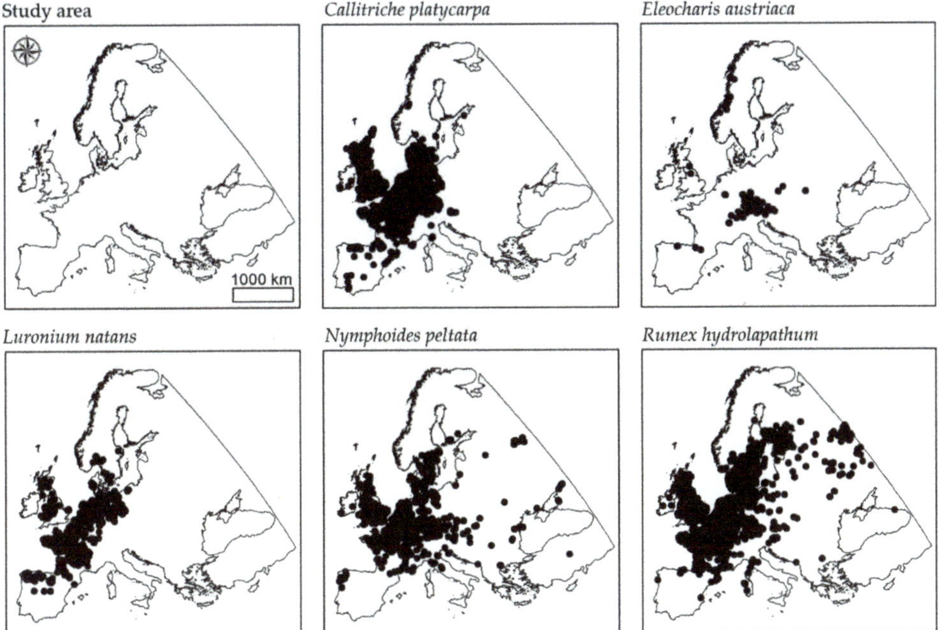

Figure 1. Occurrence records per analyzed species in the study area.

Data on the geographical distributions of the study species (Figure 1) were compiled from the Global Biodiversity Information Facility (GBIF; www.gbif.org (accessed on 3 June 2019)). The dataset was cleaned manually so that only high-quality records were used in the analysis; records conforming to these sets of conditions were retained: (i) georeferenced; (ii) with year of record; (iii) with "county" or "municipality" locality data; (iv) inland coordinates; (v) and inside the known native range of the species, as depicted by species range maps of Figure 1. We used a grid resolution of 1×1 km to remove duplicate records (i.e., only one occurrence record per grid square of 1×1 km), thereby reducing clustering (spatial bias; [28]). We omit occurrence data outside the years 1979–2013 to align with the temporal reference of the climatic variables (described below).

For our analysis, climatic variables of the current climate (representative of 1973–2013) were downloaded from the climatologies at high resolution for the Earth's land surface areas (CHELSA; http://chelsa-climate.org/ (accessed on 3 June 2019)) database at a 30 arc-second resolution (~900 m). This dataset provides improved climatic estimates in landscapes with complex topography [29]. We chose four bioclimate variables (Table 2 and Figure A2) to represent a broad range of seasonal and annual climatic patterns across the study area while minimizing redundancy (Figure A1). These variables were chosen because of their strong link with important ecological processes in plant species, such as distribution, reproduction, and phenology [30]. Furthermore, these variables are enough to explain most of the climatic variation, and other important variables (e.g., winter and summer temperatures) are strongly related to linear combinations of the four variables considered. Climate data were aggregated by averaging to 1×1 km to match the species data grid. All spatial information processing was handled using ArcGIS 10.3 [31].

Table 2. Bioclimatic variables used for this study, obtained from the CHELSA database [29], and calculated from monthly air temperature means and precipitation sums.

Variable	Calculation	Unit	Source
Annual mean temperature	$\left(\frac{\sum_{i=1}^{12} t_i}{12} \right)$	°C	CHELSA [29]
Temperature seasonality	$\left(\sqrt{\frac{1}{11} \sum_{i=1}^{12} \left(t_i - \left(\sum_{i=1}^{12} \frac{t_i}{12} \right) \right)^2} \right)$	SD	
Annual precipitation	$\sum_{i=1}^{12} p_i$	mm	
Precipitation seasonality	$\left(\sqrt{\frac{1}{11} \sum_{i=1}^{12} \left(p_i - \left(\sum_{i=1}^{12} \frac{p_i}{12} \right) \right)^2} \right) / \left(\frac{\sum_{i=1}^{12} p_i}{12} \right)$	CV	

t_i = monthly temperature, p_i = monthly precipitation, SD = standard deviation and CV = coefficient of variation.

For future climate (2050: average of 2041–2060), we used an ensemble method of two global climate models (GCMs) due to climate uncertainty [32,33]. The two GCMs used were the Community Climate System Model (CCSM4) [34] and the Hadley Global Environment Model (HadGEM2-ES) [35]. Both models have been used extensively in addressing the effects of climate change on species distributions [36]. For the future climate ensembles, we used two representative concentration pathways (RCPs) for the prescribed greenhouse gas emissions: (i) RCP 4.5, which represents medium CO_2 emissions; (ii) and RCP 8.5, which represents high CO_2 emissions [37]. The future climate data were downloaded from the CHELSA dataset at a 30 arc-second resolution and were aggregated by averaging to 1×1 km to match the species data grid (Figure A2).

In relation to projections of the effects of future climate change for European regions, the Iberian Peninsula has been identified as one of the areas that is most vulnerable to the predicted changes [38,39] and is expected to experience greater increases in temperature and aridity than other regions. In this sense, we evaluated spatial changes in the Iberian Peninsula climate characteristics through the differences between the future climate and the current climate.

2.3. Modelling Climatic Suitability

To assay whether threatened populations living on the Iberian Peninsula occur at the limit of the species' climatic tolerance we used the ecological-niche model. This approach uses known occurrence locations and spatially-explicit data on the environmental conditions (herein climatic variables) believed to restrict the geographic distribution of the target species to predict climatic suitability across the landscape [17]. To do it, we used the maximum entropy (MaxEnt) modeling [40]. MaxEnt is a machine-learning process that uses presence-only data and environmental covariates to estimate the relative suitability of one place vs. another [41]. MaxEnt has been described as especially efficient to handle complex interactions between response and covariates [41,42], and to be little sensitive to small sample sizes [43]. We used this method because it has been demonstrated to perform well in a diverse set of modeling scenarios in ecology, biogeography, and conservation, besides being widely used to fit models across many different taxa, geographical areas, and time periods [44,45].

To build the climatic-niche models for each of the species, we used the kuenm package in R [46]. This tool allows detailed calibrations of ecological-niche models in Maxent, helping to select among the complex and numerous sets of parameters those that demonstrate best performance based on significance, predictive ability, and complexity level [46]. In this study, for each species, we created 493 candidate models by combining 1 set of climatic predictors, 17 values of the regularization multiplier (0.1–1.0 at intervals of 0.1, 2–6 at intervals of 1, 8, and 10), and all 29 possible combinations of the 5 feature classes (linear = l, quadratic = q, product = p, threshold = t, and hinge = h). We evaluated the candidate model performance based on significance (partial ROC, with 500 iterations and 50% of the data for bootstrapping), omission rates (E = 5%), and model complexity (AICc). Best models for each species were selected according to the following criteria: (i) significant models with (ii) omission rates $\leq$5%. Then, from among this model set, models with delta AICc values of $\leq$2 were chosen as final models. The candidate model creation was performed using the "kuenm_cal" function and the candidate model evaluation and the best model selection was done using the "kuenm_ceval" function.

We created final models (i.e., the best fitted model for each species) using the full set of occurrences and the selected parameterizations (Table 2). We produced 10 replicates by bootstrap with logistic outputs. The final model evaluations consisted of calculations of partial ROC, omission rates, and AICc using an independent dataset. Final models were performed with the "kuenm_mod" function. Finally, we extracted habitat suitability values from the final model for the threatened species populations inhabiting the Iberian Peninsula.

2.4. Modelling Vulnerability to Climate Change

Species' vulnerability can be interpreted as a function of both extrinsic (exposure) and intrinsic (sensitivity and adaptability) traits [14,47]. Exposure is the degree to which the species will experience climate change across its distribution range [14]. Sensitivity is the degree to which the species' persistence ability is determined by the climatic conditions of its habitat, while adaptability is the ability to adapt to changes in climate through dispersal, evolutionary responses, and phenotypic plasticity [14,15]. In this study adaptability was not considered, as climatic niche evolution of species is slower than the rate of climate change [48].

We used the CNFA approach [15] to quantify vulnerability to climate change of five species of hydrophyte plants with threatened populations living on the Iberian Peninsula. This approach expands on the earlier ecological-niche factor analysis [18,49], provides spatially-explicit insight into geographic patterns of vulnerability, relies only on readily-available spatial data, and is suitable for a wide range of species and habitats [15]. One of the strengths of this approach is the ability to identify and describe aspects of climate vulnerability to climate change with relatively little information about the species itself [15]. Thus, this enables us to more proactively identify species of highest climate vulnerability and species in need of immediate conservation actions. We used this approach because it has been demonstrated to perform well in different taxa and geographical areas [15,50–52].

To quantify CNFA models we compared the species distribution in the ecological space with the global distribution of available environmental conditions [15]. We quantified two aspects of a species' niche: (i) the marginality axis (m), which is a measurement capturing the difference between the conditions used by the species and the conditions available in the global distribution and (ii) specialization axis (p), which is the ratio of size of the species niche to that of the global distribution [15]. To define the global distribution in our study, we used the combined range of the five hydrophyte species in Europe as N cells. For the distribution of each species with N cells, we used occurrence records. For the multi-dimensional ecological space composed of bioclimatic variables with C dimensions, the components of marginality and specialization are defined as the marginality factors (m_j) and the specialization factors ($p_{j1}, p_{j2}, \ldots, p_{jC-1}$), respectively. Based on the first factor, we extracted the marginality values for the threatened populations of each species in order to identify how far the climatic conditions used for such populations are from the optimal climatic conditions for the global distribution.

Following Rinnan and Lawler [15], we measured the following metrics: sensitivity (s), exposure (e), and vulnerability (v) of each species to climate change. We obtained the sensitivity factor through the marginality and specialization axis for each bioclimatic variable. We first normalized the vector ($m_j, p_{j1}, p_{j2}, \ldots, p_{jC-1}$) to ($w_{j1}, w_{j2}, \ldots, w_{jC}$). We then calculated the sensitivity factor s_j corresponding to each bioclimatic variable j as $\sum_{k=1}^{C} w_{jk}\rho_k$, where ρ_1 is the amount of specialization on the marginality axis, and $\rho_k (k > 1)$ is the amount of specialization expressed on the specialization axis. The s metric quantifies the average species specialization in each climatic variable. Thus, if a species only tolerates a narrow range of climatic conditions, we may reasonably expect it to be more sensitive to the effects of climate change. The overall sensitivity $s = \sqrt{1/C \sum_{j=1}^{C} s_j}$, can then be used to compare the sensitivity between different species.

The e metric quantifies the differences between current and future conditions (departure) inside the species range. In this sense, this metric reflects the amount of climate change a species might experience if it remains in place. The departure factor is $d_j = \sum_{i=1}^{N} p_j |\delta_{ij} - z_{ij}|$, where δ_{ij} and z_{ij} represents the value of current and future bioclimatic variable j at location i, respectively, and p_i is the habitat utilization at location i. Then, the overall exposure is $e = \sqrt{\sum_{j=1}^{C} d_j^2}$.

Finally, to calculate the species vulnerability to climate change, we combined sensitivity and exposure. To do this, we calculated the vulnerability factor v_j for each bioclimatic variable j as $\sqrt{(1 + d_j)s_j}$, and the predicted vulnerability of cell i for the global distribution as $v_{Gi} = \sqrt{\sigma_{Gi}\varepsilon_{Gi}}$. Thus, the overall vulnerability is $v = \sqrt{1/C \sum_{j=1}^{C} v_j}$. The v metric reflects the interaction between s and e to climate change. Larger values of s and e indicate higher climate sensitivity and exposure, which result in larger v values, indicating higher vulnerability in the climatic variable. See [15] for more thorough details on the CNFA process.

We implemented the CNFA method with the "cnfa" function of the package CENFA [15] in the R program [53]. We also used the "predict" function in the CENFA package to evaluate the spatial vulnerability within a potential habitat used by the threatened populations living on the Iberian Peninsula. To select the potential habitat, we used the minimum convex polygon (convex hull) produced by the full set of presence records for each species in the Iberian Peninsula. Maps of spatial vulnerability were generated with ArcGIS [31].

3. Results

3.1. Projected Changes in the Iberian Peninsula Climate Characteristics

Analysis of projected changes in the climate variables used in this study showed that the Iberian Peninsula is very likely to undergo warmer and drought events (Figure 2). These changes present similar spatial patterns for the near future (2050) under the scenarios RCPs 4.5 and 8.5. On the one hand, annual patterns of temperature are projected to increase overall on the peninsula, especially in the mountain zones. On the other hand, for the annual patterns of precipitation a decrease over the entire peninsula is projected, especially in the north and northwest. Finally, seasonal patterns of temperature and precipitation are projected to increase throughout the peninsula.

3.2. Climatic Suitability

The final best models for each species performed well in throughout the study area according to the external validations (Table 3).

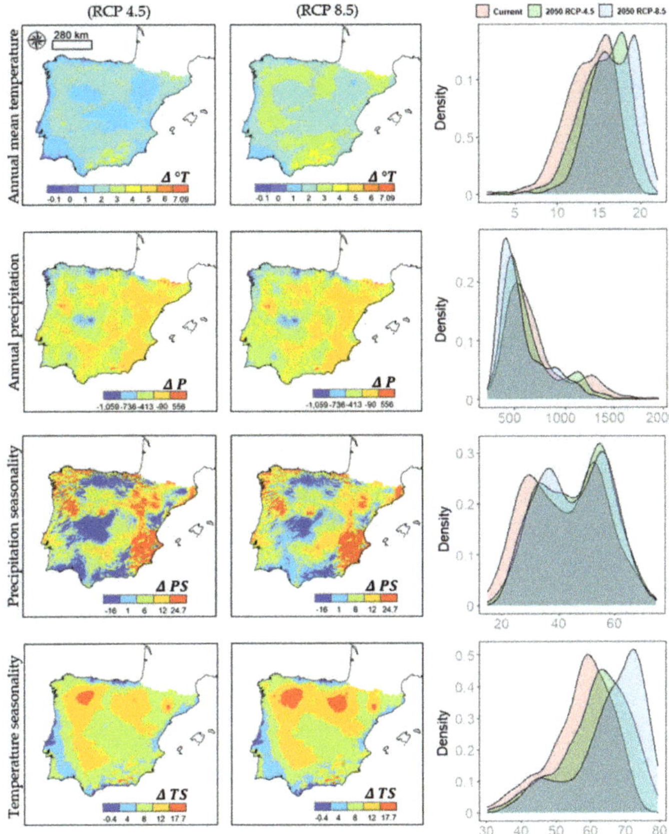

Figure 2. The first two columns are spatial differences between Iberian Peninsula climate variables for the future period (2050: average of 2041–2060) and the current period (average of 1973–2013). The third column is the distribution of climate variables throughout the Iberian Peninsula. The future climate scenario was estimated from an ensemble of two global climate models projections under the representative concentration pathways (RCPs) 4.5 and 8.5. **TS** and **PS**: temperature and precipitation seasonality, respectively.

Table 3. Final models performance for each species under optimal parameters. Feature classes (linear = l, quadratic = q, product = p, threshold = t, and hinge = h).

Species	Regularization Multiplier	Feature Classes	Mean AUC Ratio	Partial ROC	Omission Rate (5%)	AICc
Luronium natans	0.5	lqp	1.15	0.01	0.045	346.46
Rumex hydrolapathum	2	lqph	1.28	0.00	0.030	985.65
Nymphoides peltata	0.1	lqp	1.22	0.02	0.049	458.27
Callitriche platycarpa	0.1	lqp	1.19	0.01	0.023	248.13
Eleocharis austriaca	0.7	lq	1.16	0.02	0.041	148.13

These results indicate high dependence on the occurrence data and the set of climatic variables included in the analysis. In relation to the optimal climates of the analyzed species, the threatened populations of all species tend to occur in less climatic suitable habitats on the Iberian Peninsula (Figure 3).

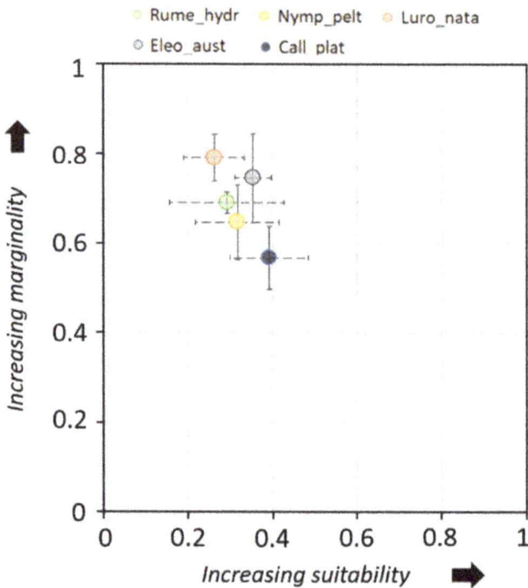

Figure 3. Mean suitability vs. mean marginality for all species. The grey bars are standard errors reflecting variation for marginality and grey dashed bars are standard errors reflecting variation for habitat suitability. Rume_hydr, *Rumex hydrolapathum*; Nymp_pelt, *Nymphoides peltate*; Luro_nata, *Luronium natans*; Eleo_aust, *Eleocharis austriaca*; Call_plat, *Callitriche platycarpa*.

3.3. Vulnerability to Climate Change

Application of the CNFA method to the calibration set indicated that only two axes (i.e., the axis of marginality and the first axis of specialization) accounted for most of the information for all species studied. The five species have high overall marginality indexes (Table 4). These results show that the niche of the species differs clearly from the mean conditions in their global distributions.

Table 4. Overall marginality (m), sensitivity (s), exposure (e), and vulnerability (v) of five hydrophyte species in the study area under future climate for the year 2050. The future climate scenarios were estimated from ensembles of two global climate models projections under the representative concentration pathways (RCPs) 4.5 and 8.5.

Species	m	s	e (RCP 4.5)	e (RCP 8.5)	v (RCP 4.5)	v (RCP 8.5)
Luronium natans	2.16	3.81	0.43	0.74	2.00	2.07
Rumex hydrolapathum	1.98	3.36	0.43	0.73	1.91	1.97
Nymphoides peltata	2.08	2.95	0.44	0.75	1.79	1.85
Callitriche platycarpa	2.06	2.86	0.43	0.68	1.74	1.79
Eleocharis austriaca	2.12	1.92	0.37	1.08	1.44	1.53

Marginality coefficients point out that species are essentially linked to wets and less seasonality environments (see m factor in Figure 4). Climatic conditions of the populations living on the Iberian Peninsula were rather different from the mean available conditions (Figure 3), indicating that these marginal populations tended to occur in less suitable climatic environments relative to each of the species' optimal climates.

The factor account for specialization (see p factor in Figure 4), mostly regarding annual mean temperature and precipitation seasonality, indicates some species' sensitivity to shift away from the optimal values of these variables. In fact, the overall sensitivity index (Table 4) shows that the ranges of species' tolerable climate conditions are quite restricted, with the greatest sensitivity to hot temperature extremes and high fluctuations in precipitation (i.e., high seasonality) (see *s* factor in Figure 4).

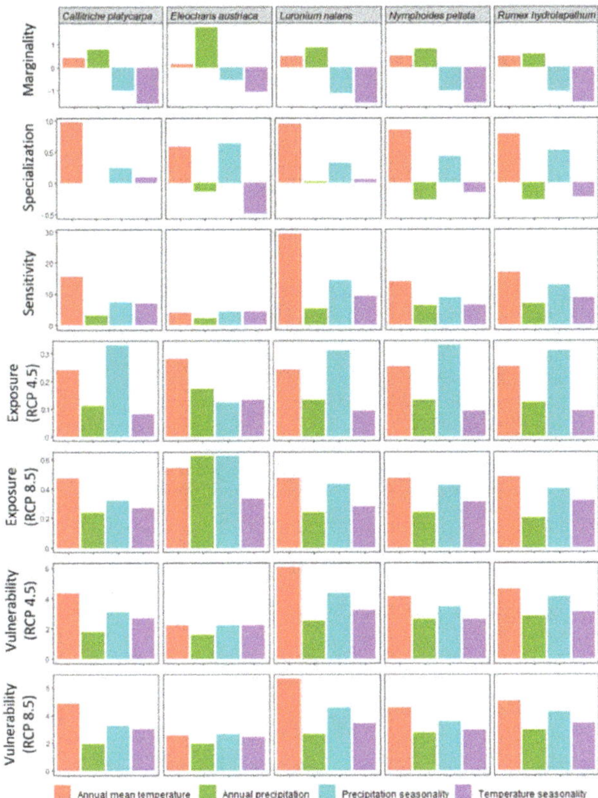

Figure 4. The marginality factor, specialization factor, sensitivity factor, exposure factor, and vulnerability factor of five hydrophytes species for four bioclimatic variables calculated under future climate for the year 2050 as estimated from an ensemble of two global climate models projections under the representative concentration pathways (RCPs) 4.5 and 8.5.

The RCP 8.5 scenario had greater departure than the RCP 4.5 scenario in almost every variable, reflecting the expected increase in climate change associated with the concentration of greenhouse gas emissions. The five species had high departure indexes, hence, this led to high overall vulnerability indexes (Table 4). In general, the species studied demonstrated high vulnerability to thermal extremes and high seasonality of precipitation and temperature (see *v* factor in Figure 4).

For the five species, the overall vulnerability index under RCP 8.5 is higher than those under RCP 4.5 (Table 4). Compared with the other hydrophyte species, *L. natans* and *R. hydrolapathum* have the highest overall vulnerability to climate change, followed by *N. peltata*, *C. platycarpa*, while *E. austriaca* shows the least overall vulnerability (Table 4). The different CO_2 emission scenarios do not change the vulnerability ranking for the five hydrophyte species.

The most vulnerable current environments for the hydrophyte threatened populations taking into account future climatic projections for the Iberian Peninsula are shown in the Figures 5 and A3. Overall, most of the examined populations occur in environments that will undergo significant climatic changes (Figures 5 and A3). The spatial patterns of climatic vulnerability under RCP 4.5 show a similar pattern for RCP 8.5 (Figures 5 and A3). However, the high CO_2 emission increases the climatic vulnerability for the studied populations (Figure 5). These results also allow us to detect areas that can serve as a refuge from climatic disturbances. Thus, the predictions suggest that the northeast zones of both hydrographic basins of the Duero and Ebro may provide potential suitable refuges from climate change for *R. hydrolapathum*, *L. natans* and *C. platycarpa*. For *N. peltata* and *E. austriaca*, climate refuges could be located in zones adjacent to the hydrographic basins of the Miño and Pais Vasco, respectively.

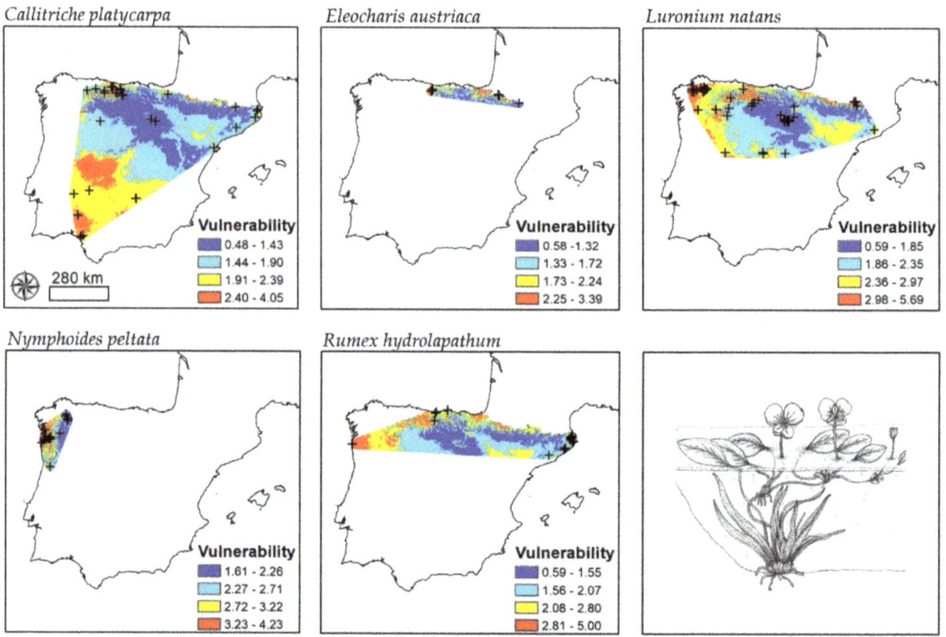

Figure 5. Predicted vulnerability across the potential habitat used by the threatened populations living on the Iberian Peninsula. The future climate scenario was estimated from an ensemble of two global climate model projections under the representative concentration pathway (RCP) 8.5 (year 2050). Predicted vulnerability for RCP 4.5, see Figure A3.

4. Discussion

4.1. Vulnerability to Climatic Change

In this study, we apply a niche-based ecological model in an attempt to explicitly assess the potential effects of the climate change on the distribution patterns of species' rear edges. In this case, hydrophytic plants were ideal due to their specific ecological requirements. In recent years, niche-based ecological models have become a strong approach for addressing this conservation issues at large scales [9,15]. This approach can provide useful information in order to lead priority conservation plans to endangered populations in a timely manner, when there is a lack of updated data about the real degree of threat for these populations, and the available resources for practical conservation actions are scarce [9].

In this study, we have found a correspondence between ecological (i.e., climatic) and geographical marginality for threatened rear edge populations of some European hydrophytes on the Iberian Peninsula. These results would support the hypothesis that marginality within the set of habitable conditions (i.e., climate niche) could represent an outstanding factor on the performance and thus persistence of isolated plant populations. Less availability of suitable environments at their rear edge distribution ranges could be translated into lower survival potential for these populations [5]. The optimal climatic habitats for our target species are associated with colder, wetter, and less seasonal environments and, for the Iberian Peninsula, these conditions are linked to high mountain habitats with high water regimes and low temperatures. However, due to the warmer, less wet, and more seasonal climates of this area in relation to Northern European climates, the suitable environments of the Iberian populations seem to be lower than the mean values in the rest of their distribution. According to these results, marginality in the Iberian populations of these species are particularly highlighted. Numerous studies provide empirical evidence of an important rear edge population decline. For example, Allen et al. [54] report range retractions and population decline (i.e., mortality) associated with elevated drought stress at species' rear edges in forest ecosystems across the globe. However, this rear edge population decline has been often questioned by empirical data [55]. These mismatches can occur, for instance, if peripheral populations are genetically isolated and adapted to conditions in border areas [56], their ecological optima are not properly established [57], their original patterns have been altered by anthropogenic land use changes [5], or their distribution patterns occur in suitable environmental conditions (i.e., microrefugium) surrounded by inhospitable regional climate [9]. On the other hand, many of the threatened populations of the analyzed species are distributed in a scattered way with few populations throughout the Iberian Peninsula. Some of them are also associated with anthropogenic activities that would make them especially vulnerable. In fact, in the Iberian territory, the distribution areas of these populations are subject to a strong anthropogenic pressure due to the conditioning of river banks, the construction of river walks, the transformation of water courses, the drainage of the water table, and livestock [24]. These factors may increase fragmentation and isolation, which can result in a significant population decline [6].

Our results indicate that the studied species have narrower climate niches, suggesting that they may be very sensitive to climate change. In fact, the two largest components of the species sensitivity factor are all associated with colder, wetter, and less seasonal environments. Analysis of projected changes in the climate variables used in this study showed that the Iberian Peninsula is very likely to undergo significant climatic changes. These changes display similar spatial patterns for the near future (2050) with increase of the temperatures and droughts in the Iberian Peninsula under scenarios RCPs 4.5 and 8.5. This climatic panorama might exacerbate the rapid decline in performance of the current threatened populations as climate warming pushes them to extirpation. Recent cases of local extinction of populations of hydrophytes in the Iberian Peninsula have already been documented [24,58]. Among the recently extinct species are *Sagittaria sagittifolia* L. and *Oenanthe aquatica* (L.) Poir., hydrophytes formerly distributed in the north of the Iberian Peninsula. The causes of their extinction are mainly associated with the loss/deterioration of their suitable habitats due to extreme climatic events such as drought and to the strong anthropogenic pressure to which the distribution areas of these species in the Iberian Peninsula are subjected [58].

It is important to note, that by neglecting other ecological processes that shape the habitat, the approach present here entails some of the issues common to niche-based ecological models [59]. For instance, our results are based on an analysis of the species' climate niches. Non-climatic constraints, such as biotic interactions, dispersal ability, and land uses are important factors that also drive the species' distribution [60,61]. Another issue is that the CNFA approach is a static method that does not capture the dynamic nature of population fluctuations over time. In addition, we must realize that the spatial scale of our approach may reduce the range of some drivers (regional/local, topographic, and microclimatic landscape features) and their effects on the study patterns [62]. However, we think that this approach is a suitable tool for a first-pass assessment of distribution patterns of species vulnerability to climate change in a spatially explicit fashion, with the potential to optimize the conservation efforts of the most vulnerable populations.

4.2. Management Implications

Climate change will have far-reaching impacts on biodiversity, including increasing extinction rates [63]. Despite that increasing impacts are expected for the future, only a few studies have aimed for a general understanding of the vulnerability of rear edge populations to climate change [38]. Herein we regard the present approach as a first step towards identifying the rear edge populations that could be potentially affected by the ongoing climate change.

Depending on each species' capacity to adapt to climate change via dispersal, rapid evolution, or other processes [63], recommended actions aimed at the conservation of these populations involve monitoring and supporting adaptive responses, prioritizing the geographic locations in which vulnerability is greatest (Figure 5). On the other hand, feasible in situ conservation measures (e.g., reduce or remove threats, maximize habitat quality, reinforcement, seed collection, translocations, enhance habitat heterogeneity, improvement of landscape connectivity [6,64]) for these populations should be concentrated in those parts that are expected to have low vulnerability to climate change. In addition, areas with low vulnerability should be considered as potential suitable climate refuges from anthropogenic climate change when planning the creation of new micro-reserves or enlarging the existing ones. Climate refugia may reduce the local extirpation risk for organisms [65].

Since there are still a lot of unanswered questions about the ecological processes involved in the persistence of these populations, it is crucial to develop several research lines to provide an effective conservation plan in the long term. The vulnerability map can help to design new studies to address knowledge gaps, for example, flowering phenology, and genetic and breeding systems of the species to assess the main factors affecting female reproductive success.

5. Conclusions

Climate change vulnerability assessments are an important tool for understanding the threat that the anthropogenic climate change poses to species and populations. In this sense, the results of our study to this methodological field show that some threatened hydrophyte populations in the Iberian Peninsula tended to occur in less suitable environments relative to each of the species' optimal climates. This result suggests a plausible explanation to the current degree of stagnancy or regression experienced by these populations. Populations of the five species showed high sensitivity and thus vulnerability to thermal extremes and droughts. Climatic predictions for 2050 displayed that most of the examined populations will tend to occur in situations of environmental risk in the Iberian Peninsula. Therefore, actions aimed at the conservation of these populations should be prioritized in the geographic locations in which vulnerability is greatest.

Author Contributions: Conceptualization, R.E.H.-L. and J.Á.S.A.; methodology, R.E.H.-L.; software, R.E.H.-L.; validation, D.R.d.l.C. and J.Á.S.A.; formal analysis, R.E.H.-L.; investigation, R.E.H.-L.; writing—original draft preparation, R.E.H.-L.; writing—review and editing, R.E.H.-L., J.Á.S.A. and D.R.d.l.C.; visualization, R.E.H.-L.; supervision, J.Á.S.A. All authors have read and agreed to the published version of the manuscript.

Funding: This research received no external funding.

Institutional Review Board Statement: Not applicable.

Informed Consent Statement: Not applicable.

Data Availability Statement: Not applicable.

Acknowledgments: R.E.H.-L. has been granted by Consejería de Educación de la Junta de Castilla y León and Fondo Social Europeo (EDU/556/2019).

Conflicts of Interest: The authors declare no conflict of interest.

Appendix A

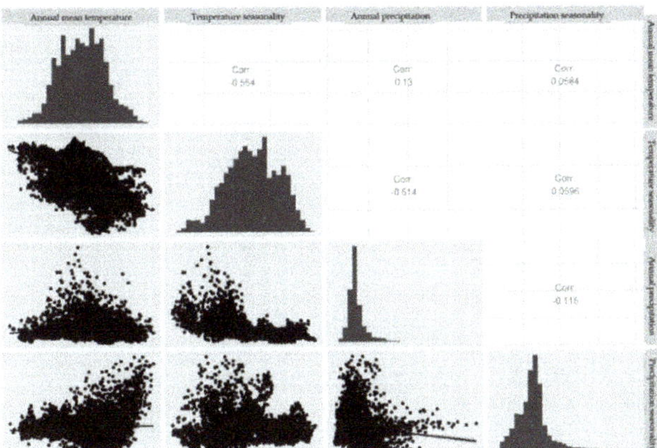

Figure A1. Pearson's correlation coefficients among the variables used to calibrate the MaxEnt and CNFA models.

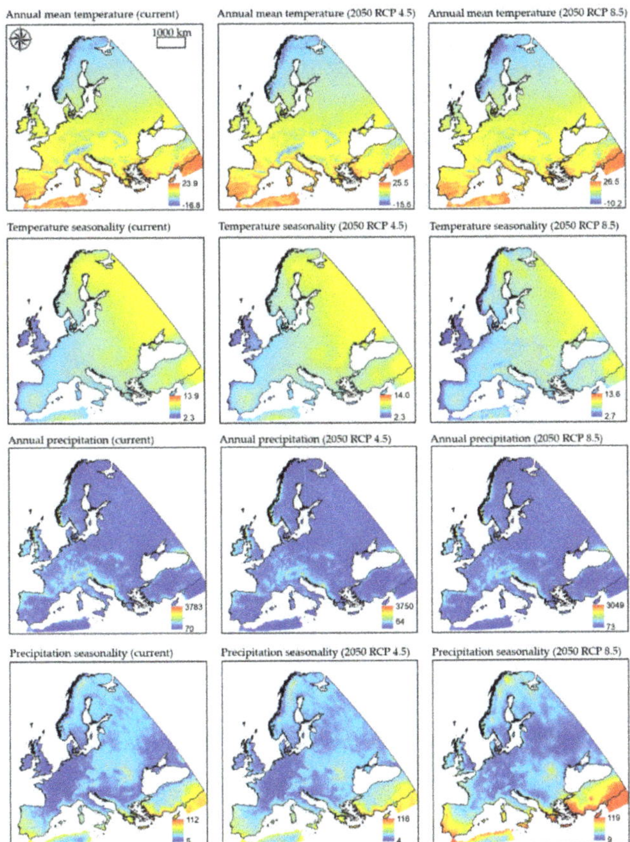

Figure A2. Spatial distribution of the climatic dataset in the study area. The future climate data were estimated from an ensemble of two global climate model projections under the representative concentration pathway (RCP) 4.5 (year 2050).

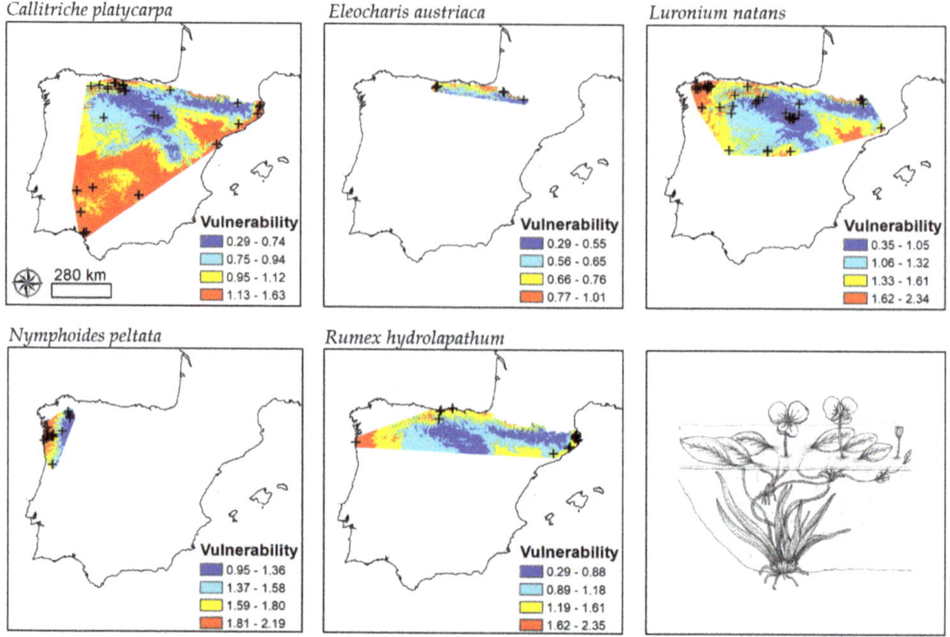

Figure A3. Predicted vulnerability across the potential habitat used by the threatened populations living on the Iberian Peninsula. The future climate scenario was estimated from an ensemble of two global climate model projections under the representative concentration pathway (RCP) 4.5 (year 2050).

References

1. Gillard, M.; Thiébaut, G.; Deleu, C.; Leroy, B. Present and future distribution of three aquatic plants taxa across the world: Decrease in native and increase in invasive ranges. *Biol. Invasions* **2017**, *19*, 2159–2170. [CrossRef]
2. Dudgeon, D. Threats to Freshwater Biodiversity in a Changing World BT. In *Global Environmental Change*; Freedman, B., Ed.; Springer: Dordecht, The Netherlands, 2014; pp. 243–253. ISBN 978-94-007-5784-4.
3. Alahuhta, J.; Heino, J.; Luoto, M. Climate change and the future distributions of aquatic macrophytes across boreal catchments. *J. Biogeogr.* **2011**, *38*, 383–393. [CrossRef]
4. Sala, O.E.; Stuart Chapin, F.; Armesto, J.J.; Berlow, E.; Bloomfield, J.; Dirzo, R.; Huber-Sanwald, E.; Huenneke, L.F.; Jackson, R.B.; Kinzig, A.; et al. Global Biodiversity Scenarios for the Year 2100. *Science* **2000**, *287*, 1770. [CrossRef] [PubMed]
5. Vilà-Cabrera, A.; Premoli, A.C.; Jump, A.S. Refining predictions of population decline at species' rear edges. *Glob. Chang. Biol.* **2019**, *25*, 1549–1560. [CrossRef] [PubMed]
6. Hampe, A.; Petit, R.J. Conserving biodiversity under climate change: The rear edge matters. *Ecol. Lett.* **2005**, *8*, 461–467. [CrossRef]
7. Abeli, T.; Ghitti, M.; Sacchi, R. Does ecological marginality reflect physiological marginality in plants? *Plant Biosyst. Int. J. Deal. Asp. Plant Biol.* **2019**, 1–9. [CrossRef]
8. Thomas, C.D.; Cameron, A.; Green, R.E.; Bakkenes, M.; Beaumont, L.J.; Collingham, Y.C.; Erasmus, B.F.N.; de Siqueira, M.F.; Grainger, A.; Hannah, L.; et al. Extinction risk from climate change. *Nature* **2004**, *427*, 145–148. [CrossRef]
9. De Medeiros-Madeira, C.; Hernández-Lambraño, R.E.; Felix-Ribeiro, K.A.; Sánchez-Agudo, J.Á. Living on the edge: Do central and marginal populations of plants differ in habitat suitability? *Plant Ecol.* **2018**, *219*, 1029–1043. [CrossRef]
10. Brown, J.H.; Mehlman, D.W.; Stevens, G.C. Spatial Variation in Abundance. *Ecology* **1995**, *76*, 2028–2043. [CrossRef]
11. Urban, M.C. Accelerating extinction risk from climate change. *Science* **2015**, *348*, 571–573. [CrossRef]
12. Pironon, S.; Villellas, J.; Morris, W.F.; Doak, D.F.; García, M.B. Do geographic, climatic or historical ranges differentiate the performance of central versus peripheral populations? *Glob. Ecol. Biogeogr.* **2015**, *24*, 611–620. [CrossRef]
13. Stanton, J.C.; Shoemaker, K.T.; Pearson, R.G.; Akçakaya, H.R. Warning times for species extinctions due to climate change. *Glob. Chang. Biol.* **2015**, *21*, 1066–1077. [CrossRef] [PubMed]
14. Williams, S.E.; Shoo, L.P.; Isaac, J.L.; Hoffmann, A.A.; Langham, G. Towards an Integrated Framework for Assessing the Vulnerability of Species to Climate Change. *PLoS Biol.* **2008**, *6*, e325. [CrossRef] [PubMed]
15. Rinnan, D.S.; Lawler, J. Climate-niche factor analysis: A spatial approach to quantifying species vulnerability to climate change. *Ecography* **2019**, *42*, 1494–1503. [CrossRef]
16. Hutchinson, G.E. Concluding Remarks. *Cold Spring Harb. Symp. Quant. Biol.* **1957**, *22*, 415–427. [CrossRef]

17. Soberón, J.; Nakamura, M. Niches and distributional areas: Concepts, methods, and assumptions. *Proc. Natl. Acad. Sci. USA* **2009**, *106*, 19644–19650. [CrossRef] [PubMed]

18. Hirzel, A.H.; Hausser, J.; Chessel, D.; Perrin, N. Ecological-Niche Factor Analysis: How to Compute Habitat-Suitability Maps Without Absence Data? *Ecology* **2002**, *83*, 2027–2036. [CrossRef]

19. Cirujano, S.; Molina, A.M.; Murillo, P.G.; Chirino, A.M. *Flora Acuática Española: Hidrófitos Vasculares*; Real Jardín Botánico, CSIC: Madrid, Spain, 2014; ISBN 8461686810.

20. Gentili, R.; Bacchetta, G.; Fenu, G.; Cogoni, D.; Abeli, T.; Rossi, G.; Salvatore, M.C.; Baroni, C.; Citterio, S. From cold to warm-stage refugia for boreo-alpine plants in southern European and Mediterranean mountains: The last chance to survive or an opportunity for speciation? *Biodiversity* **2015**, *16*, 247–261. [CrossRef]

21. Grzybowski, M.; Glińska-Lewczuk, K. Principal threats to the conservation of freshwater habitats in the continental biogeographical region of Central Europe. *Biodivers. Conserv.* **2019**, *28*, 4065–4097. [CrossRef]

22. Ortmann-Ajkai, A.; Csicsek, G.; Hollós, R.; Magyaros, V.; Wágner, L.; Lóczy, D. Twenty-Years' Changes of Wetland Vegetation: Effects of Floodplain-Level Threats. *Wetlands* **2018**, *38*, 591–604. [CrossRef]

23. Moreno, J.C. *Lista Roja 2008 de la Flora Vascular Española*; Dirección General de Medio Natural y Política Forestal (Ministerio de Medio Ambiente, y Medio Rural y Marino, y Sociedad Española de Biología de la Conservación de Plantas): Madrid, Spain, 2008; ISBN 8469173758.

24. Bañares, Á.; Blanca, G.; Gumes, J.; Moreno, J.C.; Ortiz, S. *Atlas y Libro Rojo de la Flora Vascular Amenazada de España*; Dirección General de Conservación de la Naturaleza: Madrid, Spain, 2010.

25. Carapeto, A.; Francisco, A.; Pereira, P.; Porto, M. *Lista Vermelha da Flora Vascular de Portugal Continental*; Sociedade Portuguesa de Botânica, Associação Portuguesa de Ciência da Vegetação PHYTOS e Instituto da Conservação da Natureza e das Florestas: Lisboa, Portugal, 2020; ISBN 978-972-27-2876-8.

26. IUCN The IUCN Red List of Threatened Species. Available online: http://www.iucnredlist.org (accessed on 3 June 2019).

27. Tutin, T.; Heywood, V.; Burges, N.; Valentine, D.; Walters, S.; Webb, D. *Flora Europaea: Lycopodiaceae to Platanaceae*; Tutin, T., Heywood, V., Burges, N., Valentine, D., Walters, S., Webb, D., Eds.; The Syndics of the Cambridge University Press: New York, NY, USA, 1964.

28. Hernández-Lambraño, R.E.; González-Moreno, P.; Sánchez-Agudo, J.Á. Towards the top: Niche expansion of Taraxacum officinale and Ulex europaeus in mountain regions of South America. *Australia Ecol.* **2017**, *42*, 577–589. [CrossRef]

29. Karger, D.N.; Conrad, O.; Böhner, J.; Kawohl, T.; Kreft, H.; Soria-Auza, R.W.; Zimmermann, N.E.; Linder, H.P.; Kessler, M. Climatologies at high resolution for the earth's land surface areas. *Sci. Data* **2017**, *4*, 170122. [CrossRef] [PubMed]

30. Zuckerberg, B.; Strong, C.; LaMontagne, J.M.; St. George, S.; Betancourt, J.L.; Koenig, W.D. Climate Dipoles as Continental Drivers of Plant and Animal Populations. *Trends Ecol. Evol.* **2020**, *35*, 440–453. [CrossRef] [PubMed]

31. ESRI ARCMAP 10.3.1. 2015. Available online: https://support.esri.com/en/products/desktop/arcgis-desktop/arcmap/10-3-1. (accessed on 3 June 2019).

32. Naujokaitis-Lewis, I.R.; Curtis, J.M.R.; Tischendorf, L.; Badzinski, D.; Lindsay, K.; Fortin, M.-J. Uncertainties in coupled species distribution–metapopulation dynamics models for risk assessments under climate change. *Divers. Distrib.* **2013**, *19*, 541–554. [CrossRef]

33. Buisson, L.; Thuiller, W.; Casajus, N.; Lek, S.; Grenouillet, G. Uncertainty in ensemble forecasting of species distribution. *Glob. Chang. Biol.* **2010**, *16*, 1145–1157. [CrossRef]

34. Gent, P.R.; Danabasoglu, G.; Donner, L.J.; Holland, M.M.; Hunke, E.C.; Jayne, S.R.; Lawrence, D.M.; Neale, R.B.; Rasch, P.J.; Vertenstein, M.; et al. The Community Climate System Model Version 4. *J. Clim.* **2011**, *24*, 4973–4991. [CrossRef]

35. Martin, G.M.; Bellouin, N.; Collins, W.J.; Culverwell, I.D.; Halloran, P.R.; Hardiman, S.C.; Hinton, T.J.; Jones, C.D.; McDonald, R.E.; McLaren, A.J.; et al. The HadGEM2 family of Met Office Unified Model climate configurations. *Geosci. Model Dev.* **2011**, *4*, 723–757. [CrossRef]

36. Albuquerque, F.; Benito, B.; Rodriguez, M.Á.M.; Gray, C. Potential changes in the distribution of Carnegiea gigantea under future scenarios. *PeerJ* **2018**, *6*, e5623. [CrossRef] [PubMed]

37. Harris, R.M.B.; Grose, M.R.; Lee, G.; Bindoff, N.L.; Porfirio, L.L.; Fox-Hughes, P. Climate projections for ecologists. *WIREs Clim. Chang.* **2014**, *5*, 621–637. [CrossRef]

38. Thuiller, W.; Lavorel, S.; Araújo, M.B.; Sykes, M.T.; Prentice, I.C. Climate change threats to plant diversity in Europe. *Proc. Natl. Acad. Sci. USA* **2005**, *102*, 8245–8250. [CrossRef] [PubMed]

39. Giorgi, F.; Lionello, P. Climate change projections for the Mediterranean region. *Glob. Planet. Chang.* **2008**, *63*, 90–104. [CrossRef]

40. Phillips, S.J.; Anderson, R.P.; Dudík, M.; Schapire, R.E.; Blair, M.E. Opening the black box: An open-source release of Maxent. *Ecography* **2017**, *40*, 887–893. [CrossRef]

41. Elith, J.; Phillips, S.J.; Hastie, T.; Dudík, M.; Chee, Y.E.; Yates, C.J. A statistical explanation of MaxEnt for ecologists. *Divers. Distrib.* **2011**, *17*, 43–57. [CrossRef]

42. Elith, J.; Graham, C.H.; Anderson, R.P.; Dudík, M.; Ferrier, S.; Guisan, A.; Hijmans, R.J.; Huettmann, F.; Leathwick, J.R.; Lehmann, A.; et al. Novel methods improve prediction of species' distributions from occurrence data. *Ecography* **2006**, *29*, 129–151. [CrossRef]

43. Wisz, M.S.; Hijmans, R.J.; Li, J.; Peterson, A.T.; Graham, C.H.; Guisan, A. Effects of sample size on the performance of species distribution models. *Divers. Distrib.* **2008**, *14*, 763–773. [CrossRef]

44. Elith, J.; Leathwick, J.R. Species Distribution Models: Ecological Explanation and Prediction Across Space and Time. *Annu. Rev. Ecol. Evol. Syst.* **2009**, *40*, 677–697. [CrossRef]

45. Franklin, J. *Mapping Species distributions: Spatial Inference and Prediction*; Cambridge University Press: New York, NY, USA, 2010; ISBN 1139485296.

46. Cobos, M.E.; Peterson, A.T.; Barve, N.; Osorio-Olvera, L. kuenm: An R package for detailed development of ecological niche models using Maxent. *PeerJ* **2019**, *7*, e6281. [CrossRef]

47. Pacifici, M.; Foden, W.B.; Visconti, P.; Watson, J.E.M.; Butchart, S.H.M.; Kovacs, K.M.; Scheffers, B.R.; Hole, D.G.; Martin, T.G.; Akçakaya, H.R.; et al. Assessing species vulnerability to climate change. *Nat. Clim. Chang.* **2015**, *5*, 215–224. [CrossRef]

48. Quintero, I.; Wiens, J.J. Rates of projected climate change dramatically exceed past rates of climatic niche evolution among vertebrate species. *Ecol. Lett.* **2013**, *16*, 1095–1103. [CrossRef]

49. Basille, M.; Calenge, C.; Marboutin, É.; Andersen, R.; Gaillard, J.M. Assessing habitat selection using multivariate statistics: Some refinements of the ecological-niche factor analysis. *Ecol. Model.* **2008**, *211*, 233–240. [CrossRef]

50. Raia, P.; Mondanaro, A.; Melchionna, M.; Di Febbraro, M.; Diniz-Filho, J.A.F.; Rangel, T.F.; Holden, P.B.; Carotenuto, F.; Edwards, N.R.; Lima-Ribeiro, M.S. Past extinctions of Homo species coincided with increased vulnerability to climatic change. *One Earth* **2020**, *3*, 480–490. [CrossRef]

51. Wang, W.-T.; Guo, W.-Y.; Jarvie, S.; Svenning, J.-C. The fate of Meconopsis species in the Tibeto-Himalayan region under future climate change. *Ecol. Evol.* **2021**, *11*, 887–899. [CrossRef] [PubMed]

52. Sutton, L.J.; Anderson, D.L.; Franco, M.; McClure, C.J.W.; Miranda, E.B.P.; Vargas, F.H.; Vargas González, J.D.J.V.; Puschendorf, R. Geographic range estimates and environmental requirements for the harpy eagle derived from spatial models of current and past distribution. *Ecol. Evol.* **2021**, *11*, 481–497. [CrossRef]

53. R Core Team. *R: A Language and Environment for Statistical Computing*; R Foundation for Statistical Computing: Vienna, Austria, 2019.

54. Allen, C.D.; Macalady, A.K.; Chenchouni, H.; Bachelet, D.; McDowell, N.; Vennetier, M.; Kitzberger, T.; Rigling, A.; Breshears, D.D.; Hogg, E.H.; et al. A global overview of drought and heat-induced tree mortality reveals emerging climate change risks for forests. *For. Ecol. Manag.* **2010**, *259*, 660–684. [CrossRef]

55. Pironon, S.; Papuga, G.; Villellas, J.; Angert, A.L.; García, M.B.; Thompson, J.D. Geographic variation in genetic and demographic performance: New insights from an old biogeographical paradigm. *Biol. Rev.* **2017**, *92*, 1877–1909. [CrossRef]

56. Eckert, C.G.; Samis, K.E.; Lougheed, S.C. Genetic variation across species' geographical ranges: The central-marginal hypothesis and beyond. *Mol. Ecol.* **2008**, *17*, 1170–1188. [CrossRef]

57. Sagarin, R.D.; Gaines, S.D.; Gaylord, B. Moving beyond assumptions to understand abundance distributions across the ranges of species. *Trends Ecol. Evol.* **2006**, *21*, 524–530. [CrossRef]

58. Aedo, C.; Medina, L.; Barberá, P.; Fernández-Albert, M. Extinctions of vascular plants in Spain. *Nord. J. Bot.* **2015**, *33*, 83–100. [CrossRef]

59. Jiménez-Valverde, A.; Lobo, J.M.; Hortal, J. Not as good as they seem: The importance of concepts in species distribution modelling. *Divers. Distrib.* **2008**, *14*, 885–890. [CrossRef]

60. Hernández-Lambraño, R.E.; González-Moreno, P.; Sánchez-Agudo, J.Á. Environmental factors associated with the spatial distribution of invasive plant pathogens in the Iberian Peninsula: The case of Phytophthora cinnamomi Rands. *For. Ecol. Manag.* **2018**, *419–420*, 101–109. [CrossRef]

61. Osorio-Olvera, L.; Soberón, J.; Falconi, M. On population abundance and niche structure. *Ecography* **2019**, *42*, 1415–1425. [CrossRef]

62. Oldfather, M.F.; Kling, M.M.; Sheth, S.N.; Emery, N.C.; Ackerly, D.D. Range edges in heterogeneous landscapes: Integrating geographic scale and climate complexity into range dynamics. *Glob. Chang. Biol.* **2020**, *26*, 1055–1067. [CrossRef]

63. Foden, W.B.; Butchart, S.H.M.; Stuart, S.N.; Vié, J.-C.; Akçakaya, H.R.; Angulo, A.; DeVantier, L.M.; Gutsche, A.; Turak, E.; Cao, L.; et al. Identifying the World's Most Climate Change Vulnerable Species: A Systematic Trait-Based Assessment of all Birds, Amphibians and Corals. *PLoS ONE* **2013**, *8*, e65427. [CrossRef] [PubMed]

64. Oliver, T.; Roy, D.B.; Hill, J.K.; Brereton, T.; Thomas, C.D. Heterogeneous landscapes promote population stability. *Ecol. Lett.* **2010**, *13*, 473–484. [CrossRef] [PubMed]

65. Dobrowski, S.Z. A climatic basis for microrefugia: The influence of terrain on climate. *Glob. Chang. Biol.* **2011**, *17*, 1022–1035. [CrossRef]

 *sustainability*

Article

Geomorphological Map and Quaternary Landscape Evolution of the Monfragüe Park (Cáceres, Spain)

José Luis Goy, Raquel Cruz, Antonio Martínez-Graña *, Virginia Valdés and Mariano Yenes

Department of Geology, Faculty of Sciences, Plaza de la Merced s/n, University of Salamanca, 37008 Salamanca, Spain; joselgoy@usal.es (J.L.G.); rqcruz@usal.es (R.C.); vvaldes@usal.es (V.V.); myo@usal.es (M.Y.)
* Correspondence: amgranna@usal.es; Tel.: +34-923-294546

Received: 20 September 2020; Accepted: 2 December 2020; Published: 3 December 2020

Abstract: From the geomorphological cartography, the geometric and spatial distribution of the quaternary forms and deposits are analyzed, with special relevance to the fluvial terraces that allow obtaining the chronology of the successive landscape changes of the course of the Tagus River attributed to the activity of the Fault of Alentejo-Plasencia (APF). The "Appalachian" relief of Monfragüe National Park, constituting a series of quartzitic combs with direction NW, between which they find slopes, hills and valleys following the same direction, for the dismantlement of the Cenozoic cover that was covering the substratum (still present in the central sector) and encasement of the Rivers Tagus and Tiétar. The remains of fluvial terraces inside and outside the Park stand out at different heights and so they originate from different times and show different landscapes along the routes of the Tagus river and its movement over time. In the north end (basin of the Campo Arañuelo), there are remains of ten fluvial terraces of relative importance attributed to the River Tagus (with heights relative to the thalweg between 120 and 20 m). In the south edge, there are eight levels attributed to a former fluvial drainage network, which assimilates to the River Tagus, with the more recent level reaching over 280 m on the current river. Neotectonics readjustments that rejuvenated the relief produced the elevation of the socle and cover, at the time of diversions in the path of the fluvial network, up to the structure and encasement (for supertax and/or antecedence). During the Quaternary, the activity of the Alentejo-Plasencia Fault (APF) has given rise to palaeogeographic changes in the fluvial valley of the Tagus River. During the ancient Lower Pleistocene, its course passed south of the current one (Talaván-Torrejón el Rubio basin); at the end of the Lower Pleistocene, it came out crossing the syncline through the Boquerón porthole, and the meander that bordered the town of Almaraz was abandoned; at the beginning of the Middle Pleistocene, it changes its direction, from NE–SW to SE–NW, leaving the porthole and joining the Tiétar river within the Park; later it moves somewhat to the south. These changes in the route and the anomalous fitting of the course of the Tagus River into the Paleozoic substrate, have been attributed to the APF, which, through impulses, has had a great activity from the Lower Pleistocene to the Middle Pleistocene.

Keywords: geomorphological map; Appalachian landscape; neotectonic; drainage network; superimposition-antecedence

---

1. Introduction

The Monfragüe National Park (MNP), located in the center-west of the Iberian Peninsula, represents an area to the north of the province of Cáceres, in the shape of an arch, which belongs to the geological unit of the Iberian Massif and within it to the Central-Iberian Zone. Its limits span to the north to the Campo Arañuelo Basin, to the south with the Extremeña Peneplane and the Talaván Basin, to the east with Las Villuercas and Montes de Toledo, of which it forms part, and to the W with the Neogene Basins of Cañaveral and Plasencia, originated by the Alentejo-Plasencia Fault (APF) (Figure 1).

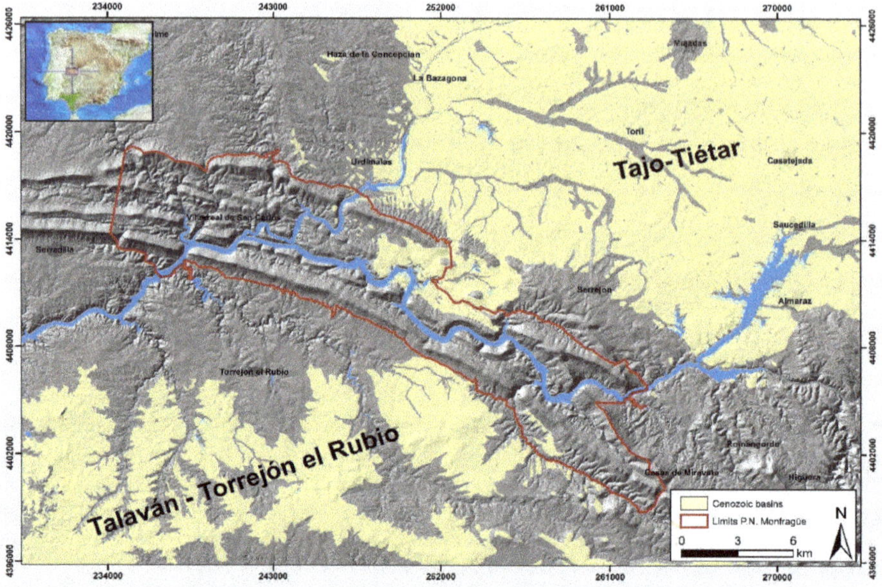

Figure 1. Location of the Monfragüe National Park.

The relief consists of a set of elongated mountain ranges and valleys with a predominant NW–SE direction giving rise to a type of "Appalachian" modeling, characterized by successive alternations of quartzites, sandstones and slates of the Paleozoic series, folded in the Variscan orogeny, devastated during the Mesozoic and rejuvenated in the alpine and post-alpine cycle.

At the level of geological formations, it falls within the geological unit of the Hesperian Massif [1], currently the Iberian Massif within the Vertical Folds Domain of the Central-Iberian Zone [2]; it is composed mainly of metamorphic rocks of the Paleozoic series (Ordovician and Silurian) folded and fractured during the Variscan orogeny, which is covered in some sectors by younger Cenozoic and Quaternary sedimentary materials. The whole complex was reactivated during the Alpine orogeny. The relief is related to differences of erodibility owing to the alternation and arrangement of its materials, the neotectonics and quaternary morphogenetic processes [3–5]. Especially noteworthy are the series of mountainous alignments, which stand out on the wide Extremadura peneplain, and the water network that runs through it, whose waters flow directly or indirectly to the Tagus River.

The environmental characteristics of this area have served for its declaration as a Natural Park (1979), Spatial protection zone for birds (ZEPA) (1998), Biosphere Reserve (2003) and finally a National Park (2006). This recognition of ecological values has not been followed by the recognition of its geological values, nor the geomorphological character of its landscapes.

The fundamental elements of the relief (ridges, hills, valleys and surfaces), are accompanied by other quaternary geomorphological units: colluviums, scree and foothills (linked to ridges and escarpments, waterfalls, portholes, river terraces, alluvial fans and glacis, related to the river network of the Park.

In the last decade, the detailed analysis of geomorphological cartography [6–10] aided by absolute and relative dating techniques, allows establishing temporal sequences that enable the interpretation of the changes in the relief and therefore the landscape generated in each temporal stage from the field study of the modeling and structures generated by external geodynamic agents (water, wind, ice, etc.) and the processes and genesis of the existing forms on the ground (erosion and deposit).

From the analysis of sequences in cross-sectional profiles of the river valleys of the rivers of the Neogenes basins of the Iberian Peninsula, mainly the Tagus basin and Duero basin, general sequences of river terraces with their topographic height in relation to the fluvial course (thalweg) are elaborated, correlating the rivers of the same basin and similar basins such as the Tajo basin. In parallel, the terraces are dated using different methods (isotopic, radiogenic, chemical-biological, geomorphological-edaphic, etc.) and correlations (paleontological with macro-microfauna and pollen, archaeological, paleomagnetic, and stable isotopes to correlate paleoenvironments). The most relevant articles, among others, that have been taken into account refer to sequences for the Iberian Peninsula [11], Tagus basin sequences [12–14] and Duero basin sequences [15–18] and sequences for Tagus-Duero basins [19,20]

The objective of this work is to carry out a detailed geomorphological cartography, which allows the interpretation of the different landscapes of the Quaternary and to determine its palaeogeographic evolution. Based on the spatial position of the quaternary fluvial deposits (terraces) and their correlation with sequences from the same Tagus Basin or with basins whose behavior is similar (Duero Basin), create a relative chronology that helps us to determine the ages of the paleogeographic changes of the river course of the Tagus river and as a consequence of the moments of activity of the Alentejo-Plasencia Fault.

The importance of this work lies, in addition to its geomorphological cartography, in that it refers to important changes in the Quaternary paleo-landscape, pointing out five major paleogeographic changes in the river valley of the Tagus River, which occurred in a reduced time interval, since the Lower (Middle) Pleistocene to the Middle Pleistocene (ancient), due to the activity in this period of time, of the APF.

2. Materials and Methods

To prepare the geomorphological map, a bibliographic review of the existing works was carried out, accompanied by field and office tasks (photointerpretation and GIS analysis), and along three stages (Figure 2):

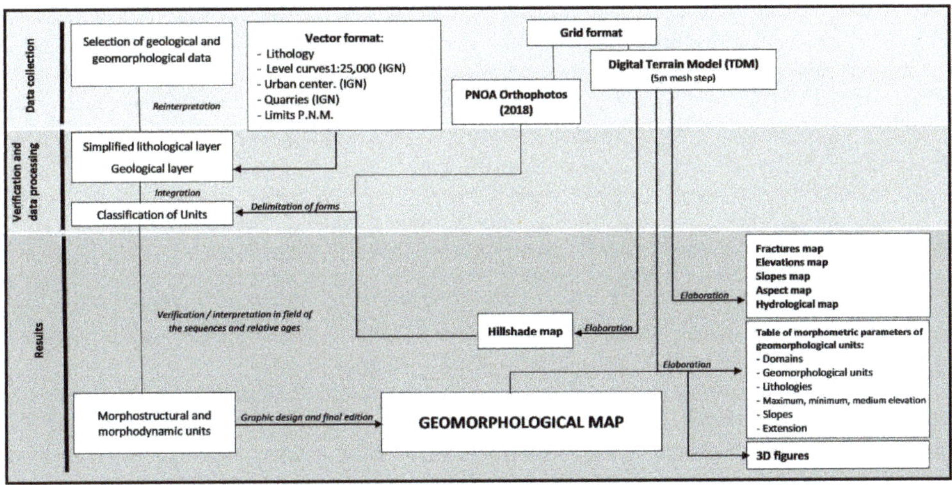

Figure 2. Methodology chart.

Stage 1: The geological framework was analyzed [21–29], with the description of the different lithologies and structures that it comprises and the elaboration of the simplified geological layer, using the geological units that are needed for the geomorphological map. From here, the geology of the area is reinterpreted, especially the part corresponding to the Quaternary deposits.

Stage 2: Preparation of the map of Geomorphological Units from existing works [30–34], especially the cartographic ones that bring us closer to the different units of the relief [4,5], then performing the geomorphological photointerpretation and the necessary field survey.

In the representation of the landforms in the legend of the Geomorphological Map, the erosive forms are separated from the depositional forms: fluvial (streams), gravitational, morphostructural, lacustrine (marsh), alteration and mixed (polygenic), giving them a geomorphological symbol with the characteristic color of each morphogenetic system common in this type of cartography. Light colors are chosen, for better contrast with the surface patterns.

Another aspect to consider is the analysis of the Quaternary deposits from the point of view of their age (chronology) based on the sequences of the fluvial terraces of the two rivers that cross the area (Tagus and Tiétar), obtaining a relative chronology comparable to those deduced in other areas of the Tagus and Duero Basins, where paleomagnetic, radiogenic (ESR, TL, OSL), isotopic (cosmogenic nuclides), paleontological dates, etc. [12,17–20,35], geomorphology (heights of the terraces, soils, cementations, etc.) and correlations based on the heights of significant scarps [36] data are available.

In the study area, these deposits have been differentiated as Lower, Middle, Upper Pleistocene and Holocene ages. To represent this chronology, the methodology of the Geological and Mining Institute of Spain has been used, which allowed us to give a relative age to both the deposits and associated forms.

Within the geomorphological map, although they are important, we have not used frames to differentiate neither the various substrates, nor the tectonic signs for the geological structures (only some faults, with a dashed line), so as not to complicate their reading, lithology and fractures maps. In this last map it can be observed how fractures favor the entrances and exits of the rivers to the Park through fluvial gorges (portholes). The other auxiliary maps (elevations, slopes and aspect) complement other aspects of geomorphology, such as characterizing the relief and including the basic toponymy. The next maps will allow us to appreciate clearly where the drainage network fits more or less, and relate it to the areas that have suffered a greater neotectonic uplift, which can be seen in the NW and SE of the Park as compared to the center. The last auxiliary map is used for analyzing recent (active) processes and their relationship with the climate; in this case, it is observed that the dominant orientations are NE and SW, which condition soils, vegetation and landscapes (Figure 3).

Stage 3: The geomorphological layer, involved the detailed digitization of the Park and its surroundings at 1:50,000 using ArcGIS. The spatial resolution of the DTM of 5 m and National Plan for Aerial Orthophotography (PNOA) Orthophotos of maximum actuality at 1:10,000, allowed a greater cartographic detail.

Process sequences are established, based on the geometric and spatial arrangement of the deposits with respect to the terrace levels, using the altimetric data from the Terrain Digital Model—TDM. Each group of forms with the same genetic origin (structural, fluvial, polygenic, gravitational, lake and alteration) is represented with a color, using the internationally recognized colors for geomorphological units. The shadow map is used as background, accompanied by a selection of contour lines, the drainage network and some toponymy elements (municipalities and quarries) extracted from the national topographic maps at a scale of 1:25,000 of the National Geographic Institute.

From the TDM, complementary maps that provide information have been generated to characterize units and interpret the evolution of the regional relief. Finally, the limit "of the area of the MNP and zone of influence" is incorporated. After the integration of all layers, the map of geomorphological units is generated, which records the presence, distribution and extension of the morphostructural and morphogenetic units, related to the neotectonic and morphodynamic processes that have operated in the region since the Late Neogene and, especially, during the Quaternary and, eventually, define the landscape evolution during time (Figure 4).

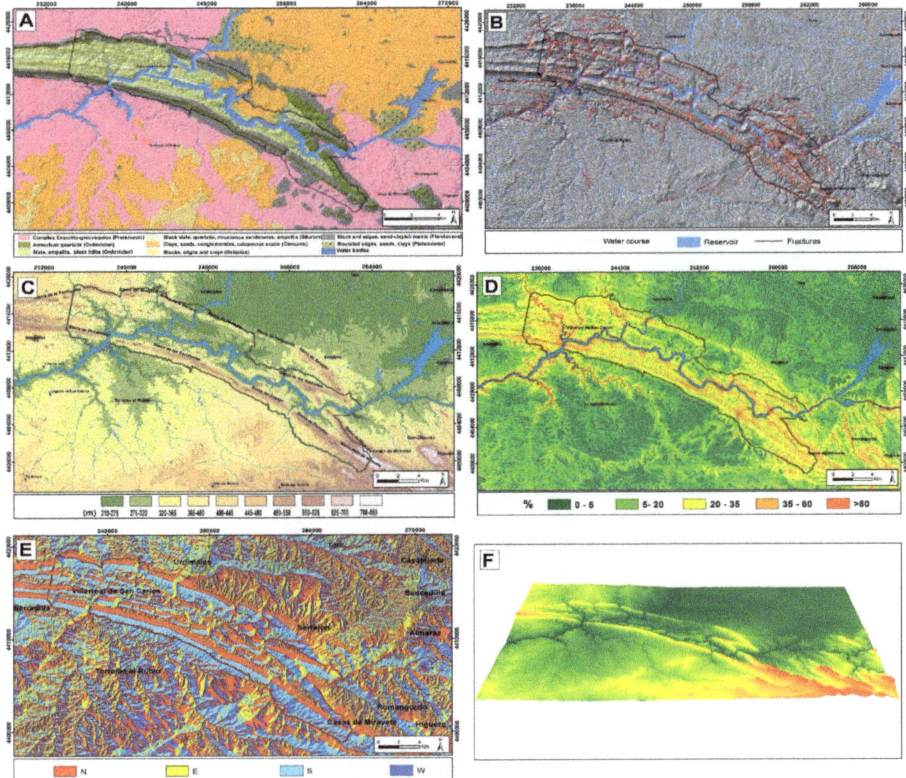

Figure 3. Auxiliary maps: Lithology (**A**) and fractures (**B**), elevations (**C**), slopes (**D**) aspect (**E**) and TDM with higher altitude zones in red, middle altitudes in yellow and lower altitude in green (**F**).

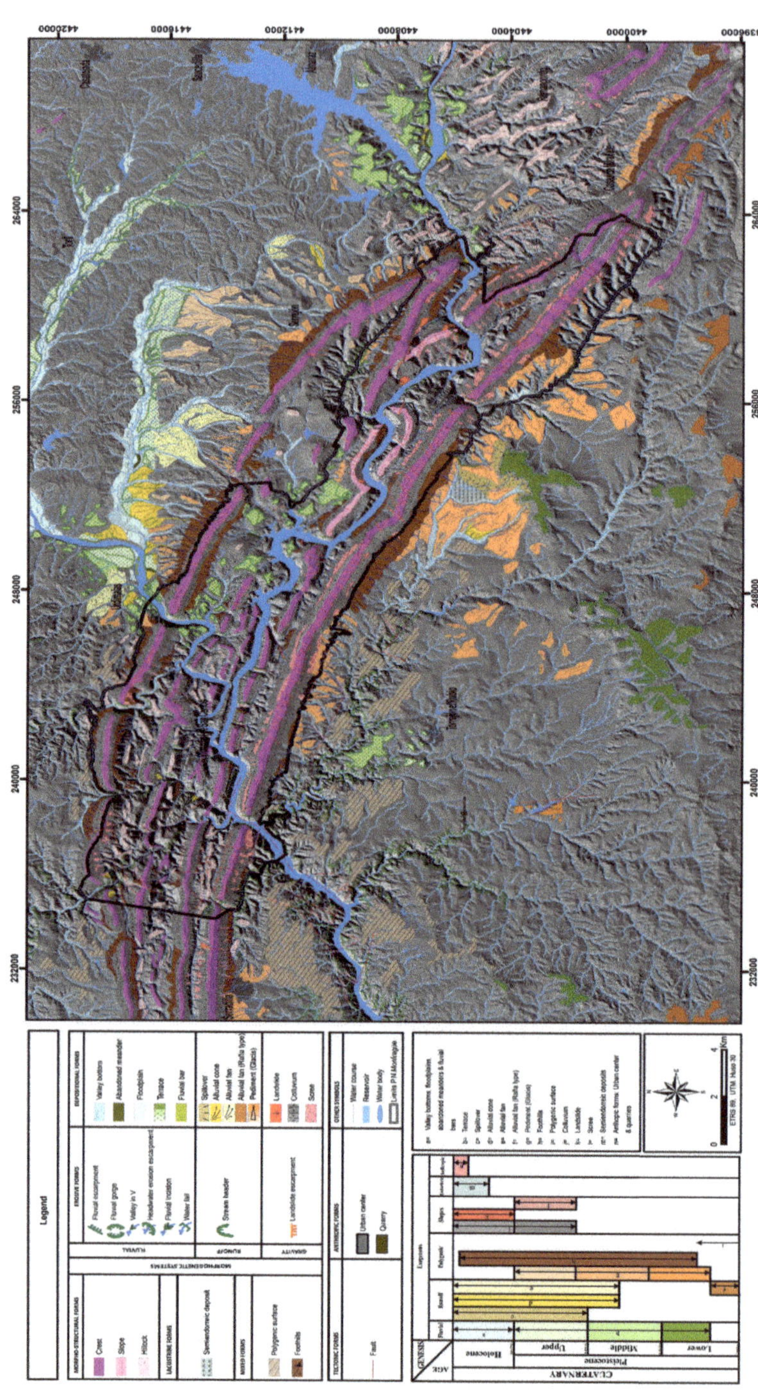

Figure 4. Geomorphological map.

3. Results

The geomorphological cartography (Figure 4) obtained from photointerpretation and field work allows an analysis of geomorphological features that can be synthesized in the following sections:

3.1. Morphostructural Modeling

This modeling is dominated by an "Appalachian" type relief that is the result of differential erosion of the ancient relief, presenting as a general characteristic three highly competent quartzite sequences, which constitute a large syncline in E–W direction with a significant convergence towards the south. These sequences they give rise to the most pronounced reliefs (ridges) with heights above 500 m for the Armorican Quartzite and lower values, around 400 m and 300 m, for the Caradoc and Criadero Quartzites, respectively. These are separated by less competent materials (sandstones and slates), which favor slopes, hills and river valleys following the same synclinal pattern (Figure 5). Partial individualizations of the general model are observed, due to bifurcations of layers, descents and/or elevations of the edges of the folds, asymmetry of these, and the existence of NE–SW faults that displace the layers (examples of Loma del Diablo and Sierra de Herguijuela), which modify the alignment of the mountains, singling out the modeling, which acquires its own pattern. The fracture systems, controlled by alpine and neotectonic movements, have affected the general curvature of the park and the development of the river network (Figures 6 and 7A).

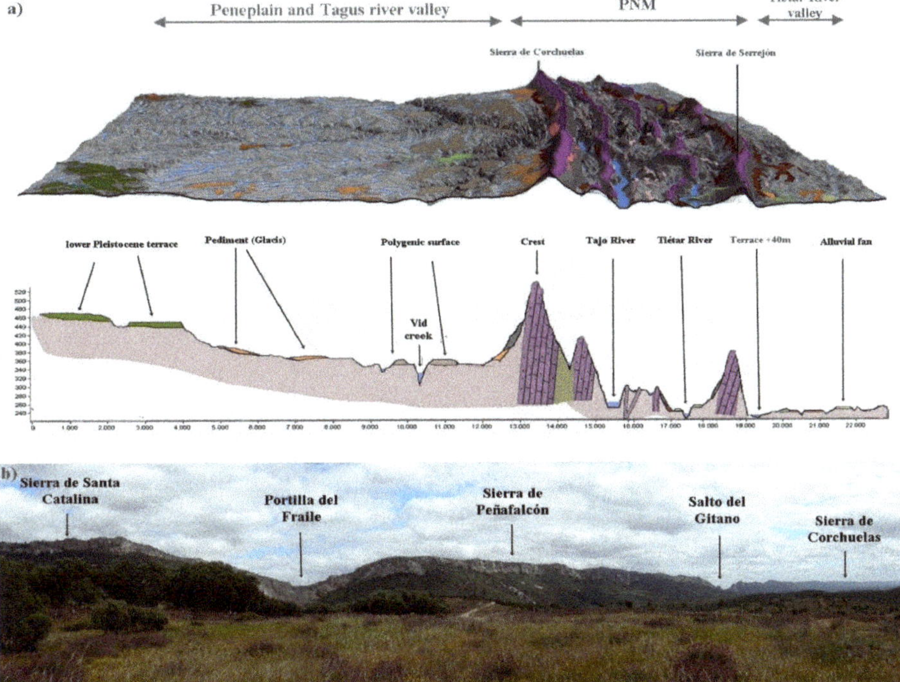

Figure 5. (a) N–S section of the Monfragüe National Park (MNP) in which the verticalized Ordovician synclinal arrangement can be seen (Tajo = Tagus). (b) Quartzite ridges of the Sierras de Santa Catalina, Peñafalcón and Corchuelas, separated by the portholes of El Fraile and Salto del Gitano, protruding from the Extremadura peneplain.

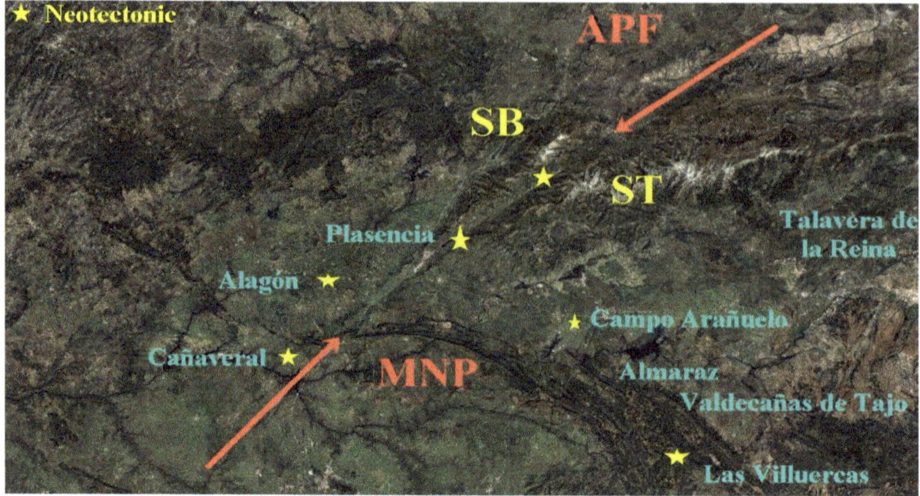

Figure 6. Satellite photo showing the location of neotectonic indicators in nearby areas. Names in yellow: SB—Mountain Range of Béjar, ST—Mountain Range of Tormantos.

Neotectonics are reflected in geomorphological indicators of nearby areas (Figure 6). The western sector of the Park is affected by the crossing of the Alentejo-Plasencia Fault (APF) that affects the Paleozoic materials, curving them convexly to the north and that Villamor (2002) [33,34] considers as a segmented synaesthetic tear failure, with evidence of movements during the Quaternary and low displacement rates, less than 0.1 (probably between 0.01 and 0.05 mm/y). Other indicators are the tilted Neogene deposits of the Plasencia and Cañaveral pull-apart basins [33], the lower and middle Pleistocene fractured river terraces in the Cañaveral basin [34]; the geomorphological anomalies indicative of recent tectonics [35,36] affecting the terraces of the Alagón River, NW of the Park, with anomalous elevations (+218 m, Cerro Marifranca), for the oldest terrace (beginning of the Lower Pleistocene); its slope: in favor (+70 m-Mesa del Val), beginning of the middle Pleistocene and counter slope (40–45 m, Los Llanos) in the middle of said Pleistocene, which promote lagoons due to poor drainage; changes in the direction of the course of said river, from NE–SW during the Lower Pleistocene and the middle of the Middle Pleistocene, to N–S; the 90° turn (from N–S to E–W) of the Alagón del Río riverbed when approaching the APF is also anomalous. The right-angle turn of the Jerte river in Plasencia (NE-SE to SE-NE), and the lack of old levels is anomalous compared to the Alagón river a few kilometers away; the tectonized terraces in the Jerte Valley and the asymmetry of the valley, a greater number of terraces on its right bank, when moving towards the fault (to the SE); the relief of Las Villuercas reactivated in the Pliocene, the deformed Paleogene and Neogene deposits in the Campo Arañuelo basin and the displacement of the Tiétar River to the NW, as a consequence of the sinking of the northern part of this basin [11]. According to this activity it has been included as an active fault in the database of active faults of the Iberian Peninsula.

From the fracture map (Figure 3B), it can be deduced that the most abundant fault systems are the NNE–SSW to NE–SW and the N–S, and to a lesser extent the E–W and NW–SE. Many of these fractures condition the drainage network by rectifying water courses, giving rise to anomalous elbows, diverting ridges and hills, causing fault scarps, faceted faces and river gorges at the entrance and exit of the rivers to the Park (Gorges or Fluvial Portholes) (Figure 8).

Figure 7. (**A**) General curvature of the synclinal of Monfrague; (**B**) Corzo porthole and entrance of the Tagus river to the synclinal; (**C**) Boqueron porthole and old outlet of the Tagus river to the Talaván basin in favor of fractures; (**D**) Fraile porthole showing incision for movement of fracture E–W; (**E**) Raña of Jaraicejo to +320 m in Talavan basin; (**F**) Tagus river terrace to +232 m in Talavan basin; (**G**) terrace of the river Tiétar of middle Pleistocene age to +80 m in Tagus-Tietar basin; (**H**) blocks by gravitational processes that show periglacial conditions in Sierra de Piatones.

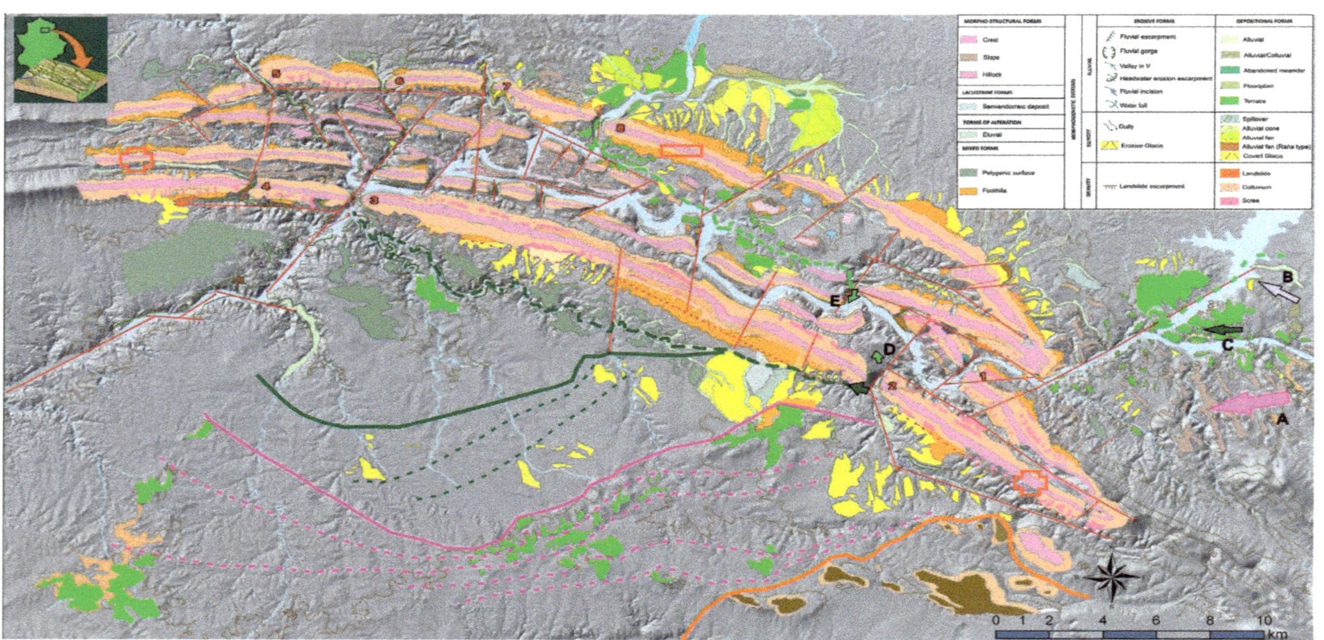

Figure 8. Quaternary morphostructure and evolution of the Tagus River channel. Colored lines and strokes represent river positions and ages (orange: Early Pleistocene; purple: Lower Pleistocene, and green: Middle Pleistocene), arrows indicate the direction of river movement. The signs + (uplift) and - (subsidence) indicate the movement of the blocks. Main portholes: 1—Salto del Corzo, 2—Boquerón Porthole, 3—Tagus Porthole (Salto del Gitano), 4—Fraile porthole, 5—Barbaón porthole, 6—Serrana porthole, 7—Calzones porthole, 8—Tiétar porthole. Stages: A. Abandonment of the Talaván-Torrejón el Rubio Basin (Early Lower Pleistocene). B. Formation of the Almaraz meander (Modern Lower Pleistocene). C. Displacement of the river towards the south and abandonment of the meander of Almaraz (Middle Ancient Pleistocene). D. Change in direction of the Tagus River within the park, from NE–SW to NW–SE (Early Middle Pleistocene). E. Abandoned channel of the Tagus River, within the Park and displacement of the course to the south: current channel (Middle Modern Pleistocene).

3.2. Quaternary Modeling

Throughout the Quaternary, fluvial dynamics played a major role, although at certain times the gravitational processes favored by periglacial environments also contributed.

3.2.1. River Modeling

The fluvial morphogenetic system is formed by the Park's water network, made up of the Tagus and Tiétar rivers, streams and tributary gorges (Malvecino, Barbaón, Calzones, de la Vid, del Cubo, etc.), clearly conditioned by the lithology and the fracture systems, presenting as a whole a "lattice drainage" pattern (Figures 3B and 4). All the courses run incised in the layers of shale and sandstone, forming V-shaped valleys, with river scarps marked in certain sectors (Figure 5).

The gorges (portholes) of La Garganta or Fraile (Figure 7D), affected by a recent fracture that has originated an important tectonic step which stands out, that of Tiétar (at the entrance of the river) and these of Corzo (Figure 7B) and Salto del Gitano (Tagus porthole) exit of the Tagus and Tietar Rivers, respectively (Figures 4 and 8). In addition, the change of direction of the La Garganta and Cubo streams, which border the park's SW flank parallel to the Armorican quartzite, incised in the Extremadura peneplain, and turn towards the interior, crossing the Armorican quartzite (Boquerón porthole) (Figure 7C) with an S-N course. These changes in direction have been favored by an NE–SW fracture and the uplift of the block located SE of the Park (Figure 8)

During the Quaternary, the rivers were deeply incised in the underlying rocks and currently there are hardly any remains of their deposits (terraces) within the synclinal megastructure (only in the central-eastern sector) where some hanging terraces have been detached from the current channels, which would imply a change in the course of the Tagus River during the Quaternary (Figure 4).

The most significant fluvial forms in the cartography are the fluvial terraces (Tagus and Tiétar rivers) (Figure 7F,G), alluvial fans and pediment (glacis). The Tagus River is the main artery that crosses the Park from SW to NE, from Salto del Corzo to the Tagus porthole (Salto del Gitano). The sequences of terraces have been analyzed before entering the MNP, inside and after leaving it, synthesizing them in 18 levels, encased in the system of ancient alluvial fans (beginning of the Quaternary), called "Rañas" (Figure 7E) that constitute the beginning of the river encasement, as Neogene basins became exoreic. The oldest terraces are found in the Talavan-Torrejon el Rubio Basin, in the southern sector, outside the Natural Park, where eight levels have been recognized (six of them recorded in this cartography and two, immediately to the west, to the S of Cañaveral) (Figures 4 and 8). Nowadays they are detached from the channel as remains of a previous course of the Tagus river during the Lower Pleistocene, before the present emplacement.

On the edges of the Neogene basin (S of Campo Arañuelo), before entering the Park, to the east and SW of Almaraz, there are remains of ten relatively large river terraces that we have related to the Tagus River (with elevations relative to the talweg between +120 and +20 m), and ages ranging from the Upper Lower Pleistocene to the Upper Pleistocene. The old terraces (on both banks) of this section do not correspond to the Tagus valley, which is an abandoned meander of the east river (Arrocampo creek)

The levels within the Park crop out between +70 and +20 m correspond to the six most recent of the previous section. Between the levels at +70 m and +60 m there is an abandoned channel (Figures 4 and 8H).

General sequence: Lower Pleistocene: Raña- +310 m; Terraces: T1- +288 m; T2- +275 m; T3- +260 m; T4- +243 m; T5- +232 m, T6- +200 m; T7- +160 m; T8- +150 m; (first change of the river course), T9- +120 m; T10- +112 m; T11- +100 m; T12- +90 m; (second change); Middle Pleistocene: T13- +70 m; (third change); T14- +60 m; T15- +50 m; T16- +40 m; Upper Pleistocene: T17- +30 m; T18- +20 m. The three sectors analyzed correspond to changes in the river course.

The Tiétar River has a sequence of 12 levels that range from late Lower Pleistocene to Holocene (Figures 4 and 8).

General sequence: Lower Pleistocene: T1- +100 m; T2- +90 m; T3- +80 m; T4- +70 m; Middle Pleistocene: T5- +60 m; T6- +50 m; T7- +40 m; T8- +30 m; Upper Pleistocene: T9- +20 m; T10- +15 m; T11- +10 m; Holocene: T12- +4 m.

The other relevant fluvial forms (mapped) are the alluvial fans and cones and the pediment (glacis) (Figures 5 and 7). They are deposits originated by non-channeled waters and among the first we must point out the "rañas" that, as said before, are dated as Early Pleistocene (Gelasian) and serve as the origin of the fluvial sequences. This deposit is found at +310 m in the eastern sector of the Talaván-Torrejón El Rubio basin, discordant on the Cenozoic arkoses and on the Proterozoic and Paleozoic substrate (outside the Park) composed of sub-rounded wedges of quartzite and quartz with a red, sandy-clay matrix. The alluvial cones and fans of Middle, Upper Pleistocene and Holocene ages are related to the Tiétar river valley to the north of the syncline, composed of sub-rounded clasts of quartz, granite, and schists in a sandy-clay matrix.

The most representative systems of cover glacis are found in the eastern part of the Talaván-Torrejón El Rubio Basin, also associated with the slopes of the Sierras de las Conchuelas, del Espejo and Pico de Miravete and to the north hills of the Sierras de Serrejón and Herguijuela. In the first case, the change in the course of the Tagus River, in Lower Pleistocene times, favored the formation of three encased glacis, associated with secondary streams (they represent connecting slopes) such as El Retuerta and La Vid. These deposits consist of a sandy-clay matrix with wedges of sub-rounded quartzite clasts.

3.2.2. Gravitational Modeling

The slopes of the quartzite ridges are covered by colluvium and scree, characteristic of the Pleistocene and Holocene periglacial environments.

The colluviums correspond to accumulations of medium and small quartzite, slate and sandstone cobbles encased in a sandy-clay matrix, downslope of the ridges which experienced mass displacement. Up to four sequences can be distinguished, related to different moments of the Quaternary (Upper Middle Pleistocene and Holocene).

The screes "Canchales term" are accumulations of medium-sized size (30 cm to 10 cm) angular blocks of quartzite, with little or no matrix, derived from the steep quartzite ridges, and arranged on colluviums during the Upper Pleistocene. They occur especially in the southern sector. The study and statistical treatment of these materials reveals a preferential orientation of major axes (NE/SW, NW/SE); suggesting that they are related to reactivations of the APF, which cause the reactivation of the stress fields [4]. Arched "lobes" and "detachment scars" are observed in the middle and lower parts of these formations, indicating post-depositional gelifluidal movements. All this points to a freeze–thaw environment typical of the cold periods of the Holocene (Figure 4).

On the surface of the peaks of the Sierra de Piatones, a "field of very angular blocks" was formed under periglacial conditions (freeze–thaw processes), by cryoclasty and gelifluction (Figure 7H).

3.2.3. Polygenic Modeling

On the slopes of the Sierras, to the north and south of the synform, there are deposits of mixed genesis partly fossilized by colluviums. Owing to their position and the type of transported material they have been called Piedmonts (Piedemontes). These are detrital deposits related to more or less steep quartzite reliefs, whose formation has been influenced by gravitational and stream processes. They appear associated with colluviums with ages ranging from late lower Pleistocene to Holocene.

In the Talaván-Torrejón el Rubio Basin, and to the NW of the Park, there are preserved remains of the Fundamental Penillanura. They consist of an old surface that erodes the Paleozoic materials of the schist-grauwaque complex, with a very gentle slope, caused by more than one morphogenetic process (Figure 7H).

4. Discussion: Paleogeographic Evolution

From the analysis of the cartography and the geomorphological features, the existence of major neotectonic movements can be deduced. These caused the displacement and uplift of the Monfragüe syncline, possibly between the end of the Neogene and middle, Middle Pleistocene (Figure 8). These explain the anomalous arrangement of terraces, the absence in the Park of ancient terraces of the Tagus river, the preservation of the Neogene cover in its central sector, the presence of abandoned channels within the Natural Park (witnessing changes in the main channel); the orientation of blocks in the scree areas and the Appalachian relief exhumed during the Quaternary could hardly be explained by climate change only. Figure 8 shows that, during the Lower Pleistocene, the Tagus River was located to the south of the present-day channel, close to the "raña" deposits, to the SE of the study area, near the current divide of the Tagus and Almonte, and that it was displacing towards the NW at that time. This figure also shows that, at the beginning of the Middle Pleistocene, the Tagus River channel was closer to the current river course, but not yet corresponding with it, producing minor changes before the accommodation in the underlying materials.

From the observation of the fracturing that affects the MNP materials, it can be deduced that the entire syncline that constitutes it has undergone bulging and curvature (Figures 4 and 8), due to the movement of the APF, originating radial-type fractures (NNE–SSW and NNW–SSE) abundant in this sector, compartmentalizing ancient geological materials into blocks, which gave rise to movements of different magnitudes (Figure 8).

Another aspect to consider is the fluvial incision rates of the Tagus River in this area, comparing them with those of Talavera de la Reina, (approximately 100 km to the East), belonging to the Campo Arañuelo sub-basin (Figure 6). The elevation of the ancient alluvial fans "rañas" in relation to the current riverbed (talweg) in both areas have been taken as a reference. In the MNP it is at +310 m (Jaraicejo) and in Talavera de la Reina (Malpica) at +220 m above the current channel. As the age attributed by us to these deposits is Early Quaternary (Lower Pleistocene-Gelasian) 2588 Ma, the calculated values are 0.119 m/Ka for the Park area and 0.085 m/Ka for the Talavera de la Reina area. This difference (3.4 cm/y) is significant, almost 30% more, especially if we take into account that the geological substractum where the river is encased is more competent in the Park area (slates, sandstones and quartzites) than in the Talavera sector (arkose), so the difference is interpreted as a neotectonic effect.

These paleogeographic changes have been due to tectonic readjustments of the variscan fractures, reactivated in the Alpine and subsequent neotectonic movements, which generated displacements, uplifts, and tilts, that have favored the changes in the course of the Tagus River. As to the origin of the embedding of the river in the Appalachian relief, we propose a generalized uplift of the base materials during the Upper Pliocene and Quaternary (antecedents) since the river is established on the Neogene materials, and, when the neotectonic uplift occurs, it forces the materials to fit into said socket (superimposition), at the same time that erosion dismantles the Cenozoic materials, exposing those of the underlying Paleozoic.

The main changes that the Tagus river has undergone are: (Figures 4 and 8)

1. During the Lower-Ancient Pleistocene, the river flowed further south, in a general south-western direction, depositing a succession of terraces (T1- +288 m to T8- +150 m), in the Talaván-Torrejón el Rubio Basin, which are unrelated of the current course and river dynamics.

2. In the Lower-Modern Pleistocene, the river course bordered the town of Almazán, forming a wide meander, currently occupied by the Arrocampo creek, with four levels of terraces on both margins (T9 + 120 m to T12 + 90 m). The river abandoned the meander between the Lower Pleistocene and the beginning of the Middle Pleistocene moving to the north of Valdecañas de Tajo, and rectified its course to an east–west direction.

3. At this time, the river crossed the Park leaving it through the Boquerón porthole, attached to the relief and headed towards Torrejón el Rubio, where remains of this system of terraces are found

near the town, to go through the current Retuerta river to converge with the Tiétar river that left the MNP through the Salto del Gitano. This old course had a path parallel to the current one.

4. Between the Lower Pleistocene and the beginning of the Middle Pleistocene (as marked by terraces T12-90 m and T13-70 m), the Tagus River changed direction from NE–SW to SE–NW, in the Boquerón porthole, as a consequence of the uplift and tipping towards the north of the block formed by the Sierras de Piatones and Pico de Miravete (to the south) and Miravete (to the north), to join the Tiétar to the north of its current position. This can be deduced from the layout of the terraces, located north of its channel, and further confirmed by observing Figure 5a, which shows, for this section of the Park, a greater fit (age) of the Tiétar river bed than that of the Tagus river. This implies that the Tiétar River has maintained its course within the Park since the Lower Pleistocene and exited the Park area through the Salto del Gitano. In contrast, the Tagus River occupied its place around the Middle Pleistocene.

5. During the Middle–Late Pleistocene (T13- +70 m-T14- +60 m), new movements forced the abandonment of the channel and the Tagus river moved to the south.

5. Conclusions

The elaborated geomorphological cartography, together with the auxiliary maps, has allowed a global vision of the evolution of the Quaternary landscape, from the analysis of the different forms, their geometric and spatial distribution, as well as the morphogenetic processes that have originated them, especially fluvial and morphostructural morphogenesis. This analysis makes it possible to elaborate a relative sequence of Quaternary deposits and forms.

During the Neogene, arcosic materials covered the structure of the Monfragüe synclinal that can still be seen in the north-central sector of the interior of the synclinal. This indicates a later dismantling of these materials due to the erosion caused by the fluvial networks of the Tagus and Tietar rivers, favored by the uplifting of the structure during the Quaternary.

The origin of this uplift is attributed to the reactivation of variscal fractures during the Alpine orogeny and by neotectonics (Pliocene and Quaternary) as well as the activity during all this time of the APF. The movement of this fracture has given rise to the northward curvature of the synclinal structure, its uplift and tipping of blocks, favoring the incision of the current river network and the displacement of the Tagus River.

The landscape of the current relief in the MNP has its origin in the lifting of the Paleozoic materials during the Pliocene and Quaternary, the installation of the drainage network (antecedent process) on the Neogenes materials and its incision on said materials to generate the superimposition of the river in the Paleozoic materials.

The spatial distribution of the terraces of the river Tagus (before and after the syncline) allows us to locate (in time) the changes in the course of the river's fluvial course during the Pleistocene. There have been five major changes in the route of the River Tagus between the Upper and Middle Pleistocene.

The incision rates of the river course of the Tagus River and its large incision in the Paleozoic substrate (30% higher than the incision of the same river as it passes through the town of Talavera de la Reina 100 km to the east and upstream in the same basin) allow us to deduce a rise by means of impulses by the activity of the APF during the Quaternary.

The time of greatest neotectonic activity in the park occurred between the early Middle Pleistocene and the early Middle Modern Pleistocene.

Author Contributions: Conceptualization, J.L.G. and R.C.; methodology, J.L.G., R.C. and A.M.-G.; software, V.V.; validation, J.L.G., R.C. and A.M.-G.; formal analysis, J.L.G.; investigation, J.L.G., R.C. and A.M.-G.; resources, V.V.; data curation, J.L.G., R.C. and A.M.-G.; writing—original draft preparation, J.L.G., R.C. and A.M.-G.; writing—review and editing, J.L.G., R.C. and A.M.-G.; visualization, J.L.G., R.C. and A.M.-G.; supervision, A.M.-G. and M.Y.; project administration, A.M.-G.; funding acquisition, A.M.-G. All authors have read and agreed to the published version of the manuscript.

Funding: This research received no external funding.

Acknowledgments: This research was methodologically helped by project Junta Castilla y León SA044G18.

Conflicts of Interest: The authors declare no conflict of interest.

References

1. Hernández, P.E. *Síntesis Fisiográfica y Geológica de España*; Trabajos del Museo Nacional de Ciencias Naturales: Madrid, Spain, 1934; p. 584. (In Spanish)
2. Diez Balda, M.A.; Vegas, R.; González-Lodeiro, F. Structure of the Central Iberian Zone. In *Pre-Mesozoic Geology of Iberia*; Dallmeyer, R.D., Martínez García, E., Eds.; Springer: Berlin/Heidelberg, Germany, 1990; pp. 172–188.
3. Gumiel, P.; Campos, R.; Segura, M.; Monteserín, V. *Guía Didáctica del Parque Natural de Monfragüe*; Junta de Extremadura: Mérida, Spain, 2003. (In Spanish)
4. Soto Alonso, S. *Cartografía Geomorfológica del Parque Nacional de Monfragüe, (Proyecto Fin de Carrera (Inédito))*; University Alcalá de Henares: Madrid, Spain, 2006. (In Spanish)
5. Goy, J.L.; Cruz, R.; Martínez-Graña, A.; Zazo, C. Geomorfología del Parque Nacional de Monfragüe: Cartografía y evolución Cuaternaria. In Proceedings of the XIII Reunión Nacional de Geomorfología, Cáceres, Spain, 12 September 2014; pp. 299–302. (In Spanish).
6. Benito-Calvo, A.; Pérez-González, A. Erosion surfaces and Neogene landscape evolution in the NE Duero Basin (north-central Spain). *Geomorphology* **2007**, *88*, 226–241. [CrossRef]
7. Forte, F.; Pennetta, L. Geomorphological Map of the Salento Peninsula (southern Italy). *J. Maps* **2012**, *3*, 173–180. [CrossRef]
8. Karymbalis, E.; Papanastassiou, D.; Gaki-Papanastassiou, K.; Tsanakas, K.; Maroukian, H. Geomorphological study of Cephalonia Island, Ionian Sea, Western Greece. *J. Maps* **2013**, *9*, 121–134. [CrossRef]
9. Migiros, G.; Bathrellos, G.D.; Skilodimou, H.D.; Theodoros, K. Pinios (Peneus) River (Central Greece): Hydrological-Geomorphological elements and changes during the Quaternary. *Cent. Eur. J. Geosci.* **2011**, *3*, 215–228. [CrossRef]
10. Pérez-González, A.; Gallardo-Millan, J.L.; Uribelarrea del Val, D.; Panera, J.; Rubio-Jara, S. La inversión Matuyama-Brunhes en la secuencia de terrazas del río Jarama entre Velilla de San Antonio y Altos de la Mejorada, al SE de Madrid (España). *Estud. Geológicos* **2013**, *69*, 35–46. (In Spanish) [CrossRef]
11. Santiesteban, J.; Schulte, L. Fluvial networks of the Iberian Peninsula: A chronological framework. *Quat. Sci. Rev.* **2007**, *26*, 2738–2757. [CrossRef]
12. Cunha, P.P.; Almeida, N.; Aubry, T.; Martins, A.A.; Murray, A.S.; Buylaert, J.-P.; Sohbati, R.; Raposo, L.; Rocha, L. Records of human occupation from Pleistoceneriver terrace and aeolian sediments in the Arneiro depression (Lower Tejo River, central eastern Portugal). *Geomorphology* **2012**, *165*, 78–90. [CrossRef]
13. Pérez-González, A.; Silva, P.G.; Calvo, J.P.; de Vicente, G.; González Casado, J.M. *Mapa Geológico de España, E. 1:50,000 (2ª Serie)*; Talavera de La Reina n° 627; Instituto Geológico y Minero de España: Madrid, Spain, 2009. (In Spanish)
14. Rosina, P.; Voinchet, P.; Bahain, J.J.; Cristanao, J.; Falgueres, C.H. Dating the onset of Lower Tagus River terraces formation using electron spin resonance. *J. Quat. Sci.* **2014**, *29*, 153–162. [CrossRef]
15. Moreno, D. Datation par ESR de Quartz Optiquement Blanchis(ESR-OB) de la Région de Atapuerca (Burgos, Espagne). Applicationau Site préhistorique de Gran Dolina (Contexte Karstique) et Auxsyste'mes Fluviatiles Quaternaires de l'Arlanzon et l'Arlanza. Ph.D. Thesis, Universitat Rovira i Virgili, Tarragone, Spain, 2011.
16. Benito-Calvo, A.; Ortega, A.I.; Navazo, M.; Moreno, D.; Pérez-González, A.; Parés, J.M.; Bermúdez de Castro, J.M.; Carbonell, E. Evolución geodinámica pleistocena del valle del río Arlanzón. Implicaciones en la formación del sistema endokásrtico y los yacimientos al aire libre de la Sierra de Atapuerca (Burgos, España). *Boletín Geológico y Min.* **2018**, *129*, 59–82. (In Spanish) [CrossRef]
17. Goy, J.L.; Rodríguez López, G.; Martínez-Graña, A.M.; Cruz, R.; Valdés, V. Geomorphological Analysis Applied to the Evolution of the Quaternary Landscape of the Tormes River (Salamanca, Spain). *Sustainability* **2019**, *11*, 7255. [CrossRef]
18. Martín-Martín, I.; Gabriel-Silva, P.; Martínez-Graña, A.M. Geomorphological and Geochronological Analysis Applied to the Quaternary Landscape Evolution of the Yeltes River (Salamanca, Spain). *Sustainability* **2020**, *12*, 7869. [CrossRef]

19. Roquero, E.; Silva, P.G.; Zazo, C.; Goy, J.L.; Masana, J. Soil evolution indices in fluvial terrace chronosequences of Central Spain (Tagus and Duero fluvial basins). *Quat. Int.* **2015**, *376*, 101–113. [CrossRef]
20. Silva, P.G.; Roquero, E.; López-Recio, M.; Huerta, P.; Martínez-Graña, A.M. Chronology of fluvial terrace sequences for large Atlantic Rivers in the Iberian Peninsula (Upper Tagus and Duero drainage basins, Central Spain). *Quat. Sci. Rev.* **2016**, *166*, 188–203. [CrossRef]
21. Martin, D.; Bascones, L.; Corretge, L.G. *Cartografía y Memoria del Mapa Geológico de España E: 1/50,000 (2ª Serie). Cañaveral (650)*; Instituto Geológico y Minero de España (IGME): Madrid, Spain, 1987. (In Spanish)
22. Contreras, E.; Roldán, F.J.; Sánchez, R. *Cartografía y Memoria del Mapa Geológico de España E:1/50,000 (2ª Serie) de Navalmoral de la Mata (624)*; Instituto Geológico y Minero de España (IGME): Madrid, Spain, 2006. (In Spanish)
23. Gumiel, P.; Arias, M.; Monteserín, V.; Segura, M. Modelo geológico 3D de la estructura en sinforme de Monfragüe: Un valor añadido al patrimonio geológico del Parque Nacional. *Boletín Geológico y Min.* **2010**, *121*, 15–28. (In Spanish)
24. Gumiel, P.; Campos, R.; Muñoz Barco, P.; Martínez, E. Sinforme de Monfragüe. In *Patrimonio Geológico de Extremadura: Geodiversidad y Lugares de Interés Geológico*; Muñoz Barco, P., Martínez, E., Eds.; Consejería de Medio Ambiente: Seville, Spain; Junta de Extremadura: Madrid, Spain, 2005; 478p. (In Spanish)
25. Duque Macias, J. Un paseo por la geología del sinclinal de Monfragüe (Cáceres). *Meridies* **1999**, *3*, 31–48. (In Spanish)
26. Martín-Serrano, A.; Molina, E. *Montes de Toledo y Extremadura. En: Memoria del Mapa de Cuaternario de España a Escala 1:1,000,000*; IGME: Madrid, Spain, 1989. (In Spanish)
27. Martin-Serrano, A. La definición y el encajamiento de la red fluvial actual sobre el Macizo Hespérico Peninsular en el marco de su geodinámica alpina. *Rev. Soc. Geol. España* **1991**, *4*, 337–351. (In Spanish)
28. Martín-Serrano, A. El paisaje del área fuente cenozoica: Evolución e implicaciones; correlación con el registro sedimentario de las cuencas. *Ciências Terra. Earth Sci. J.* **2000**, *14*, 25–38. (In Spanish)
29. Martín-Serrano, A.; y Molina, E. *El Macizo Ibérico. Memoria del Mapa Geomorfológico de España a Escala 1:1,100,000*; Martín-Serrano, A., Ed.; IGME: Madrid, Spain, 2005; pp. 65–85. (In Spanish)
30. Fernández Macarro, B.; Blanco, J.A. Evolución morfológica de la depresión de Talaván Torrejón el Rubio (Cáceres, España). In Proceedings of the Actas de la I Reunión Geomorfológica de España, Teruel, Spain, 17–20 September 1990; pp. 753–762. (In Spanish).
31. Martínez-Graña, A.M.; Silva, P.G.; Goy, J.L.; Elez, J.; Valdés, V.; Zazo, C. Geomorphology applied to landscape analysis for planning and management of natural spaces. Case study: Las Batuecas-S. de Francia and Quilamas natural parks, (Salamanca, Spain). *Sci. Total Environ.* **2017**, *584–585*, 175–188. [CrossRef]
32. Martínez-Graña, A.M.; Goy, J.L.; Zazo, C.; Silva, P.G.; Santos-Francés, F. Configuration and evolution of the landscape from the geomorphological map in the Natural Parks Batuecas-Quilamas (Central System, SW Salamanca, Spain). *Sustainability* **2017**, *9*, 1458. [CrossRef]
33. Villamor, M.P. Cinemática Terciaria y Cuaternaria de la Falla de Alentejo-Plasencia y su Influencia en la Peligrosidad Sísmica del Interior de la Península Ibérica. Ph.D. Thesis, University Complutense, Madrid, Spain, 2002; 343p. (In Spanish).
34. Villamor, P.; Capote, R.; Stirling, M.W.; Tsige, M.; Berryman, K.R.; Martinez Diaz, J.J.; Martín-González, F. Contribution of active faults in the intraplate area of Iberia to seismic hazard: The Alentejo-Plasencia Fault. *J. Iber. Geol.* **2012**, *38*, 85–111. [CrossRef]
35. Villamor, P.; Capote, R.; Tsige, M. Actividad neotectónica de la falla de Alentejo Plasencia en Extremadura (macizo Hespérico). *Geogaceta* **1996**, *20*, 925–928. (In Spanish)
36. Goy, J.L.; Zazo, C. Cuaternario y Geomorfología. In *Cartografía y Memoria del Mapa Geológico de España E: 1/50,000 (2ª Serie). Torrejoncillo (622)*; Bascones, L., Martin, D., Eds.; Instituto Geológico y Minero de España (IGME): Madrid, Spain, 1987. (In Spanish)

Publisher's Note: MDPI stays neutral with regard to jurisdictional claims in published maps and institutional affiliations.

 *sustainability*

Article

The Vega Alta of Segura River (Southeast of Spain): A Wetland of International Importance

Gustavo Ballesteros-Pelegrín, Daniel Ibarra-Marinas and Ramón García-Marín *

Department of Geography, University of Murcia, Campus La Merced, 30001 Murcia, Spain; gabp1@um.es (G.B.-P.); adaniel.ibarra@um.es (D.I.-M.)
* Correspondence: ramongm@um.es

Abstract: The Ramsar Convention is an intergovernmental treaty for the conservation and wise use of wetlands, which establishes nine criteria related to natural values and a cultural one that wetlands must meet to be included in the list of wetlands of international importance. We aim to evaluate if the wetlands of the Vega Alta of the Segura River (southeast of Spain) meet the requirements to fulfil this agreement. Thanks to meticulous fieldwork and a bibliographic review related to the stated objective, we collected information on the existing environmental and cultural values. The results show that this set of wetlands is home to 11 species of threatened vertebrates in Spain, two priority habitats in the European Union, as well as cultural values related to their origin, conservation, and ecological functioning. Likewise, in the area there are archaeological sites, traditional uses of water associated with the cultivation of rice, and religious manifestations. Effective wetland inventories and rigorous analyzes of their ecological and environmental characteristics, as well as their socioeconomic functions, need to be carried out in order to improve their management and protection. This wetland certainly meets three criteria to be included in the Ramsar Convention list of wetlands.

Keywords: Ramsar Site; types of wetlands; endangered species; priority habitats; cultural values

 check for **updates**

Citation: Ballesteros-Pelegrín, G.; Ibarra-Marinas, D.; García-Marín, R. The Vega Alta of Segura River (Southeast of Spain): A Wetland of International Importance. *Sustainability* **2021**, *13*, 3145. https://doi.org/10.3390/su13063145

Academic Editor: Antonio Miguel Martínez-Graña

Received: 21 January 2021
Accepted: 9 March 2021
Published: 12 March 2021

Publisher's Note: MDPI stays neutral with regard to jurisdictional claims in published maps and institutional affiliations.

1. Introduction

The Convention on Wetlands of International Importance (Ramsar Convention) is the oldest of the modern intergovernmental agreements on the environment. The treaty was negotiated in the 1960s between countries and non-governmental organizations concerned about the loss and degradation of wetland habitats for migratory waterbirds [1,2], as it has been estimated that more than 50% of humid areas have disappeared over the last 150 years, currently occupying 6% of the surface land area [3–5].

This convention was adopted in the Iranian city of Ramsar in 1971 and entered into effect in 1975 with one main objective: "The conservation and wise use of wetlands, through national action and through international cooperation, in order to contribute towards achieving sustainable development throughout the world." The convention makes a broad definition of wetlands, which covers all "lakes and rivers, underground aquifers, swamps and marshes, wet grasslands, peatlands, oases, estuaries, deltas and tidal flats, mangroves and other coastal areas, coral reefs, and artificial sites such as fish ponds, rice fields, reservoirs, and salt flats" [1].

The convention highlights the importance of wetlands for the important ecosystem services they provide [6], among which the regulation of water regimes and their usefulness as reservoirs of biodiversity stand out. It also recognizes that they generate a resource of great economic, scientific, cultural, and recreational value, in addition to playing a fundamental role in adapting and mitigating climate change.

A requirement for those countries that adhere to the convention is that they designate at least one wetland of international importance, and therefore, the Ramsar List of Wetlands is a prestigious list, since it integrates the most important wetlands in the world from the point of view of its ecological interest and for the conservation of biodiversity.

The List of Wetlands declared as Ramsar Sites is the most extensive network of protected areas on the planet. As of December 2020, it had more than 2400 sites, covering more than 2.5 million square kilometers in the 171 countries that have signed the convention [7].

The Ramsar Convention has a set of tools for national wetland inventories [8], and goal number 6 of the Sustainable Development Goals is the protection and restoration of aquatic ecosystems such as wetlands. However, recent trends in the global assessment of wetlands indicate that between 1970 and 2015 the area of natural wetlands hasdecreased by approximately 35%, while constructed wetlands, mostly rice fields and reservoirs, almost doubled during this period and now constitute 12% of catalogued wetlands. Nevertheless, increases in constructed wetlands have not compensated for the loss of natural wetlands [9–11].

To halt the global deterioration and disappearance of wetlands, it is necessary to improve the identification, planning, and management of wetlands that have relevant ecological and cultural values [12], and implement conservation and sustainable development measures [13–16].

The inclusion of a wetland as a wetland of international importance requires that it meet at least some relevant requirements related to the presence of threatened species or ecological communities, to which cultural values related to their origin, conservation, and/or ecological functioning can be added [17–19].

Spain has been a party to the convention since 1982 and has declared a total of 75 sites as wetland of international importance, covering an area of 304,564 hectares, in such a way that it recognizes its commitment to contribute to sustainable development, actively working at local, regional and state level for the conservation and wise use of wetlands [2,20].

The government of Spain created the Wetlands Committee in 1994 as a consultative and cooperative body between public administrations (Royal Decree 1424/2008) in order to provide the necessary mechanisms to apply and disseminate the philosophy of protection and wise use of wetlands. In addition to supporting the processing, approval, and application of national wetland policies, the management of wetlands, in particular Ramsar sites, the inclusion of new sites on the list, the control of the application of Resolutions and Recommendations of the COP (Conference of the Contracting Parties) at the national level and the writing of National Reports, etc.

Within the framework of this committee, in 2000, the Autonomous Organization of National Parks prepared the "Spanish Strategic Plan for the Conservation and Rational Use of Wetlands, within the framework of the aquatic ecosystems on which they depend", with the aim of laying the foundations to "guarantee the conservation and wise use of wetlands, including the restoration or rehabilitation of those that have been destroyed or degraded" [21].

As a consequence, the Autonomous Community of the Region of Murcia (southeast of Spain) has 3 Ramsar Sites: the Mar Menor Lagoon (designated in 1994) and the Campotéjar and Las Moreras lagoons (both chosen in 2011). It has also identified two other wetlands that meet the criteria to be included as wetland of international importanceunder the Ramsar Convention: the Ajauque and RamblaSalada Wetlands, in the municipalities of Santomera and Fortuna, and the Vega Alta of the Segura River in the municipalities of Calasparra, Moratalla, and Hellín (the latter administratively belonging to the province of Albacete and the Autonomous Community of Castilla-La Mancha).

Most of the wetlands of the Vega Alta of the Segura River are included within the Natura Network 2000 (Red Natura 2000) through the body of Site of Community Interest. They approved a Comprehensive Management Plan (Decree No. 55/2015, BORM No. 1 of 17 April) [22], whereall wetlands are in the Inventory of Wetlands of Spain (BOE No. 139, of 11 June 2019) [23].

The environmental and cultural values have been studied by various authors [24–32]. It is proposed, as a starting hypothesis, that the Vega Alta of the Segura River meets the requirements to be included as a Ramsar Site. In this way, the objective of this study has been to compile, organize, and evaluate the available information on the types of wetlands and natural/cultural values, as well as to verify if, indeed, the Vega Alta of the Segura River meets the necessary criteria to be included in the list of wetland of international importance.

2. Materials and Methods

2.1. Study Area

The wetlands included in the Vega Alta of the Segura River cover an area of approximately 2500 hectares located largely within the administrative area of the Region of Murcia, in the municipalities of Cieza, Calasparra, and Moratalla, as well asa smaller part in the province of Albacete, within the municipality of Hellín (Figure 1).

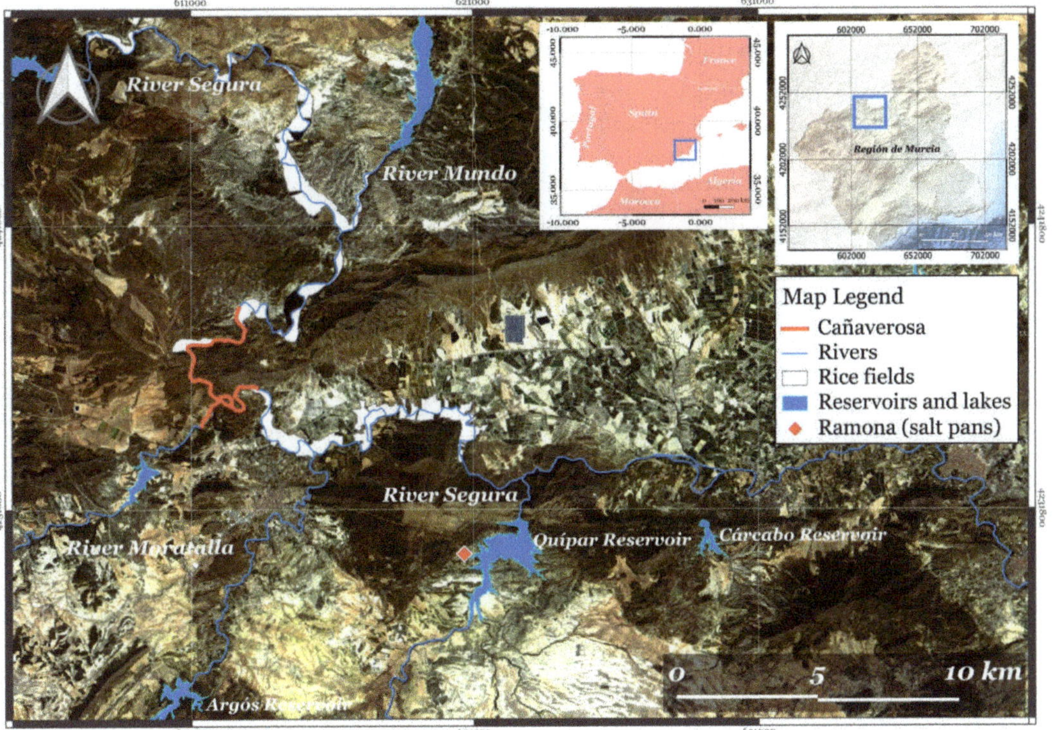

Figure 1. Location of the Vega Alta of the Segura River wetlands (southeast of Spain). Source: self-made. Cañaverosa: Natural reserve composed of groves and riverside forest.

The geographical position of the Vega Alta of the Segura River determines the presence of a Mediterranean climate, with average annual temperatures ranging between 13 and 18 °C, but with marked fluctuations, ranging from temperatures below 0 °C in winter and up to 40 °C in the summer period. Average annual rainfall varies between 200 and 350 mm per year, which increases to 300–500 mm in the westernmost part of the Moratalla River [32].

The Segura River is the main watercourse that crosses the territory on its northwest flank, where it acts as a natural boundary between the Province of Albacete and the Autonomous Community of Murcia, to which must be added other tributary rivers of the Segura River: Benamor, Argos, Quípar, Alhárabe, and Moratalla [33]. It is located in the outer part of the Betic Mountain Ranges, where the Prebetic and Subbetic Zones are represented. The most abundant geological substratum are limestone and dolomite, which are harder, older materials and typical of the base, located in the central nuclei of the mountains. Among these, there are softer, Neogene, and alluvial quaternary materials, which must be added, and to the south of the Cañaverosa area the river is embedded in a wide area of conglomerates [27].

The study area is located in the central area of the Segura River basin, crossed and supported by the river itself and the main tributaries in this area, the Rambla de Cañaverosa and the Moratalla, Argos and Quípar rivers, all of them on its right bank [34]. In this area, the nine groups of wetlands are located in places mostly protected by the Special Conservation Area of the Sierras and the Vega Alta of the Segura River, as well as the Alhárabe and Moratalla rivers, the Special Conservation Area of the Sierra del Molino, Embalse del Quípar, and Llanos del Cagitán, which are connected by the Segura River and its tributaries: Mundo, Moratalla, Argos and Quípar rivers. Moreover, its surrounded by around a 1000 hectares of land area dedicated to rice cultivation that are currently not covered by any environmental protection body [35].

2.2. Methodology

In the first place, the types of wetlands that exist in the Vega Alta of the Segura River were demarcated, taking as a starting point the cataloguing of wetlands included in the inventory of wetlands approved by the Resolution of 21 May 2019, of the General Directorate of Biodiversity and Environmental Quality, by which 53 new wetlands of the Autonomous Community of the Region of Murcia wereincluded in the Spanish Inventory of wetlands (BOE No. 139 of 11 June 2019) [23]. Likewise, the limits of some wetlands were revised after carrying out specific field samplings for this research. Wetlands were managed using the Ramsar Convention's international classification system for wetlands, which groupedthem broadly, allowing the most significant and representative types of wetlands to be identified [17].

The procedure was proposed by the protocol for the inclusion of Spanish wetlands in the list of international importance (Ramsar Convention) and technical annex, prepared by the Wetlands Committee of the Ministry of Environment of the government of Spain in 2011, and approved by the National Commission for the Protection of Nature on 4 December 2007, which is an adaptation to the Spanish case of Manual 17. Designation of Ramsar sites include: strategic framework and guidelines for the future development of the list of wetlands of international importance (2010), a document that is part of the set of Ramsar Manuals (21 Manuals), which the Ramsar Secretariat made available to the public. It establishes some related criteria on the representativeness of the site, i.e., rare or unique wetland in a biogeographic region, sites of importance to conserve biodiversity, and an additional criterion related to cultural values (Table 1).

Table 1. Ramsar criteria for the identification of wetlands of international importance.

Sites Comprising Representative, Rare, or Unique Wetland Types
Criterion 1: A wetland may be considered of international importance if it contains a representative, rare or unique example of a natural or near-natural wetland type in a biogeographic region.

Criteria based on species and ecological communities
Criterion 2: A wetland may be considered of international importance if it supports vulnerable, endangered or critically endangered species, or threatened ecological communities.
Criterion 3: A wetland may be considered of international importance if it supports populations of plant and/or animal species important to maintain the diversity of a biogeographic region.
Criterion 4: A wetland may be considered of international importance if it supports plant and/or animal species when they are in a critical stage of their biological cycle.

Specific criteria based on waterfowl
Criterion 5: A wetland may be considered of international importance if it regularly supports a population of 20,000 or more waterfowl birds.
Criterion 6: A wetland may be considered of international importance if it regularly supports 1% of the individuals in a population of a species or subspecies of waterfowl.

Table 1. *Cont.*

Sites Comprising Representative, Rare, or Unique Wetland Types
Specific criteria based on fish
<u>Criterion 7</u>: A wetland may be considered of international importance if it supports a significant proportion of indigenous fish subspecies, species or families, life cycle stages, species and/or population interactions that are representative of the benefits and/or values of wetlands, and in this way it contributes to the world's biological diversity.   <u>Criterion 8</u>: A wetland may be considered of international importance if it is an important food source for fish, is a spawning area, an area of development and growth, and/or a migration route on which fish stocks within or outside the wetland depend.
Specific criteria based on other taxonomic groups
<u>Criterion 9</u>: A wetland may be considered of international importance if it regularly supports 1% of the individuals in the population of a subspecies or species that is a non-aviary animal.
Additionalcriteria
<u>Cultural value</u>: A wetland can also be considered of international importance when, in addition to ecological criteria, it has examples that illustrate important cultural value, whether material or not, related to its origin, conservation, and/or ecological functioning.

Source: Ramsar (2010) [17] and Committee on Wetlands (2011) [36].

The information on natural values has been prepared from specific field samples carried out in the study area, as well as a bibliographic compilation of the publications available on the related habitats [24,33] and vertebrates associated with aquatic environments [29,32,35,37], classified as threatened on a national and international scale, taking into account the threat criteria of the International Union for Conservation of Nature (IUCN) [38].

2.3. Field Work

Sampling of the different groups of vertebrates has been carried out periodically or recurrently between 2002 and 2021, which have covered the wetlands and river courses included in this study, as well as adjacent areas. For the ichthyofauna inventory, the channels and other bodies of water were sampled using a wide variety of fishing methodologies: active (electric fishing, landing net) or passive (minnow traps, traps, trammel nets, gill nets) [39]. To make the list of threatened chiroptera, the natural or artificial underground cavities were sampled (especially between the months of May to October). A visual inspection of the interior and/or trapping of the animals was carried out using fog nets and harp-traps [40]. These inspections were supplemented with ultrasound detectors (SONY model DCR-TRV33E video camera with infrared focus from Wildlife Engineering Inc.; SONY model WM6-DC recorder; PetterssonElektronik model D-90A ultrasound detectors) [41,42].

For the bird inventory, censuses were carried out by observing and identifying birds with a telescope or binoculars in the different wetlands, generally first thing in the morning, in the absence of wind or rain, taking note of all the species observed [43]. The amphibians, given the high extension of the area to be examined, were detected by quantitative sampling of larvae, a method that allows detecting species when the abundances of adult specimens are low and of nocturnal species during the day [44].

3. Results

In the Vega Alta of the Segura River, three different types of wetlands were described. They are formed by natural wetlands, artificial reservoirs of ecological interest, salt pans, and rice fields flooded cyclically through a system of crop rotation, and that together encompass 10humid areas and aquatic environments that are connected to each other by the Segura River and some of its tributaries (Table 2).

Table 2. Types of wetlands in the Vega Alta of the Segura River.

Zone	Area (ha)	Type of Wetland	Ramsar Code
Natural Wetlands			
Segura, Mundo, Moratalla, Argos, Quípar and rambla del CárcaboRivers	-	Natural stretches of permanent waterways (includes riverbanks)	M
"Sotos and Bosques de Ribera de Cañaverosa" Nature Reserve	56.7	Natural stretches of permanent waterways (includes riverbanks)	M
"Cañón deAlmadenes" Nature Reserve	40.1	Natural stretches of permanent waterways (includes riverbanks)	M
Artificial or modified wetlands with ecological interest			
La RiscaReservoir	15.3	Reservoirs of ecological interest and that function as wetlands	6
Argos Reservoir	126.6		
MoratallaReservoir	1.3		
CárcaboReservoir	16.0		
QuíparReservoir	423.6		
Salinas de la Ramona	2.1	Salt flats	5
Rice fields of Calasparra, Salmerón, and Mundo	875	Flooded land of ecological interest	5

Source: prepared from Ramsar 2010 [17] and Directorate General for the Natural Environment (2015) [45].

The analysis of the natural values of the wetlands and aquatic environments allows us to conclude that these wetlands meet, as a whole, 3of the 10criteria defined by Ramsar (2010) for this area to be declared a Wetland of International Importance:

i. Criterion 2: in the Vega Alta of the Segura River, 11 threatened species have been found in Spain, to which we must add two species of fish included in Annex II of the Habitat Directive of the European Union (Table 3). Among them, it is worth highlighting the populations of chub (*Squaliuspyrenaicus*), a species of fish that is considered in danger of extinction in the Segura Riverbasin, since it has a distribution that is restricted to the upper part of the River Mundo and some enclaves isolated from other tributaries of the Segura River [29]. It is alsonoteworthy that populations of bigeye buzzard (*Myotiscapaccinii*) are in danger of extinction and are distributed in the Region of Murcia, mainly around the courses and bodies of water of the Vega or the Segura River [30,33].

The Quípar river is considered important for the conservation of aquatic invertebrates, with the presence of 39 species of aquatic and semiaquatic coleopterans, among which the presence of 5 Iberian endemism's stand out (Table 4).

Two habitats classified as priority in the context of the European Union were located in the humid areas of the Vega Alta of the Segura River (Table 5): habitat 7210: Calcareous bogs of *Cladiummariscus* and with *Caricondavallanae* species, whose associations 621012 Cladiomarisci-Caricetumhispidae O. Bolòs 1926 and 621123 Typho-Schoenoplectetumglauci Br.-Bl. andO. Bolòs 1958 present a wide distribution along these sections of the Segura riverbed and its main tributaries; habitat 7220*: Petrifying springs with tuff formation (*Cratoneurion*), constituted by the association 622027 *Tracheliocauruleae-Adiantetumcapilli-veneris* O. Bolòs 1957 [24,33].

Table 3. Threatened species in the Vega Alta of the Segura River.

CommonName	Scientific Name	Red Books of Spain	Spanish Catalogue of Species in Danger of Extinction (RD131/2011)	Habitat Directive (92/43/CEE)	Birds Directive (2009/147/CE)
			Fishes		
Chub	*Squaliuspyrenaicus*			A.II	
Iberiannase	*Chondrostomapolylepis*			A.II	
			Mammals		
Cave bats	*Myniopterusschereibersii*		VU	A.II	
Long-fingeredbat	*Myotiscapaccinii*		EPE	A.II	
Lesser mouse-earedbat	*Myotisblythii*		VU	A.II	
Greater mouse-earedbat	*Myotismyotis*		VU	A.II	
			Birds		
Squacco Heron	*Ardeolaralloides*		VU		A.I
Western marshharrier	*Circusaeroginosus*	VU	VU		A.I
Montagu'sharrier	*Circuspygargus*	VU	VU		A.I
Western osprey	*Pandionhaliaetus*	EPCr	VU		A.I
Eurasian teal	*Anas crecca*	VU			
Commonreedbunting	*Emberizaschoeniclus*	VU			A.I
			Amphibians		
Beticmidwifetoad	*Alytesdickhilleni*	VU	VU		

Legend: A.I: Annex I; A.II: Annex II; EPCr: critically endangered; EPE: in danger of extinction; VU: vulnerable. Source: prepared from the information provided by the Directorate General for the Environment (2010) [33]; Doadrio (2002) [46]; and Madroño et al. (2004) [47].

Table 4. Endemic and vulnerable aquatic beetles.

Species	Iberian Endemism	Regional Vulnerability
Agabusramblae	X	
Hydrochusnooreinus	X	High
Helophorusbrevipalpis		High
Newbrioporusbaeticus	X	
Ochthebiusdelgadoi	X	
Stictonectesepipleuricus	X	

Source: prepared from Sánchez-Fernández et al. (2004) [31] andGeneral Directorate of the Natural Environment (2015) [45].

Table 5. Habitats of priority interest in the European Union in the Vega Alta of the Segura River.

Code	Habitat
7210*	Calcareous bogs of *Cladiummariscus* and with *Caricondavallanae* species
7220*	Petrifying springs with tuff formation (*Cratoneurion*)

Source: own elaboration from Alcaráz et al. (2008) [24].

ii.   Criterion 3: in the Vega Alta of the Segura River, in addition to the two habitats included in criterion 2, nine habitats associated with humid environments of Annex I of the Habitats Directive of the European Union were described with Global Assessment A [24,33]. They are as follows:

(a)   Habitat 1410. Mediterranean saline grasslands (*Juntaliamaritimi*) occupies 1.10 hectares and is distributed in narrow strips around the Segura riverbed and in brackish areas around the Cárcabo reservoir.

(b)   Habitat 3150. Natural eutrophic lakes and lagoons, with *Magnopotamion or Hydrocharition* vegetation (underwater grasslands), andan area of 20.18 hectares.

(c)   Habitat 3250. Mediterranean rivers of permanent flow with *Glauciumflavum*, with an extension of 2.35 hectares.

(d)   Habitat 3280. Mediterranean rivers of permanent flow of the Paspalo-Agrostidion with riparian plant curtains of *Salix and Populusalba*, which occupy a total of 2.26 hectares, and which colonize surfaces with frequent floods along the banks of rivers, streams, and irrigation canals.

  (e) Habitat 3290. Mediterranean rivers of intermittent flow of the *Paspalo-Agrost idion*. These are reed beds that develop in non-permanent waters.

  (f) Habitat 6420. Mediterranean hygrophilous herbaceous communities (reed beds), with an area of 60.88 hectares.

  (g) Habitat 6430. Hygrophilous eutrophic megaphorbs of the plain fringes and mountainous to alpine steeps, with a total area of 23.54 hectares.

  (h) Habitat 92AO. Gallery forests of *Salix alba* and *Populus alba* (poplars and willows), which develop in the upper and middle sections of the Segura river basin, with a layout in two parallel bands located on both sides of the riverbed, together occupying an expanse of 84.49 hectares. These formations are usually accompanied by elms, ash trees, willows, and Magnoliopsida, in addition to other shrub species, such as oleanders.

  (i) Habitat 92DO. Thermo-Mediterranean riparian thickets and galleries (*Nerio-Tamaricetea* and *Securinegiontinctoriae*) (Magnoliopsida and oleanders), with a vegetation formed mainly by poplars, tars, Magnoliopsida, and oleanders, which are distributed along the riverbeds and occupy an area of 128.5 hectares.

iii. Cultural values: associated with the river terraces of the Segura River and the Mundo River are rice fields, which have been cultivated in a traditional way since the 18th century. It is a type of humid area of an artificial nature, both due to its origin and its use, which is characterized by the significant returns from irrigation that periodically and naturally recharge the underlying aquifer. They constitute shallow aquatic ecosystems (15–20 cm), which have a dry phase during winter and another wet phase that lasts from April or May to November or December [33]. Rice cultivation occupied a potential area of 2463 hectares at the beginning of the 20th century, which was made official and protected by the Royal Decree of February 1, 1908, the date on which the delimitation of the CotoArrocero (Rice farms) located in the municipal terms of Hellín (Castilla-La Mancha), Moratalla, and Calasparra (Region of Murcia), with the latter municipality having a larger area (approximately 1000 hectares), roots, and tradition (Sánchez-Lorente, 1999). In 1986, the Protected Designation of Origin (PDO) was approved for the Bomba varieties (cultivated since the 19th century) and BalillaXSollana (hybrid variety obtained in 1948). This certificate guarantees the quality of the product and its origin, so that Calasparra rice became the first to have this endorsement in the world, with a Regulatory Council that must ensure compliance with the regulations during the process and methods cultivation that ensure its quality. There are two main factors that influence the origin, uniqueness, and quality of the product [48,49], and that make it of cultural value closely interrelated with the natural values of the environment.

  (a) The rice fields are located in small valleys, with slopes that make mechanization difficult, located at an average altitude of 450 m above sea level, in a fertile valley surrounded by mountains and fed by the clean moving waters of the Segura River, with a system of ditches and floodgates that take the water from the river and that generate a renewing current, flooding the rice field. This is met with a structure supported at different levels and plots separated by small boundaries. The excess water after irrigation returns to the main river.

  (b) Use of traditional cultivation methods that, in part, are maintained over time: adaptation of the land, sowing, weeding, and harvesting of the produce. The sowing is done with seeds that have been previously hydrated, to prevent them from floating and can be spread over the flooded boxes, which have been previously conditioned and nourished. The cultivated varieties (Bomba and BalillaXSollana) are adapted to the special characteristics of this environment.

On the other hand, there is also another rich cultural heritage in the study area, characterized by the presence of traces of the first settlers of the Iberian southeast: cave paintings, tools remains, ceramics, and utensils located in the cave-chasm of La Serreta, known as "Abrigo de losMonigotes" or "Abrigo del Pozo", "caveand spring of las Enredaderas",

"Abrigos del Laberinto", "Abrigo de las Escalerillas", "Cueva de losPucheros", or "Cueva del Arco" [33]. Finally, it is also necessary to mention the combinations of the heritage generated from the different uses of water, which have conceived an extensive network of ditches and dams, i.e., the presence of hydraulic heritage closely related to the culture of water, which in turn was once closely linked to the traditional uses of rice cultivation, with rice mills and other buildings associated with religious worship, which also constitute elements of cultural and ethnographic interest in the Vega Alta of the Segura River, such as the Sanctuary of Our Lady of Hope (place of pilgrimage), located in the town of Calasparra [50–52].

4. Discussion

According to Scott andJones [53], the classification of wetlands is extremely problematic, and the definition of the term wetland is a difficult and controversial starting point. Nonetheless, considerable effort has been made in developing national and regional wetland classifications, although the only attempt to establish a global system has been under the auspices of the Ramsar Convention [54]. The incorporation of a wetland to the Ramsar List requires meeting some of the requirements that demonstrate that it has natural values of international importance [17,19]. As a consequence of the requirement that countries that adhere to the Convention must include at least one wetland, the Ramsar List of Wetlands has become the most extensive network of protected areas on the planet and continues to grow every year, as it is a prestigious list that favors the interest and worldwide projection of wetlands that are included on the list [55–60], which predisposes populations and countries that have humid areas to study them to see if they meet the criteria in order to achieve their incorporation (as is the case of the study at hand).

However, the protection of a wetland within the framework of the Ramsar Convention should be the starting point for developing plans and strategies that generate the necessary instruments to guarantee its conservation and sustainable use, as well as the application of integrated wetland management policies [56,61,62]. In the case of the Vega Alta of the Segura River, it is necessary to integrate the management of natural wetlands, artificial wetlands, which make up structures of water regulation (reservoirs), and the different surrounding land uses, such as rice fields, which maintain extensive areas flooded with rotation systems and which preserve high natural values, compatible with agricultural and social or recreational use.

In Costa Rica, for example, there are several studies on different wetlands that conclude the relevance of these reservoirs for the maintenance of the ecological dynamics of the surrounding territory, and how changes in land use in the areas adjacent to this protected wild areas generates alterations in this delicate ecosystems. Among others, the wetlands of the Térraba-Sierpe Delta [63–66], the high Andean wetlands of the Chirripó National Park [67], and the San Vito wetland [68] have been analyzed. Performing analyzes similar to those mentioned would be of great interest to complete the information of the wetland in this analyzed work.

South America has a large proportion of wetlands compared to other continents. While most of these wetlands remained in relatively good condition until a few decades ago, pressures from land use and climate change have threatened their integrity in recent years [69].

On the other hand, the wetlands of Saudi Arabia, located in a region with water stress, vulnerable to climate change and other global changes (changes in nearby land uses, for example), as in our case study, present significant challenges for long-term maintenance [70].

In India, the reasons for the loss of wetlands are urbanization, changes in land use, and pollution. There is no adequate regulatory framework for their conservation [71] and, in this sense, future research should also focus on the institutional factors that influence their condition and evolution. The same is happening in China, and some authors call for improvements in legislation and management systems for wetlands [72]. According to Hu et al. [73], the greatest loss of wetlands occurred in Asia, but the most serious situation currently resides in Europe.

We are seeing that numerous ecosystems of environmental, patrimonial, social, and economic interest, such as wetlands, are in decline, suffering the consequences of poor land use planning. Extensive restoration of these critical ecosystems is increasingly seen as critical to conserving biodiversity and stabilizing the Earth's climate [74]. Support to biodiversity, improving water quality, reducing floods, and sequestering carbon are key functions that are impacted when wetlands are lost or degraded.Restoration techniques are improving, although the recovery of lost biodiversity is challenged by invasive species, which thrive under disturbance and displace the natives [75]. The Ramsar Convention has helped many nations protect their wetlands, but effective wetland inventories need to be rigorously carried out.

5. Conclusions

The wetlands of the Vega Alta of the Segura River, which together have a surface area of more than 1500 hectares, joined together by the Segura Riverand its tributaries, meet three of the 10 criteria indicated by the Ramsar Convention to be classified, as a whole, as a wetland of international importance. The presence of 11 species of vertebrates have been classified in Spain as vulnerable or in danger of extinction, two endemic aquatic beetles, another two habitats considered as a priority in the European Union, and another nine habitats associated with humid environments that are included in Annex I of the Habitats Directive are considered important to maintaining the high biological diversity. This is also true for relevant cultural representations related to the existence of the wetland.

Finally, the cultural values associated with the ecological functioning of the Segura River must also be taken into consideration, such as the traditional cultivation of rice introduced in the 18th century, of great quality, uniqueness, and originality, together with other manifestations of traditional uses of water and assets of cultural interest, with relevant paleontological and archaeological sites, as well as buildings closely associated with religious worship.

Author Contributions: Conceptualization, G.B.-P., D.I.-M., and R.G.-M.; methodology, G.B.-P. and D.I.-M.; formal analysis, G.B.-P., D.I.-M., and R.G.-M.; investigation, G.B.-P., D.I.-M., and R.G.-M.; data curation, G.B.-P., D.I.-M., and R.G.-M.; writing—original draft preparation, G.B.-P., D.I.-M., and R.G.-M.; writing—review and editing, G.B.-P. and R.G.-M.; supervision, R.G.-M. All authors have read and agreed to the published version of the manuscript.

Funding: This research received no external funding.

Institutional Review Board Statement: Not applicable.

Informed Consent Statement: Not applicable.

Data Availability Statement: Data sharing not applicable.

Acknowledgments: Authors want to thank anonymous reviewers for their suggestions, which have helped to significantly improve the manuscript.

Conflicts of Interest: The authors declare no conflict of interest.

References

1. Vélez, F. *ImpactosSobre Zonas Húmedas Naturales*; Instituto Nacional para la Conservación de la Naturaleza (Ministerio de Agricultura); Servicio de Publicacionesagrarias: Madrid, Spain, 1979.
2. Ramsar. The Convention on Wetlands and Its Mission. Gland. Ramsar Convention on Wetlands. Available online: https://ramsar.org/about/the-convention-on-wetlands-and-its-mission (accessed on 28 November 2020).
3. Dugan, P. *Wetlands in Danger. A World Conservation Atlas*; Oxford University Press: Oxford, UK, 1993.
4. O´Connel, M.J. Detecting, measuring and reversing changes to wetlands. *Wetl. Ecol. Manag.* **2003**, *11*, 197–401.
5. Engle, V.D. Estimating the Provision of Ecosystem Services by Gulf of Mexico Coastal Wetlands. *Wetlands* **2011**, *31*, 179–193. [CrossRef]

6.   Maltby, E.; Acreman, M.C. Ecosystem services of wetlands: Pathfinder for a new paradigm. *Hydrol. Sci. J.* **2011**, *56*, 1341–1359. [CrossRef]

7.   Ramsar. Wetlands of International Importance. Gland. Ramsar Convention on Wetlands. Available online: https://www.ramsar. org/sites-countries/wetlands-of-international-importance (accessed on 7 December 2020).

8.   Convention on Wetlands. A New Toolkit for National Wetlands Inventories. Gland, Switzerland: Convention on Wetlands Se-Cretariat. Available online: https://www.ramsar.org/sites/default/files/documents/library/nwi_toolkit_2020_e.pdf (accessed on 10 December 2020).

9.   Davidson, N.C. How much wetland has the world lost? Long-term and recent trends in global wetland area. *Mar. Freshw. Res.* **2014**, *65*, 934–941. [CrossRef]

10. Darrah, S.E.; Shennan-Farpón, Y.; Loh, J.; Davidson, N.C.; Finlayson, C.M.; Gardner, R.C.; Walpole, M.J. Improvements to the Wetland Extent Trends (WET) index as a tool for monitoring natural and human-made wetlands. *Ecol. Indic.* **2019**, *99*, 294–298. [CrossRef]

11. Ramsar. Global Wetland Outlook: State of the World's Wetlands and Their Services to People. Gland: Ramsar Convention Se-Cretariat. Available online: https://www.ramsar.org/sites/default/files/flipbooks/ramsar_gwo_english_web.pdf (accessed on 10 December 2020).

12. Finlayson, C.M.; Milton, G.R.; Prentice, R.C.; Davidson, N.C. (Eds.) *The Wetland Book-II*; Springer: Dordrecht, The Netherlands, 2016. Available online: https://link.springer.com/referenceworkentry/10.1007%2F978-94-007-6173-5_186-1 (accessed on 10 December 2020).

13. Bennett, E.M.; Cramer, W.; Begossi, A.; Cundill, G.; Díaz, S.; Egoh, B.N.; Geijzendorffer, I.R.; Krug, C.B.; Lavorel, S.; Lazos, E.; et al. Linking biodiversity, ecosystem services, and human well-being: Three challenges for designing research for sustainability. *Curr. Opin. Environ. Sustain.* **2015**, *14*, 76–85. [CrossRef]

14. Mauerhofer, V.; Kim, R.E.; Stevens, C. When implementation works: A comparison of Ramsar convention implementatiton in different continents. *Environ. Sci. Policy* **2015**, *51*, 95–105. [CrossRef]

15. Worboys, G.L.; Lockwood, M.; Kothari, A.; Feary, S.; Pulsford, I. *Protected Area Governance and Management*; Anu Press: Canberra, Australia, 2015. Available online: https://www.jstor.org/stable/pdf/j.ctt1657v5d.13.pdf (accessed on 10 December 2020).

16. Gell, P.A.; Finlayson, C.M.; Davidson, N.C. Understanding Change in the Ecological Character of Ramsar Wetlands: Pers-Pectives from a Deeper Time—Synthesis. *Mar. Freshw. Res.* **2016**, *67*, 869. Available online: https://www.publish.csiro.au/MF/mf16075 (accessed on 7 December 2020). [CrossRef]

17. Ramsar. *Designating Ramsar Sites: Strategic Framework and Guidelines for the Future Development of the List of Wetlands of Inter-National Importance, Ramsar Handbooks for the Wise Use of Wetlands*, 4th ed.; Ramsar Convention Secretariat: Gland, Switzerland. Available online: https://www.ramsar.org/sites/default/files/documents/pdf/lib/hbk4-17.pdf (accessed on 10 December 2020).

18. Tittensor, D.P.; Walpole, M.; Hill, S.L.L.; Boyce, D.G.; Britten, G.L.; Burgess, N.D.; Butchart, S.H.M.; Leadley, P.W.; Regan, E.C.; Alkemade, R.; et al. A mid-term analysis of progress toward international biodiversity targets. *Science* **2014**, *346*, 241–244. [CrossRef]

19. Asaad, I.; Lundquist, C.J.; Erdmann, M.V.; Costello, M.J. Ecological criteria to identify areas for biodiversity conservation. *Biol. Conserv.* **2017**, *213*, 309–316. [CrossRef]

20. BOE (Official Gazette of the State of Spain). No. 199, of August 20, 1982, Pages 22472 to Instrument of March 18, 1982 of Accession by Spain to the Convention on Wetlands of International Importance, Especially as Habitat for Waterfowl, Made in Ramsar on February 2, 1971. Available online: https://www.boe.es/buscar/doc.php?id=BOE-A-1982-21179 (accessed on 21 January 2021).

21. Dirección General de Biodiversidad. Plan Estratégico Español Para la Conservación y Uso Racional de los Humedales. Madrid, Ministerio para la Transición Ecológica y el Reto Demográfico. Available online: https://www.miteco.gob.es/es/biodiversidad/temas/ecosistemas-y-conectividad/conservacion-de-humedales/ch_estratg_plan_estrategico_curh.aspx (accessed on 7 December 2020).

22. BORM (Official Gazette of the Region of Murcia). Decree No. 55/2015, of April 17, on the Declaration of Special Conservation Zones and Approval of the Comprehensive Management Plan of the Natura 2000 Network Protected Areas of the Northwest Region of Murcia. Available online: http://www.murcianatural.carm.es/web/guest/visor-novedades/-/asset_publisher/xK0B/content/3492119;jsessionid=B8D7B5128FF5FCDACA86AA076ADD0E43?artId=3492119 (accessed on 28 November 2020).

23. BOE.Núm. 139, de 11 de junio de 2019, páginas 61138 a Resolution of May 21, 2019, of the General Directorate of Biodiversity and Environmental Quality, by Which 53 New Wetlands of the Autonomous Community of the Region of Murcia Are Included in the Spanish Inventory of Wetlands. Available online: https://www.boe.es/diario_boe/txt.php?id=BOE-A-2019-8692 (accessed on 10 December 2020).

24. Alcaráz, F.; Barreña, J.A.; Clemente, C.M.; González, J.L.; Ribera, D.; Ríos, S. *Manual de interpretación de los Hábitats naturales y seminaturales de la Región de Murcia. TomoConsejería de Agricultura, Agua y Medio Ambiente de la Región de Murcia*; Ayuntamiento de Murcia: Murcia, Spain, 2008.

25. Ballesteros, G.A.; Ruzafa, A. *Contrastes naturales en la regiónbioclimática del mediterráneo. Museo de la Ciencia y el Agua del Ayuntamiento de Murcia*; Ayuntamiento de Murcia: Murcia, Spain, 2006.

26. Calvo, J.F.; Esteve, M.A.; López, F. *Biodiversidad: Contribución a suConocimiento y Conservaciónen la Región de Murcia*; Servicio de Publicaciones de la Universidad de Murcia: Murcia, Spain, 2000.

27. Conesa, C. *El Medio Físico de la Región de Murcia*; Universidad de Murcia: Murcia, Spain, 2006.

28. Díez de Revenga, E.; Ballesteros, G.A.; Castillo, V.; Falcó, M.D.; García, J.A.; González, G.; Giménez, M.; Gómez, R.; Picazo, H.; Rodier, A.; et al. Estrategia Regional para la Conservación y el UsoSostenible de la Diversidad Bio-lógica. Consejería de Agricultura, Agua y Medio Ambiente de la Región de Murcia. Available online: http://www.murcianatural.carm.es/web/guest/estrategias/-/journal_content/56_INSTANCE_9GoI/14/84596 (accessed on 21 December 2020).
29. Oliva, F.J.; Zamora, J.M.; Franco, J.M.; Zamora, A.; Sánchez, A.; Amat, A.; Guillén, A.; Guerrero, A.; Torralva, M. Pecesdulceacuícolas de la cuenca del río Segura. Murcia: Asociación de Naturalistas del Sureste. Available online: https://www.asociacionanse.org/descarga-guia-peces-y-atlas-odonatos/20190131/ (accessed on 28 November 2020).
30. Robledano, F.; Calvo, J.F.; Hernández, V. *Libro Rojo de los Vertebrados de la Región de Murcia*; ComunidadAutónoma de la Región de Murcia: Murcia, Spain, 2006.
31. Sánchez-Fernández, D.; Abellán, P.; Velasco, J.; Millán, A. Áreasprioritarias de conservaciónen la cuenca del río Segura utilizando los coleópterosacuáticoscomoindicadores. *Limnetica* **2004**, *23*, 209–228.
32. Ribas-Martínez, S.; Asensi, A.; Costa, M.; Fernández-González, F.; Llorens, L.; Masalles, R.; Molero Mesa, J.; Penas, A.; Pérez de Paz, P.I. El Proyecto de Cartografía e Inventariación de los Tipos de hábitats de la Directiva 92/43/CEE enEspaña. Colloques Phytosociologiques XXII. Available online: http://www.jolube.es/pdf/Rivas-Mart_al_1993_Habitats_Espana_Coll_Phytos_22.pdf (accessed on 21 December 2020).
33. General Directorate of the Environment (Dirección General de Medio Ambiente). Natura 2000-Standard Data form Sierras y Vega Alta del Segura y Ríos Alhárabe y Moratalla. *ComunidadAutónoma de la Región de Murcia*. Available online: http://www.murcianatural.carm.es/c/document_library/get_file?uuid=dbb98483-9f7b-48e4-bad9-bdc8936ca1dd&groupId=14 (accessed on 28 November 2020).
34. Rodríguez, T.; López, F. InvestigaciónInterdisciplinarSobre las DeformacionesRecientesen el Sector Meridional de la Vega Alta del Segura (Murcia). CriteriosHidrogeológicosAplicables al Estudio de la Neotectónicaen el SuresteEspañol. Primeras Jornadas SobreNeotectónica y suAplicación al Análisis de Riesgos de EmplazamientoEnergíaIndustrias. *Energía Nucl.* **1984**, *149*, 259–266.
35. Ballesteros, G.A.; Ruiz, V.; Espín, D.; Ibarra, D. El noroestemurciano, tierra de contrastes. In *GeografíaAplicadaen la Región de Murcia. Guía de las Salidas de Campo*; García, R., Alonso, F., Belmonte, F., Eds.; Asociación de GeógrafosEspañoles: Madrid, Spain, 2016; pp. 123–150.
36. Comité de Humedales. *Protocolo de Inclusión de Humedalesespañolesen la Lista de ImportanciaInternacional (Convenio de Ramsar) y Anexo Técnico*; Ministerio de Medio Ambiente de España: Madrid, Spain, 2011.
37. Yelo, D.; Calvo, J.F. Aproximación a la distribución y estatus de los mamíferoscarnívorosen la Región de Murcia. *Galemys* **2004**, *16*, 21–37.
38. IUCN. The IUCN Red List of Threatened Species. Available online: http://www.iucnredlist.org/ (accessed on 15 September 2008).
39. Torralva, M.; Oliva, F.J.; Andreu, A.; Verdiell, D.; Miñano, P.A.; Egea, A. *Atlas de distribución de los pecesepicontinentales de la Región de Murcia*; Consejería de Industria y Medio Ambiente de la Región de Murcia: Murcia, Spain, 2005.
40. Guardiola, A.; Fernández, M.P.; Olivares, E. Evaluación de las coloniasmurcianas de quirópterosincluidosen el Anexo II de la DirectivaHábitats. In *Actas del III Congreso de la Naturaleza de la Región de Murcia (Octubre 2004)*; Asociación de Natu-ralistas del Sureste: Murcia, Spain, 2004; pp. 147–154.
41. Kunz, T.H.; Thomas, D.; Richards, G.C.; Tidemann, C.R.; Pierson, E.D.; Racey, P.A. *Observational Techniques for Batsw*; de Wilson, D.E., Cole, F.R., Nicchols, J.D., Rudran, R., Foster, M.S., Eds.; Measuring and Monitoring Biological Diversity. Standard Methods for Mammals; Smihsonian Institution Press: Washington, DC, USA; London, UK, 1996; pp. 105–114.
42. Barlow, K. *Expedition Field Techiques. Bats*; Expedition Advisory Centre-RGS: London, UK, 1999.
43. Tellería, J.L. *Objetivos y Métodos del Seguimiento de Poblaciones de Aves*; Sánchez, E.A., Ed.; Actas de las XV Jornadas Orni-tológicasEspañolas, 25-SEO/BirdLife: Madrid, Spain, 2000.
44. Atlas de distribución de los anfíbios de la Región de Murcia. Available online: http://www.murcianatural.carm.es/web/guest/visor-contenidos-dinamicos?artId=107483&groupId=14&version=1.0 (accessed on 21 September 2009).
45. Directorate General for the Natural Environment (Dirección General de Medio Natural). *Plan de Gestión Integral de los espaciosprotegidos de la Red Natura 2000 del Noroeste de la Región de Murcia*; Consejería de Agricultura y Agua de la Región de Murcia: Murcia, Spain, 2015.
46. Doadrio, I. *Atlas y Libro Rojo de los pecescontinentales de España*; Ministerio de Medio Ambiente de España: Madrid, Spain, 2001.
47. Madroño, A.; González, C.; Atienza, J.C. Libro Rojo de las aves de España. Madrid: Ministerio de Medio Ambiente de España. Available online: https://www.seo.org/wp-content/uploads/2012/04/Libro_Rojo_Aves.pdf (accessed on 21 December 2020).
48. Sánchez-Llorente, J.G. Historia del arroz enCalasparra. Archivo Municipal de Calasparra. Available online: http://www.arrozdecalasparra.com/historia.pdf (accessed on 28 November 2020).
49. Bernal, J.A. ProtecciónagroambientalenCalasparra gracias al arroz. *Desarro. Rural Sosten.* **2014**, *22*, 26–27.
50. Gil Meseguer, E. Paisajesculturales del regadíotradicional e históricoen la Vega Alta de Segura. *Proc. Irrig. Soc. Landsc. Tribut. Thomas F. Glick* **2014**, 1–12. [CrossRef]
51. Espín, J.M.G. La construcción y ampliación de los regadíostradicionales e históricosen la Vega Alta de Segura: Sucesión de azudes y acequias, artilugioshidráulicosescalonados y motores de elevación de aguas. *Proc. Irrig. Soc. Landsc. Tribut. Thomas F. Glick* **2014**, 402–417. [CrossRef]
52. López, M.; Andrés, M. Estudio de la capacidad de acogida y planificación de las áreasrecreativas de Calasparra (Mur-cia). *Cuad. Tur.* **2000**, *6*, 103–121.

53. Scott, D.A.; Jones, T.A. Classification and inventory of wetlands: A global overview. *Vegetatio* **1995**, *118*, 3–16. [CrossRef]
54. Stroud, D.A. *Selecting Ramsar Sites: The Development of the Criteria for the Designation of Wetlands of International Importan-ce—1971*; Joint Nature Conservation Committee: Peterborough, UK, 2006.
55. Mitchell, A.H. Wise use of wetlands, the Ramsar Convention on wetlands, and the need for an Asian Regional Wetlands Training Initiative. *Aquat. Ecosyst. Health Manag.* **2001**, *4*, 235–242. [CrossRef]
56. Zalidis, G.C.; Takavakoglou, V.; Panoras, A.; Bilas, G.; Katsavouni, S. Re-Establishing a Sustainable Wetland at Former Lake Karla, Greece, Using Ramsar Restoration Guidelines. *Environ. Manag.* **2004**, *34*, 875–886. [CrossRef]
57. Prahalad, V.N.; Kriwoken, L. Implementation of the Ramsar Convention on Wetlands in Tasmania, Australia. *J. Int. Wildl. Law Policy* **2010**, *13*, 205–239. [CrossRef]
58. Aminu, M.; Ludin, A.N.B.M.; Matori, A.-N.; Yusof, K.W.; Dano, L.U.; Chandio, I.A. A spatial decision support system (SDSS) for sustainable tourism planning in Johor Ramsar sites, Malaysia. *Environ. Earth Sci.* **2013**, *70*, 1113–1124. [CrossRef]
59. Wittmann, F.; Householder, E.; Lopes, A.; Wittmann, A.D.O.; Junk, W.J.; Piedade, M.T. Implementation of the Ramsar Convention on South American wetlands: An update. *Res. Rep. Biodivers. Stud.* **2015**, *4*, 47–58. [CrossRef]
60. Marín, V.H.; Delgado, L.E.; Tironi-Silva, A.; Finlayson, C.M. Exploring Social-Ecological Complexities of Wetlands of International Importance (Ramsar Sites): The Carlos Anwandter Sanctuary (Valdivia, Chile) as a Case Study. *Wetlands* **2017**, *38*, 1171–1182. [CrossRef]
61. Nhuan, M.T.; Ngoc, N.T.M.; Huong, N.Q.; Hue, N.T.H.; Tue, N.T.; Ngoc, P.B. Assessment of Vietnam Coastal Wetland Vulnerability for Sustainable Use (Case Study in Xuanthuy Ramsar Site, Vietnam). *J. Wetl. Ecol.* **2009**, *2*, 1–16. [CrossRef]
62. Cherkaoui, S.I.; Magri, N.; Hanane, S. Factors predicting Ramsar site occupancy by threatened waterfowl: The case of the Marbled Teal Marmaronettaangustirostris and Ferruginous Duck Aythya nyroca in Morocco. *Ardeola* **2016**, *63*, 295–309. [CrossRef]
63. Acuña-Piedra, J.F.; Quesada-Román, A. Evolucióngeomorfológica entre 1948 y 2012 del delta Térraba—Sierpe, Costa Rica. *Cuatern. Geomorfol.* **2016**, *30*, 49. [CrossRef]
64. Quesada-Román, A.; Acuña-Piedra, J.F. Efectosclimáticos y antrópicosen la morfogénesis de islaGuarumal, Humedal Nacional Térraba-Sierpe, Costa Rica. *Rev. Cienc. Ambient.* **2017**, *51*, 169–180. [CrossRef]
65. Acuña-Piedra, J.F.; Quesada-Román, A. Cambiosen el uso y cobertura de la tierra entre 1948 y 2012 en el Humedal Nacional Térraba-Sierpe, Costa Rica. *Rev. Cienc. Mar. Costeras* **2017**, *9*, 9. [CrossRef]
66. Acuña-Piedra, J.; Quesada-Román, A.; Vargas-Bolãnos, C. Coverage and Distribution of the Mangrove Species in the Térraba-Sierpe National Wetland, Costa Rica. *Anuário Inst. Geociências* **2018**, *41*, 120–129. [CrossRef]
67. Veas-Ayala, N.; Quesada-Román, A.; Hidalgo, H.G.; Alfaro, E.J. Humedales del Parque Nacional Chirripó, Costa Rica: Características, relacionesgeomorfológicas y escenarios de cambioclimático. *Rev. Biol. Trop.* **2018**, *66*, 1436. [CrossRef]
68. Quesada-Román, A.; Mora-Vega, A. Impactosambientales y variabilidadclimáticaen el humedal de San Vito, Coto Brus, Costa Rica. *Rev. Cienc. Ambient.* **2017**, *51*, 16. [CrossRef]
69. Kandus, P.; Minotti, P.G.; Morandeira, N.S.; Grimson, R.; Trilla, G.G.; González, E.B.; Martín, L.S.; Gayol, M.P. Remote sensing of wetlands in South America: Status and challenges. *Int. J. Remote. Sens.* **2017**, *39*, 993–1016. [CrossRef]
70. Al-Obaid, S.; Samraoui, B.; Thomas, J.; El-Serehy, H.A.; Alfarhan, A.H.; Schneider, W.; O'Connell, M. An overview of wetlands of Saudi Arabia: Values, threats, and perspectives. *Ambio* **2016**, *46*, 98–108. [CrossRef] [PubMed]
71. Bassi, N.; Kumar, M.D.; Sharma, A.; Pardha-Saradhi, P. Status of wetlands in India: A review of extent, ecosystem benefits, threats and management strategies. *J. Hydrol. Reg. Stud.* **2014**, *2*, 1–19. [CrossRef]
72. Meng, W.; He, M.; Hu, B.; Mo, X.; Li, H.; Liu, B.; Wang, Z. Status of wetlands in China: A review of extent, degradation, issues and recommendations for improvement. *Ocean Coast. Manag.* **2017**, *146*, 50–59. [CrossRef]
73. Hu, S.; Niu, Z.; Chen, Y.; Li, L.; Zhang, H. Global wetlands: Potential distribution, wetland loss, and status. *Sci. Total. Environ.* **2017**, *586*, 319–327. [CrossRef] [PubMed]
74. Strassburg, B.B.N.; Iribarrem, A.; Beyer, H.L.; Cordeiro, C.L.; Crouzeilles, R.; Jakovac, C.C.; Junqueira, A.B.; Lacerda, E.; Latawiec, A.E.; Balmford, A.; et al. Global priority areas for ecosystem restoration. *Nature* **2020**, *586*, 724–729. [CrossRef] [PubMed]
75. Zedler, J.B.; Kercher, S. WETLAND RESOURCES: Status, Trends, Ecosystem Services, and Restorability. *Annu. Rev. Environ. Resour.* **2005**, *30*, 39–74. [CrossRef]

sustainability

Article

Connectivity Predicts Presence but Not Population Density in the Habitat-Specific Mountain Lizard *Iberolacerta martinezricai*

Diego Lizana-Ciudad [1], Víctor J. Colino-Rabanal [1,*] , Óscar J. Arribas [2] and Miguel Lizana [1]

[1] Section of Zoology, Department of Animal Biology, Parasitology, Ecology, Edaphology and Agronomic Chemistry, Campus Miguel de Unamuno, University of Salamanca, 37071 Salamanca, Spain; helgait@usal.es (D.L.-C.); lizana@usal.es (M.L.)

[2] Ntra. Sra. de Calatañazor 17 b, 42004 Soria, Spain; oarribas@xtec.cat

* Correspondence: vcolino@usal.es; Tel.: +34-676-643-770

Abstract: The Batuecan lizard *Iberolacerta martinezricai* is a critically endangered species due to its significantly reduced distribution, which is restricted to the scree slopes (SS) of a few mountain peaks within the Batuecas-Sierra de Francia Natural Park (western Spain). Given its high specialisation in this type of discontinuous habitat, the long-term conservation of the species requires maintaining the connectivity between populations. This study analyses the contribution of connectivity, as well as other patch-related factors, in the distribution and density patterns of the species. With this aim, 67 SS were sampled by line transects from May to October 2018. Each SS was characterised using variables indicative of the microhabitat conditions for the lizard. Inter-SS connectivity was quantified using graph theory for seven distances. Generalised linear models (GLMs) were performed for both presence and density. Model results showed that while connectivity was a relevant factor in the presence of lizards, density only involved patch-related variables. Discrepancies probably occurred because the factors influencing presence operate on a wider scale than those of abundance. In view of the results, the best-connected SS, but also those where the lizard is most abundant and from which more dispersed individuals are likely to depart, seem to be the essential patches in any conservation strategy. The results may also be relevant to other species with habitat-specific requirements.

Keywords: Batuecan lizard; connectivity; endangered species; graph theory; *Iberolacerta martinezricai*; mountain lizard; network analysis

Citation: Lizana-Ciudad, D.; Colino-Rabanal, V.J.; Arribas, Ó.J.; Lizana, M. Connectivity Predicts Presence but Not Population Density in the Habitat-Specific Mountain Lizard *Iberolacerta martinezricai*. *Sustainability* **2021**, *13*, 2647. https://doi.org/10.3390/su13052647

Academic Editors: Antonio Martinez Graña, José Ángel Sánchez Agudo, Gioele Capillo and C. Ronald Caroll

Received: 9 January 2021
Accepted: 25 February 2021
Published: 2 March 2021

Publisher's Note: MDPI stays neutral with regard to jurisdictional claims in published maps and institutional affiliations.

1. Introduction

Although climate warming poses a major threat to the terrestrial vertebrate species confined to high mountain areas, not all species are equally vulnerable [1–3]. Ectotherms are particularly sensitive because they depend on external heat sources to maintain body temperature [4]. In these species, an increase in temperature is likely to induce fast growth and thermal stress, which can accelerate the rate of ageing [5]. To cope with the variation in environmental conditions, species can either try to adapt their physiology or behaviour to extend their thermal niche to these new conditions, or move to new areas where their thermal niche is now located [6,7]. The first option, acclimatisation, seems limited as mountain species tend to be cold-specialists [8], experiencing a greater decline in fitness if body temperatures exceed the optimum [9–12]. The second one, elevational shifts, are not an option for mountaintop species, as there are no colder areas available to migrate to [13,14]. In addition, mountain species may compete with and be displaced by generalist species that are present in the surrounding lowlands and that may move into higher altitudes with increasing temperature [1,15]. All of these factors make mountain ectotherms, such as the Batuecan lizard *Iberolacerta martinezricai*, among the most vulnerable animals in the world [2].

Iberolacerta martinezricai is a medium-sized, globally threatened rock lizard of very reduced spatial distribution that lives in less than 20 km^2 of discontinuous habitat within

67

the limits of the Batuecas-Sierra de Francia Natural Park (Castile and Leon, western Spain) [16]. *I. martinezricai* is the most restricted, and probably one of the rarest, threatened reptile species in continental Europe.

Due to the projected scenarios of climate warming and drought for the Mediterranean mountain region [17,18], the predicted changes in the altitudinal and latitudinal distribution for the Spanish herpetofauna indicate a significant decline in suitable habitat for the species [19–21]. In addition, two other Mediterranean lizard species are increasingly found near the summit of the Peña de Francia mountain: *Podarcis guadarramae* and *Timon lepidus*. The presence of these lizards could indicate the advance of Mediterranean forms toward the summit of the mountain, corresponding to a progressive cornering of the Batuecan lizard in the highest areas by competition and climatic causes [22].

Nevertheless, climate warming and probable interspecific competition are not the only threats that the Batuecan lizard faces. Although the perceived remoteness and inaccessibility of these mountain areas give them some natural protection from anthropogenic impacts, human pressure on these ecosystems has increased markedly in recent decades [23]. Thus, recreational activities in mountains are associated with reduced vertebrate richness and abundance [24,25], particularly in lizards [26,27]. In the case of *I. martinezricai*, within its area of distribution there is a mountain monastery frequented by tourists, which makes human pressure in the area relevant. This impact is particularly important during summer months, which coincides with the active period of this species [22]. The dirt track to the monastery was constructed in 1920 and asphalted in 1961, facilitating massive access for motor vehicles. The highly visible, characteristic antenna was installed in the 1970s. These landmarks show the increasing human pressure upon the area over the past hundred years [28].

Access roads to the summit and unpaved roads in the vicinity may also affect the species. Even though small lizards seem not to be the group most affected by road-kill [29,30], other impacts of the roads, such as the barrier effect or fragmentation, can influence their populations [31]. Moreover, the asphalting of the road at the summit caused, at least, necrosis of the fingers of adult specimens, which could be due to sunbathing in contact with the pasty asphalt. These specimens had asphalt-coated toes and had lost parts of those toes, with naked phalanges poking out [32]. Human presence in the area can also favour the occurrence of wildfires [33], which can irremediably change the microclimate of the areas concerned.

All of these pressures and threats to the species are compounded by the fact that *I. martinezricai* is a strictly saxicolous lizard specialising in a particular type of discontinuous and naturally fragmented habitat, such as scree slopes (SS), which are interspersed with grass and broom mountain vegetation [16,22]. Although the Batuecan lizard has evolved in this naturally fragmented habitat, current human pressures may compromise the dynamics necessary to maintain its viability in the medium and long term. The negative effects of fragmentation on lizards have been widely documented. Fragmentation can alter home ranges and movements [34], demographic structure [35], or community disassembly [36]. Not all species are equally susceptible, and those with the strongest habitat preferences show the greatest response to landscape fragmentation [35,37,38].

Thus, as it is a lizard highly specialised in a specific habitat and that most probably responds to a classical model of metapopulation dynamics, it is crucial to study the importance of the connectivity between the favourable patches for the species. Previous studies have shown that some patch-related variables shape the distribution and the abundance of lizard populations in the SS [16], but the spatial structure of the SS network and the importance of each SS within this network have not been evaluated. With this objective, this study aims to assess whether connectivity between SS, together with other intra-SS factors, contributes to defining the patterns of presence and density of the Batuecan lizard in the Batuecas-Peña de Francia Natural Park. It is hypothesised that connectivity will be more related to presence than to lizard density, as the latter is more likely to be closely linked to intra-patch availability of resources. If such discrepancies exist, it is likely that certain regionalisation within the species' own range will emerge. Understanding these dynamics will help guide management and conservation plans for this threatened species.

2. Materials and Methods

2.1. Study Area

The study area is in the Las Batuecas-Sierra de Francia Natural Park (BSFNP), Province of Salamanca (Central Spain). BSFNP has been recognised as a biosphere reserve by UNESCO as well as a Special Protection Area and Special Area of Conservation within the Natura 2000 Network. BSFNP is part of the western area of the Sistema Central, with the Pico Hastiala (1735 m) and the crest of the Peña de Francia (1723 m) as its highest peaks (Figure 1). The area is characterised by a Mediterranean mountain climate, with hot dry summers and cold winters in which, along with in spring, the maximum rainfall occurs [39].

Figure 1. (**a**). Location of the study site within the Iberian Peninsula and silhouette of the Batuecas-Sierra de Francia Natural Park. The approximate distribution area of the Batuecan lizard *Iberolacerta martinezricai* is indicated by a red line; (**b**). Three-dimensional image of the mountainous area where the specie is located; (**c**). Scree slope, the Batuecan lizard habitat, interspersed with broom mountain vegetation; (**d**). Specimen of male *Iberolacerta martinezricai*. Photos (**c**,**d**): D. Lizana Ciudad.

Regarding the geology, the area is dominated by Palaeozoic metamorphic materials: quartzites, schists, and slates. The differential erosion of these materials has formed a rugged relief with peaks comprising the hardest Armorican quartzite (Hastiala, Peña de Francia, Mesa del Francés, Rongiero), and deep river valleys in the areas with less resistant rocks. The peaks (hanging synclines) form the watershed divide between the basins of the Tagus and the Duero, with a marked contrast between the northern and southern parts due to the difference in altitude between the Castilian and Extremaduran plateaus. The slopes of the quartzite crests are covered by colluvium and screes that have been partly phytostabilised because of processes that are still active and linked to a periglacial morphogenetic context. These SS are the characteristic habitat of the *I. martinezricai* lizard.

Due to the differences in altitude, up to three bioclimatic levels are represented, each with a characteristic climate and vegetation, The Oro-Mediterranean forms the highest areas (altitude above 1600 m.a.s.l.), where mountainous shrub vegetation dominates, mainly *Cytisus oromediterraneus*. The Supra-Mediterranean level (1000–1600 m.a.s.l.) is where the potential vegetation comprises Pyrenean oak *Quercus pyrenaica* forests and a shrub stratum of heather *Erica arborea* and *Erica australis*; some scattered plantations of *Pinus nigra* and

Pinus sylvestris are also present. The lower parts belong to the Meso-Mediterranean level (400–1000 m.a.s.l.), which is characterised by wood pastures of holm oak *Quercus ilex* and cork oaks *Quercus suber*.

2.2. Focal Species

The Batuecan lizard *I. martinezricai* is one of the 8 species of the genus distributed in different mountain ranges of the Iberian Peninsula and the Alps. The speciation of the genus *Iberolacerta* Arribas, 1997 is likely to have occurred by allopathic divergence during the late Tertiary because the complex relief of southern Europe provided isolated habitats with limited possibilities of dispersion between them [40]. This may have been helped by competition from wall lizards (genus *Podarcis*) present in the lower areas [41]. Mountains acted as refuges during the glacial oscillations of the Quaternary [40]. Genetic analyses show that the Batuecan lizard has been present in the mountainous area it currently inhabits for at least 2 ± 0.3 Ma [41].

The area of distribution for *I. martinezricai* is restricted to an area of 20 km^2 of discontinuous habitat within the boundaries of the Batuecas-Sierra de Francia Natural Park. Specifically, the lizard is contained in a polygon formed by the peaks of Peña de Francia, Hastiala, Rongiero, and Peña Orconera, on whose steep SS the habitat available for the species is concentrated [16,28]. Although it can be found from 800 m.a.s.l., it mainly occupies medium and high altitudes, above 1300 m.a.s.l. These mountains are surrounded by low-altitude areas with unsuitable habitats for the lizard. In terms of abundance, maximum densities of 50 individuals per hectare are reached [16], although this is lower on most SS. It prefers habitats with a certain amount of humidity and with the presence of mosses and lichens [16,28]. The total number of specimens for the species has been estimated to be between 1200 and 1500 individuals [16], distributed in several fragmented populations with a different degree of connectivity between them. Due to this reduced distribution, it has considered critically endangered since 2006 (CR B2ab(v), C2a(ii) in the IUCN Red List) [32,42,43]. Since 2019, it is also listed as Endangered in the Spanish National Catalogue of Threatened Species. Thus, the species is one of the European reptiles in most urgent need of conservation [40].

The species is a medium-sized lizard with a dark dorsal colouration, usually brown or grey with a black, sometimes greenish, reticulate, and bright blue axillary ocelli (between one and three) in most individuals. Sexual dimorphism is less pronounced than in other rock lizards. Sexual maturation is probably reached when the lizards are three years old, and the longest living recorded specimens were 8 years old [44]. The temperature range within which the species remains active is relatively wide (22.8–39.2 °C). It is a good thermoregulator in relation to substrate temperature, but not so good with respect to air temperature [28]. Regarding reproduction, copulation begins at the end of April and oviposition takes place during the summer months (especially at the end of July). The size of the clutches ranges from two to six eggs, with an average of 4.23. The incubation period lasts between 33 and 42 days [45]. There are still no data on species movement, its capacity to disperse, or its willingness to cross different habitats of the landscape matrix in which the SS are located.

2.3. Field Surveys and Density Estimation

Sampling was carried out from May to October 2018, covering the activity period of the lizard. During the spring, the Batuecan lizard shows a unimodal pattern of activity. In summer, the pattern is more bimodal, with a main peak during the first hours of sunlight and a marked reduction afterwards as the temperature increases. Evening activity is much less [28]. Patches of the characteristic habitat of these lizards (mountain SS) were positively selected as the preferred study sites. Less favourable areas for the species were a priori excluded from the surveys according to our current knowledge of its habitat preferences, namely low-elevation areas with warmer temperatures. At each SS, surveys were performed between 08:00 h and 19:00 h GMT along linear transects of about 100 m [46] (average of 96.89 m; differences are due to accessibility difficulties in some SS, given the abrupt topography of the terrain). At least one transect per 10 hectares of scree surface was carried out in each SS (average of 1 transect per 3.58 ha). For SS with more than one transect, sampling was conducted on the same day and transects were separated by at least 50 m to avoid double counting. All individuals seen in a 3 metre wide band on either side of the transect were recorded. The distance at which individuals were observed was measured perpendicularly to the transect line with a Bosch DLE50 Professional laser distance meter. Whenever possible, the sex and age of the individuals reported were identified.

Distance 6.0 software [47–49] was used to estimate the density and abundance of lizards from the observations collected during the fieldwork. The number of data were sufficient to achieve an adequate adjustment of the detection function [50]. Using the lowest value of the Akaike Information Criterion and the smallest proportion of the χ^2 goodness-of-fit divided by its degree of freedom, the detection function model that best fit the study data was selected [48,49]. In this case, a half-normal key function with cosine adjustment was adopted.

2.4. Connectivity Analysis for I. martinezricai

Spatial connectivity is defined as the ability of the landscape to facilitate or impede the movement of individuals of a species between fragments of habitat. It is usually divided into two components: structural and functional [51]. The structural component is determined by the spatial connection of different types of habitat in the landscape, and the functional component refers to the response of individuals and species to the physical structure of the landscape. The latter is influenced by the species' habitat requirements, tolerance of altered habitats, and life stage. In this sense, species, although living in the same habitat, have different behavioural responses and therefore experience different levels of connectivity [52].

The study of ecological networks and connectivity is often carried out using graph theory [53–55]. In terms of landscape ecology, a network consists of a set of favourable habitat patches, the nodes (in this case the SS, the habitat of *I. martinericai*), embedded in a non-favourable habitat matrix [56]. If any ecological flow occurs between any two nodes of the network, both nodes are said to be "connected" [57]. These analyses are useful for the design of conservation strategies [58,59].

For the connectivity analysis of the metapopulations of lizards within the SS network, the probability of connectivity index (PC), based on this graph theory, was used [60]. As the Batuecan lizard is strongly saxicolous and SS are discrete geographical units, the definition of the nodes was relatively simple. The analyses were carried out using the CONEFOR software [61]. This index describes the probability that two organisms situated at random in the landscape are in patches (in this case, SS) that are interconnected. Therefore, PC indicates the importance of each SS within the network. The mathematical expression for PC is:

$$PC = \frac{\sum_{i=1}^{n} \sum_{j=1}^{n} a_i a_j \times p_{ij}^*}{A_L^2} = \frac{PC_{num}}{A_L^2}$$

in which n, in this case, is the number of SS in the study zone (not only those sampled in the study, but all of the mapped SS, a total of 156); a_i, a_j are the areas of the SS i and j; A_L is the area of the study zone, including the SS and the landscape matrix located between them; and p_{ij}^* is the maximum probability of dispersion between i and j, which is determined by the following negative exponential function [60]:

$$p_{ij} = e^{-kd_{ij}}$$

where the p_{ij} value is usually calculated from d_{ij} distances obtained from mark-recapture data or genetic analysis. The software adjusts the previous function for these values, obtaining a constant k, from which the dispersion probabilities are calculated and, therefore, the connectivity probability for each SS. In this case, as there is no information on the capacity of dispersion and mobility between SS, connectivity was simulated for different distances. Thus, connectivity probabilities of 50% for seven distances were tested: 25, 50, 100, 250, 500, 1000 and 10,000 (practically unlimited connectivity) metres. Considering the characteristics of the species, we believe that this distance interval includes both the most frequent events of short distance dispersal and events of colonisation of distant areas.

Saura and Rubio [62] showed that PC can be divided into three fractions quantifying the different ways in which a habitat patch (in this case, a SS) can contribute to connectivity: dPC_{intra} shows the contribution of a given SS in terms of intrapatch connectivity, and it is related to the habitat availability; dPC_{flux} corresponds to the attribute-weighted dispersion flow (usually area-weighted) through the maximum-likelihood paths of a given SS with all other SS in the network when it is the point of origin or destination of that flow; and $dPC_{connector}$ for a given SS quantifies the contribution of that SS to the connectivity between the rest of the network. A SS can only contribute to this fraction if it is part of the maximum-likelihood paths between SS other than itself. dPC_{flux} and $dPC_{connector}$ are measured in the same units and can be directly compared to each other.

All of the potential SS habitats (not only those sampled) for the species in its small natural range were included in the connectivity analysis. SS cartography was obtained from the Spanish Land Occupancy Information System (SIOSE) developed by the National Geographic Institute (IGN), which is the most detailed database of land occupation for all of Spain, at a reference scale 1:25,000. Although SIOSE shows the main SS in the area with remarkable precision, some tesserae are mapped as mixed units that include smaller SS, which are probably also interesting for the species. The well-defined SS in the orthophotos within these mixed units were mapped and joined with the rest of the SS for analysis.

2.5. Presence and Density Models for I. martinezricai

A species inhabits those areas where abiotic conditions are favourable and where the community of species allows its coexistence, and in accessible places that could be colonised in both evolutionary and ecological terms [63]. All of these factors interact dynamically to produce a geographic range that is unique to each species [64]. These distribution ranges may shift, expand, or contract in response to changes in environmental conditions [65]. On the other hand, population density is also related to environmental characteristics, individual traits, and intra- and interspecific interactions [66,67]. However, the factors involved and/or their intensity do not necessarily have to be the same to determine presence and abundance.

Together with the probability of connectivity index (PC), the presence and density models included as explanatory variables those already identified as relevant for the species in previous studies [16]. Although climatic data from sources such as WorldCLIM2 [68] are often used in presence/abundance models, the resolution of these sources of information is not valid to account for the different microclimates that contribute to explaining the specific and reduced geographical range of *I. martinezricai*. For the same reason, climatic data from one or more nearby weather stations would not have been useful either. To overcome these limitations, a set of variables related to topography that could well reveal microclimatic aspects relevant to the species presence or density was selected (altitude ALT and orientation ORI). As it seems that the species positively selects those SS under which there is more humidity, the Topographic Humidity Index (TWI) was introduced into the analyses. TWI describes the tendency of a "cell" to accumulate surface water [69]. The abundance of lichens and bryophytes in SS rocks (LIC) (related to humidity levels, the slope (SLP), and rock size (ROS)) were also considered. The description of all of the variables considered is included in Table 1.

Table 1. Description of the variables used in the presence and density models for *I. martinezricai* in the Batuecas-Sierra de Francia Natural Park.

Variable	Abbrev.	Description
Altitude	ALT	Obtained from a 5 metre resolution DEM, developed by the Spanish National Geographic Institute through interpolation from LIDAR data. The minimum (ALTmin), average (ALTave), maximum elevation (ALTmax), and altitudinal range (ALTran) were considered.
Slope	SLP	Obtained from the same DEM, expressed in degrees.
Orientation	ORI	From the 5 metre DEM using the ArcGIS 10.6 "aspect" tool. Five orientations were defined: North (N), South (S), East (E), West (W), and flat.
Probability of connectivity	PC	An indicator of landscape connectivity based on habitat availability, dispersal probabilities between habitat patches and graph structures (see Section Section 2.4). Obtained from SS cartography and CONEFOR software.
Topographic wetness index	TWI	Describes the tendency of an area to accumulate water. Obtained from the 5 m DEM using ArcGIS geoprocesses.
Lichen cover	LIC	Average coverage of lichens and bryophytes on the rocks.
Rock size	ROS	Each SS was classified in relation to the average size of the rock fragments. Three categories were considered: 1 (<50 cm), 2 (50–100 cm), 3 (>100 cm).

To investigate the relative importance of each variable in species presence and density, generalised linear models (GLMs) were performed using R [70]. A preliminary statistical comparison between SS with and without the lizard was carried out using non-parametric Mann–Whitney U tests for the quantitative continuous variables and the Pearson Chi-Square test for the categorical one (ORI). GLM with a logit-link function and binomial distribution were used to model presence/absence data and GLM with a log-link function gamma distribution to model species density. To avoid introducing noise into the GLMs, potential explanatory variables that did not provide information were previously eliminated. Collinearity among explanatory variables was explored using tolerance levels, in addition to Goodman and Kruskal's gamma, to test for categorical variables against other variables. Based on this, the variables that were highly correlated were removed. Tolerance levels were sufficiently high (>0.1), and Goodman and Kruskal's gamma coefficients were low enough (<0.4) to allow the inclusion of the variables included here.

For both presence and density, a full model with all predictors was tested first. We then removed variables that were clearly insignificant. To select from the multiple models, the Akaike's information criterion (AICc) was used as a measure of overall model fit [71]. For the best fit GLM models (ΔICc = 0), the calc.relimp procedure was implemented in R-package "relaimpo" [72] to decompose the variance of the final models among the different predictors and interactions.

3. Results

Sampling Results

A total of 67 SS were sampled, with the species present in 30 of them (44.7%). The total number of individuals observed was 154, including adults (49%, 23% males and 26% females), sub-adults (22%), and juveniles (13%). No proper identification could be made in 16% of the cases. In general, the specimens were detected when the temperature was not very high (around 22–25 °C) and in the early morning hours. The average density of lizards for the SS with confirmed species presence was 41.44 ± 27.81 individuals/ha, with a maximum of 99.39 individuals/ha. Figure 2 shows the SS sampled and those where the presence of the species was detected. Table S1 (see Supplementary Material) shows the results for each SS including area, number of transects, mean length, number of individuals located, and density.

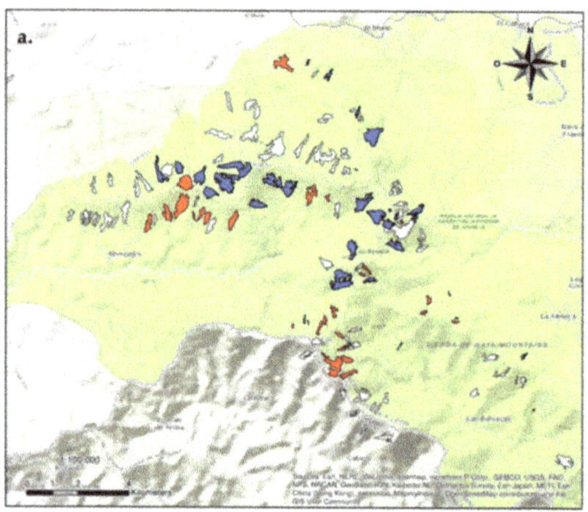

Figure 2. Map of the scree slopes with a presence of *I. martinezricai* (in blue) and those where it was not found during the fieldwork (in red), both in two dimensions (**a**) and in three dimensions (**b**). The scree slopes (SS) that appear without colour were not sampled in this study but are potential habitats for the species.

The results of the Mann–Whitney U and Pearson Chi-Square tests are shown in Table 2. The SS with lizards present showed a higher maximum, average and minimum altitude, lower slope, and higher PC (at all distances considered) than SS without the species. The three elements that compound the PC also showed significant differences. Rock size and lichen coverage were greater in the positive SS. No differences were found for the range of altitudes and TWI (Table 1), nor for the orientation according to the Pearson Chi-Square test.

Table 2. Average values in the SS with and without the presence of the species and the results of the Mann–Whitney U (quantitative variables) and Pearson Chi-Square (categorical variables) tests comparing both groups. The significance level was 0.05. Statistically significant differences are shown in bold.

	SS Positive (Average)	SS Negative (Average)	U Mann Whitney	Z	*p*-Value
ALT_{min}	1330.1	1180.5	241.5	−3.612	**<0.001**
ALT_{max}	1510.7	1344.8	189	−4.319	**<0.001**
ALT_{ave}	1421.6	1263.9	198	−4.197	**<0.001**
ALT_{ran}	180.7	164.3	463	−0.632	0.527
SLP	24.77	26.50	340	−2.287	**0.022**
TWI	9.41	9.03	408	−1.372	0.170
PC_{10000}	2.05	1.13	318	−2.583	**0.010**
PC_{1000}	3.46	1.31	264	−3.310	**0.001**
PC_{500}	3.61	1.30	268	−3.256	**0.001**
PC_{250}	3.43	1.26	287	−3.001	**0.003**
PC_{100}	3.39	1.29	228	−3.311	**0.001**
PC_{50}	3.53	1.26	253	−3.331	**0.001**
PC_{25}	3.49	1.29	228	−3.211	**0.001**
PC_{intra} *	0.13	0.06	318	−2.583	**0.010**
PC_{flux} *	2.50	0.92	262	−3.336	**0.001**
$PC_{connector}$ *	0.98	0.32	327	−2.464	**0.014**
LIC			275.5	−2.542	**0.011**
ROS			205.5	−3.334	**0.001**
			χ^2	d.f.	*p*-value
ORI			3.871	3	0.276

* The PC_{intra}, PC_{flux}, $PC_{connector}$ values shown in the table are those obtained for the PC_{50}, the distance for which a better adjustment of the presence models was obtained (see Table 3).

The results of presence and density models are shown in Table 3. The most parsimonious models included four variables (ALTave, PC_{50}, ROS, and SLP) for the presence model and 2 (LIC and ALTave) for the density model. The best presence model showed a total deviance of 48.29%. ALTave contributed 38.1%, PC_{50} 24.4%, ROS 27.8%, and SLP 9.7%. The total deviance explained by the best density model was 30.63%. LIC contributed 85.7% and ALTave 14.2%.

Table 3. Ranking of generalised linear models for Batuecan lizard presence and density, using Akaike's Information Criterion corrected for small sample sizes (AICc). Total deviance explained (%$\sum$dev) for the best models, and % explained by the main factor are also shown.

Candidate Presence Models	ΔAIC_c	%$\sum$dev	Candidate Density Models	ΔAIC_c	%$\sum$dev
ALTave+PC_{50}+ROS+SLP	0	48.29	LIC+ALTave	0	30.63
ALTave+PC_{50}+ROS+SLP+LIC	1.63	ALT_{ave} (38.1)	LIC	0.50	LIC (85.7)
ALTave+PC_{50}+ROS	3.64		LIC+ALTave+PC_{25}	2.00	
ALTave+PC_{50}	10.35		LIC+ALTave+PC_{25}+SLP	3.50	
ALTave	20.46		LIC+ALTave+PC_{25}+SLP+ROS	5.50	
PC_{50}	22.91		ALTave	7.00	
PC_{25}	24.05		PC_{25}	7.00	
PC_{100}	24.29		SLP	7.10	
PC_{1000}	25.85		ORI	7.10	
PC_{500}	26.12		PC_{500}	7.40	
PC_{250}	26.54		PC_{250}	7.40	
PC_{10000}	26.87		ROS	7.40	
ROS	27.11		PC_{10000}	7.50	
SLP	33.14		PC_{100}	7.50	
LIC	34.33		PC_{1000}	7.60	
ORI	38.90		PC_{50}	7.60	

4. Discussion

According to the results, it seems that the species presence involves variables related to connectivity as well as SS characteristics. Species density is only influenced by the microhabitat conditions present in each SS. Thus, the probability of lizard presence in an SS is correlated with the importance of that SS for the connectivity of the SS network, higher altitudes, less pronounced slopes, and large rock sizes. The variable that explained the highest percentage of the deviance was the altitude of the SS. A large rock size favours the presence of cavities that serve as refuges for the lizards, storing fresh air and generating a more humid environment, probably with a greater density of prey [16]. As mass movements are gravitational phenomena, less pronounced slopes favour larger rock sizes in the SS. The probability of occurrence for the species is minimal for SS in low and thermophilic areas, far from and poorly connected to the distribution centre. On the contrary, lizard density depends not on the position of the SS within the network but on the characteristics of each SS. The populations of *I. martinezricai* are more abundant in the SS with greater coverage of lichens and mosses (largely responsible for the explained deviance), indicators of greater humidity and possibly greater availability of prey. This abundance and presence respond to different factors, or at least there are variations in the weight of their contribution, which has already been described for other species of lizards [73]. The discrepancies are probably due to the fact that the factors influencing presence operate on a wider scale than those of abundance [74]. This seems to be the case for the Batuecan lizard, where the presence is shaped by both inter-SS spatial relations and intra-SS characteristics, and density only by the latter. However, it depends on the species, since, for example, in forest lizards, it has been found that habitat quality is also the most relevant variable in their spatial distribution [75].

The importance of the connectivity indicates that the Batuecan lizard probably acts as a metapopulation in which each SS constitutes a small subpopulation, reaching an equilibrium over time by compensating for local extinctions through recolonisations from other SS [76]. The best fit of the models was obtained for PCs with small dispersal distances, which could indicate a limited capacity of the species to reach distant SS. This limited capacity, together with the importance of intra-SS characteristics in lizard density, is likely to lead to the formation of clusters of interacting individuals, called neighbourhoods, that are regionally organised into continuous networks within the total range of the species [77,78]. In addition, connectivity was shown to be important for all distances tested, which may be indicating population flows operating at different scales. Structural connectivity shapes the spatial distribution of the lizard. The three fractions in which PC can be divided showed differences between SS with and without lizard presence. This means that both habitat availability and dispersal flows are important for species presence in a given SS.

The results for lizard presence are maintained over time, since the current distribution is very similar to that obtained 10 years ago [16], which shows a remarkable stability. The scarce changes between the two periods are mainly located in SS with low density for the species in the low altitude peripheral areas of its distribution area. These differences could be explained by the difficulty of detection, conditioned by the meteorology of the sampling days. In contrast, the comparison for density does not show such a sustained pattern over time. This may be due to several causes. First, the sampling in this study was carried out using transects rather than plots in order to cover a greater number of SS. Secondly, this is a species present in very low densities, so changes in the detection of one or two individuals, strongly influenced by weather conditions, can lead to significant changes in the densities of lizards obtained. These uncertainties mean that the results for the density models should be taken with caution.

Connectivity is especially relevant in the context of climate change. With increasing temperatures in summer and a reduction in humidity in the SS, it is possible that the species may experience an elevational shift, abandoning the patches located at lower altitudes where habitat conditions seem better for thermophilic generalists [15]. Individuals need to be given the opportunity to reach other, more suitable SS, under new environmental conditions. In addition, mountain species may be preyed on, compete with, and be displaced by generalist species present in the surrounding lowlands that may go to higher altitudes. For example, during the fieldwork, several species that can compete or prey on *I. martinezricai* were found within the same altitudinal range, such as the smooth snake *Coronella austriaca*, the Lataste's Viper *Vipera latastei*, the Lusitanian lizard *Podarcis guadarramae*, or the Algerian psammodromus *Psammodromus algirus*. *P. guadarramae* is likely to be the main candidate to outcompete or substitute *I. martinezricai*, as it is also saxicolous and occupies the same habitat. However, at present, there are not enough data available to describe the potential impact of these species on the Batuecan lizard populations.

Furthermore, considering connectivity in the specific case of the Batuecan lizard is important because, unlike other mountain species that inhabit remote areas with low human pressure, the presence of the monastery makes the mountains that constitute the small area of distribution of the lizard very popular for tourism. Although tourists do not usually have access to SS, tourism pressure on the Peña de Francia can affect the quality of habitat for *I. martinezricai* (i.e., with food remains) and attract opportunistic species that can predate upon lizards (foxes, stone martens, and rats, among others). Perhaps more importantly, this human presence can favour the occurrence of forest fires, which are frequent in the area. The population consequences of fragmentation due to human pressures on different lizard species have been widely documented [36,79,80].

The results, therefore, highlight the need to include, with particular emphasis, connectivity criteria in the management and conservation plans of threatened species that are highly specialised in discontinuous habitats, to maintain their populations in the long term. Active or passive strategies need to be developed to promote accessibility among different SS, ensuring movement of individuals and gene flow between patches [81–83]. Connectivity analyses also allow the identification of key patches [84]. However, not only the best-connected SS, but also those where the lizard is most abundant and from which more dispersed individuals are likely to depart, seem essential in any conservation strategy for the species.

This connectivity analysis should be complemented and extended by other detailed studies based on genetic analysis [85] and movement resistance in relation to the different land uses between the SS, quantifying the effect of the configuration and quality of the matrix in gene flow and genetic variation [55,86]. Furthermore, future studies on the effects of climate change, thermal ecology, predation, and competition with other lizards are essential for the design of appropriate management tools for the species.

5. Conclusions

Connectivity, together with patch-related variables, are important factors in explaining the presence of the critically endangered Batuecan lizard *I. martinezricai* in the SS of the Batuecas-Sierra de Francia Natural Park. In contrast, lizard density in each SS is not related to the spatial configuration of the SS network, but only patch-related variables have an influence, especially those indicative of a higher degree of humidity. Considering the metapopulation dynamics that emerge in these discontinuous habitats, the discrepancies in the variables related to lizard presence and density mean that, in terms of conservation, it is necessary to give special importance to two types of SS. On the one hand, those SS that act as the main connectors within the network favour the recolonisation processes (which counteracts local extinctions). On the other hand, those SS whose internal characteristics favour abundant populations that feed the stochastic processes are involved in the long-term maintenance of all these metapopulation dynamics. Based on these two groups of SS, we would expect to see the formation of "neighbourhoods" of interacting individuals, organised regionally within the total range of the lizard. These spatial dynamics should be taken into account when designing conservation plans for the species. Moreover, the results may also be relevant to other species with habitat-specific requirements.

Supplementary Materials: The following are available online at https://www.mdpi.com/2071-105 0/13/5/2647/s1, Table S1: Observed presence and absence, and density (lizards/Ha.) of *Iberolacerta martinezricai* at the 67 scree slopes.

Author Contributions: Conceptualization, D.L.-C., V.J.C.-R. and M.L.; methodology, D.L.-C., V.J.C.-R., Ó.J.A. and M.L.; formal analysis, D.L.-C., V.J.C.-R. and M.L.; investigation, D.L.-C., V.J.C.-R. and M.L.; resources, V.J.C.-R.; data curation, D.L.-C. and M.L.; writing—original draft preparation, V.J.C.-R., D.L.-C., Ó.J.A. and M.L.; writing—review and editing, V.J.C.-R.; visualization, D.L.-C., V.J.C.-R. and M.L.; supervision, M.L.; project administration, M.L.; funding acquisition, M.L. All authors have read and agreed to the published version of the manuscript.

Funding: The study was financed by the Regional Government of the Junta de Castilla y León, Spain. "SA26/18-Monitoring the state of conservation of the populations of Iberolacerta martinezricai in the Natural Park of Las Batuecas-Sierra de Francia (Salamanca)".

Institutional Review Board Statement: Not applicable.

Informed Consent Statement: Not applicable.

Data Availability Statement: The data presented in this study are available on request from the corresponding author.

Acknowledgments: The authors wish to acknowledge the support provided by the Director of the Natural Park of Batuecas-Sierra de Francia, Juan Carlos Velasco, the Head of Section for Natural Areas and Protected Species of the Regional Ministry of the Environment of the Junta de Castilla y León in the Province of Salamanca, Roberto Carbonell, and many other environmental workers from the Junta de Castilla y León and the Fundación Patrimonio Natural in Salamanca and in the Natural Park of Batuecas-Sierra de Francia.

Conflicts of Interest: The authors declare no conflict of interest. The funders had no role in the design of the study; in the collection, analyses, or interpretation of data; in the writing of the manuscript, or in the decision to publish the results.

References

1. Araújo, M.B.; Thuiller, W.; Pearson, R.G. Climate warming and the decline of amphibians and reptiles in Europe. *J. Biogeogr.* **2006**, *33*, 1712–1728. [CrossRef]
2. Sinervo, B.; Mendez De La Cruz, F.; Miles, D.B.; Heulin, B.; Bastiaans, E.; Villagran Santa Cruz, M.; Lara Resendiz, R.; Martinez Mendez, N.; Calderon Espinosa, M.L.; Meza Lazaro, R.N.; et al. Erosion of lizard diversity by climate change and altered thermal niches. *Science* **2010**, *328*, 894–899. [CrossRef]
3. Foden, W.B.; Young, B.E.; Akçakaya, H.R.; Garcia, R.A.; Hoffmann, A.A.; Stein, B.A.; Thomas, C.D.; Wheatley, C.J.; Bickford, D.; Carr, J.A.; et al. Climate change vulnerability assessment of species. *Wiley Interdisc. Rev. Clim. Chang.* **2019**, *10*, e551. [CrossRef]
4. Aragón, P.; Rodríguez, M.A.; Olalla-Tárraga, M.A.; Lobo, J.M. Predicted impact of climate change on threatened terrestrial vertebrates in central Spain highlights differences between endotherms and ectotherms. *Anim. Conserv.* **2010**, *13*, 363–373. [CrossRef]
5. Burraco, P.; Orizaola, G.; Monaghan, P.; Metcalfe, N.B. Climate change and ageing in ectotherms. *Glob. Chang. Biol.* **2020**, *26*, 5371–5381. [CrossRef]
6. Maggini, R.; Lehmann, A.; Kéry, M.; Schmid, H.; Beniston, M.; Jenni, L.; Zbinden, N. Are Swiss birds tracking climate change? Detecting elevational shifts using response curve shapes. *Ecol. Modell.* **2011**, *222*, 21–32. [CrossRef]
7. Donelson, J.M.; Sunday, J.M.; Figueira, W.F.; Gaitán-Espitia, J.D.; Hobday, A.J.; Johnson, C.R.; Leis, J.M.; Ling, S.D.; Marshall, D.; Pandolfi, J.M.; et al. Understanding interactions between plasticity, adaptation, and range shifts in response to marine environmental change. *Phil. Trans. R. Soc. B* **2019**, *374*, 20180186. [CrossRef]
8. Aguado, S.; Braña, F. Thermoregulation in a cold-adapted species (cyren's rocklizard, *Iberolacerta cyreni*): Influence of thermal environment and associated costs. *Can. J. Zool.* **2014**, *92*, 955–964. [CrossRef]
9. Huey, R.B.; Kearney, M.R.; Krockenberger, A.; Holtum, J.A.; Jess, M.; Williams, S.E. Predicting organismal vulnerability to climate warming: Roles of behaviour, physiology, and adaptation. *Phil. Trans. R. Soc. B* **2012**, *367*, 1665–1679. [CrossRef]
10. Muñoz, M.M.; Stimola, M.A.; Algar, A.C.; Conover, A.; Rodriguez, A.J.; Landestoy, M.A.; Bakken, G.S.; Losos, J.B. Evolutionary stasis and lability in thermal physiology in a group of tropical lizards. *Phil. Trans. R. Soc. B* **2014**, *281*, 20132433. [CrossRef] [PubMed]
11. Gunderson, A.R.; Stillman, J.H. Plasticity in thermal tolerance has limited potential to buffer ectotherms from global warming. *Phil. Trans. R. Soc. B* **2015**, *282*, 20150401. [CrossRef] [PubMed]
12. Buckley, L.B.; Ehrenberger, J.C.; Angilletta, M.J. Thermoregulatory behaviour limits local adaptation of thermal niches and confers sensitivity to climate change. *Funct. Ecol.* **2015**, *29*, 1038–1047. [CrossRef]
13. Berg, M.P.; Kiers, E.T.; Driessen, G.; Van Der Heijden, M.; Kooi, B.W.; Kuenen, F.; Liefting, M.; Verhoef, H.A.; Ellers, J. Adapt or disperse: Understanding species persistence in a changing world. *Glob. Chang. Biol.* **2010**, *16*, 587–598. [CrossRef]
14. McCain, C.M. Global analysis of reptile elevational diversity. *Glob. Ecol. Biogeogr.* **2010**, *19*, 541–553. [CrossRef]
15. Ortega, Z.; Mencía, A.; Pérez-Mellado, V. Are mountain habitats becoming more suitable for generalist than cold-adapted lizards thermoregulation? *PeerJ* **2016**, *4*, e2085. [CrossRef] [PubMed]
16. Carbonero, J.; García-Díaz, P.; Ávila, C.; Arribas, O.; Lizana, M. Distribution, habitat characterization and conservation status of Iberolacerta martinezricai (ARRIBAS, 1996), in the Sierra de Francia, Salamanca. Spain (Squamata: Sauria: Lacertilidae). *Herpetozoa* **2016**, *28*, 149–165.
17. Araújo, M.B.; Guilhaumon, F.; Neto, D.R.; Pozo, I.; Calmaestra, R. *Impactos, Vulnerabilidad y Adaptación al Cambio Climático de la Biodiversidad Española. Fauna de Vertebrados*; Dirección General de Medio Natural y Política Forestal: Madrid, Spain, 2011.
18. Maiorano, L.; Amori, G.; Capula, M.; Falcucci, A.; Masi, M.; Montemaggiori, A.; Pottier, J.; Psomas, A.; Rondinini, C.; Russo, D.; et al. Threats from climate change to terrestrial vertebrate hotspots in Europe. *PLoS ONE* **2013**, *8*, e2085. [CrossRef] [PubMed]
19. Nogués-Bravo, D.; Araújo, M.B.; Lasanta, T.; Moreno, J.L. Climate change in Mediterranean mountains during the 21st century. *Ambio* **2008**, *37*, 280–285. [CrossRef]
20. Moreno-Rueda, G.; Pleguezuelos, J.M.; Pizarro, M.; Montori, A. Northward shifts of the distributions of Spanish reptiles in association with climate change. *Conserv. Biol.* **2012**, *26*, 278–283. [CrossRef] [PubMed]
21. Pleguezuelos, J.M. Vulnerabilidad de los reptiles ibéricos al cambio climático. In *Los Bosques y la Biodiversidad Frente al Cambio Climático: Impactos, Vulnerabilidad y Adaptación en España*; Herrero, A., Zavala, M.A., Eds.; Ministerio de Agricultura, Alimentación y Medio Ambiente: Madrid, Spain, 2015; pp. 143–151.
22. Arribas, O. New data on the Peña de Francia Mountain Lizard 'Lacerta' cyreni martinezricai. Arribas, 1996. *Herpetozoa* **1999**, *12*, 119–128.
23. Elsen, P.R.; Monahan, W.B.; Merenlender, A.M. Topography and human pressure in mountain ranges alter expected species responses to climate change. *Nat. Commun.* **2020**, *11*, 1974. [CrossRef] [PubMed]
24. Sato, C.F.; Wood, J.T.; Lindenmayer, D.B. The effects of winter recreation on alpine and subalpine fauna: A systematic review and meta-analysis. *PLoS ONE* **2013**, *8*, e64282. [CrossRef]
25. Courtney, L.; Sarah, E.; Reed, S.E.; Merenlender, A.M.; Crooks, K.R. A meta-analysis of recreation effects on vertebrate species richness and abundance. *Conserv. Sci. Pract.* **2019**, *1*, e93.
26. Amo, L.; Lopez, P.; Martin, J. Habitat deterioration affects body condition of lizards: A behavioral approach with *Iberolacerta cyreni* lizards inhabiting ski resorts. *Biol. Conserv.* **2007**, *135*, 77–85. [CrossRef]

27. Sato, C.F.; Wood, J.T.; Schroder, M.; Green, K.; Osborne, W.S.; Michael, D.R.; Lindenmayer, D.B. An experiment to test key hypotheses of the drivers of reptile distribution in subalpine ski resorts. *J. Appl. Ecol.* **2014**, *51*, 13–22. [CrossRef]

28. Arribas, O.J. Thermoregulation, activity and microhabitat selection in the rare and endangered Batuecan Rock Lizard, *Iberolacerta martinezricai* (Arribas, 1996) (Squamata: Sauria: Lacertidae). *Herpetozoa* **2013**, *26*, 77–90.

29. Meek, P. Patterns of reptile road-kills in the Vendée region of western France. *Herpetol. J.* **2009**, *19*, 135–142.

30. Gonçalves, L.O.; Alvares., D.J.; Teixeira, F.Z.; Schuck, G.G.; Coelho, I.P.; Esperandio, I.B.; Anza, J.; Beduschi, J.; Bastazini, V.A.G.; Kindel, A. Reptile road-kills in Southern Brazil: Composition, hot moments and hotspots. *Sci. Total Environ.* **2018**, *615*, 1438–1445. [CrossRef] [PubMed]

31. Colino-Rabanal, V.J.; Lizana, M. Herpetofauna and roads: A review. *Basic Appl. Herpetol.* **2012**, *26*, 5–31. [CrossRef]

32. Arribas, O. Lagartija batueca *Iberolacerta martinezricai*. In *Enciclopedia Virtual de los Vertebrados Españoles*; Salvador, A., Marco, A., Eds.; Museo Nacional de Ciencias Naturales: Madrid, Spain, 2015; Available online: http://www.vertebradosibericos.org/ (accessed on 15 April 2020).

33. Green, K.; Sanecki, G. Immediate and short-term responses of bird and mammal assemblages to a subalpine wildfire in the Snowy Mountains, Australia. *Austral Ecol.* **2006**, *31*, 673–681. [CrossRef]

34. Young, M.E.; Ryberg, W.A.; Fitzgerald, L.A.; Hibbitts, T.J. Fragmentation alters home range and movements of the Dunes Sagebrush Lizard (*Sceloporus arenicolus*). *Can. J. Zool.* **2018**, *96*, 905–912. [CrossRef]

35. Walkup, D.K.; Leavitt, D.J.; Fitzgerald, L.A. Effects of habitat fragmentation on population structure of dune-dwelling lizards. *Ecosphere* **2017**, *8*, e01729. [CrossRef]

36. Leavitt, D.J.; Fitzgerald, L.A. Disassembly of a dune-dwelling lizard community due to landscape fragmentation. *Ecosphere* **2013**, *4*, 97. [CrossRef]

37. Vega, L.E.; Bellagamba, P.J.; Fitzgerald, L.A. Long-term effects of anthropogenic habitat disturbance on a lizard assemblage inhabiting coastal dunes in Argentina. *Can. J. Zool.* **2000**, *78*, 1653–1660. [CrossRef]

38. Henle, K.; Davies, K.F.; Kleyer, M.; Margules, C.; Settele, J. Predictors of species sensitivity to fragmentation. *Biodivers. Conserv.* **2004**, *13*, 207–251. [CrossRef]

39. Rivas-Martínez, S.; Fernandez-González, F.; Sánchez, D. El Sistema Central Español. De la sierra de Ayllón a la Serra da Estrela. In *La vegetación de España*; Peinado, M., Rivas-Martínez, S., Eds.; Universidad de Alcalá: Alcalá de Henares, Madrid, Spain, 1987; pp. 419–452.

40. Crochet, P.A.; Chaline, O.; Surget-Groba, Y.; Debain, C.; Cheylan, M. Speciation in mountains: Phylogeography and phylogeny of the rock lizards genus *Iberolacerta* (Reptilia: Lacertidae). *Mol. Phylogenet. Evol.* **2004**, *30*, 860–886. [CrossRef] [PubMed]

41. Carranza, S.; Arnold, E.N.; Amat, F. DNA phylogeny of Lacerta (*Iberolacerta*) and other lacertine lizards (Reptilia: Lacertidae): Did competition cause long-term mountain restriction? *Syst. Biodivers.* **2004**, *2*, 57–77. [CrossRef]

42. Carbonero, J.; Lizana, M.; García, P.; Arribas, O. *Distribución, Estado de Conservación y Medidas de Gestión para la Lagartija Serrana de la Peña de Francia (Iberolacerta martinezricai) en el Parque Natural de Batuecas-sierra de Francia (Unpublished Report)*; Fundación Patrimonio Natural-Junta de Castilla y León: Salamanca, Spain, 2008.

43. Pérez-Mellado, V.; Márquez, R.; Martínez-Solano, I. *Iberolacerta martinezricai*. IUCN Red List of Threatened Species. 2009: E.T61516A12499291. Available online: www.iucnredlist.org (accessed on 12 March 2020).

44. Arribas, O.J. Growth, sex-dimorphism and predation pressure in the Batuecan Lizard, *Iberolacerta martinezricai* (Arribas, 1996). *Butllet. Soc. Catal. Herpetol.* **2014**, *21*, 147–173.

45. Arribas, O.J. Reproductive characteristics of the Batuecan Lizard, *Iberolacerta martinezricai* (ARRIBAS, 1996) (Squamata: Sauria: Lacertidae). *Herpetozoa* **2018**, *30*, 187–202.

46. Sutherland, W.J. *Ecological Census Techniques: A Handbook*; Cambridge University Press: Cambridge, UK, 2006.

47. Buckland, S.T.; Anderson, D.R.; Burnham, K.P.; Laake, J.L.; Borchers, D.L.; Thomas, L. *Introduction to Distance Sampling. Estimating Abundance of Biological Populations*; Oxford University Press: Oxford, UK, 2001.

48. Buckland, S.T.; Anderson, D.R.; Burnham, K.P.; Laake, J.L.; Borchers, D.L.; Thomas, L. *Advanced Distance Sampling. Estimating Abundance of Biological Populations*; Oxford University Press: Oxford, UK, 2004.

49. Thomas, L.; Buckland, S.T.; Rexstad, E.A.; Laake, J.L.; Strindberg, S.; Hedley, S.L.; Bishop, J.R.B.; Marques, T.A.; Burnham, K.P. Distance software: Design and analysis of distance sampling surveys for estimating population size. *J. Appl. Ecol.* **2010**, *47*, 5–14. [CrossRef] [PubMed]

50. Thomas, L.; Laake, J.L.; Rexstad, E.A.; Strindberg, S.; Marques, F.; Buckland, S.; Borchers, D.; Anderson, D.; Burnham, K.; Burt, M. *User's Guide Distance 6.0 Release 2*; Research Unit for Wildlife Population Assessment, University of St. Andrews: St. Andrews, UK, 2009.

51. Taylor, P.D.; Fahrig, L.; With, K.A. Landscape connectivity: A return to the basics. In *Connectivity Conservation*; Crooks, K.R., Sanjayan, M., Eds.; Cambridge University Press: Cambridge, UK, 2006.

52. Bennett, G. *Integrating Biodiversity Conservation and Sustainable Use: Lessons Learned from Ecological Networks*; IUCN: Gland, Switzerland; Cambridge, UK, 2004.

53. Urban, D.L.; Minor, M.S.; Treml, E.A.; Schick, R.S. Graph models of habitat mosaics. *Ecol. Lett.* **2009**, *12*, 260–273. [CrossRef]

54. Kool, J.T.; Moilanen, A.; Treml, E.A. Population connectivity: Recent advances and new perspectives. *Landsc. Ecol.* **2013**, *28*, 165–185. [CrossRef]

55. Harris, K.M.; Dickinson, K.J.M.; Whigham, P.A. Functional connectivity and matrix quality: Network analysis for a critically endangered New Zealand lizard. *Landsc. Ecol.* **2014**, *29*, 41–53. [CrossRef]
56. Urban, D.; Keitt, T. Landscape connectivity: A graph theoretic perspective. *Ecology* **2001**, *82*, 1205–1218. [CrossRef]
57. Pascual-Hortal, L.; Saura, S. Comparison and development of new graph-based landscape connectivity indices: Towards the priorization of habitat patches and corridors for conservation. *Landsc. Ecol.* **2006**, *21*, 959–967. [CrossRef]
58. Decout, S.; Manel, S.; Miaud, C.; Luque, S. Integrative approach for landscape-based graph connectivity analysis: A case study with the common frog (*Rana temporaria*) in human-dominated landscapes. *Landsc. Ecol.* **2012**, *27*, 267–279. [CrossRef]
59. Clauzel, C.; Bannwarth, C.; Foltete, J.C. Integrating regional-scale connectivity in habitat restoration: An application for amphibian conservation in eastern France. *J. Nat. Conserv.* **2015**, *23*, 98–107. [CrossRef]
60. Saura, S.; Pascual-Hortal, L. A new habitat availability index to integrate connectivity in landscape conservation planning: Comparising with existing indices and application to a case study. *Landsc. Urban Plan.* **2007**, *83*, 91–103. [CrossRef]
61. Saura, S.; Torne, J. Conefor Sensinode 2.2: A software package for quantifying the importance of habitat patches for landscape connectivity. *Environ. Model. Softw.* **2009**, *24*, 135–139. [CrossRef]
62. Saura, S.; Rubio, L. A common currency for the different ways in which patches and links can contribute to habitat availability and connectivity in the landscape. *Ecography* **2010**, *33*, 523–537. [CrossRef]
63. Soberón, J.; Peterson, A.T. Interpretation of models of fundamental ecological niches and species' distributional areas. *Biodivers. Inform.* **2005**, *2*, 1–10. [CrossRef]
64. Brown, J.; Stevens, G.G.; Kaufman, D.M. The geographic range: Size, shape, boundaries, and internal structure. *Annu. Rev. Ecol. Syst.* **1996**, *27*, 597–623. [CrossRef]
65. Hampe, A.; Jump, A.S. Climate Relicts Past, Present, Future. *Annu. Rev. Ecol. Evol. Syst.* **2011**, *42*, 313–333. [CrossRef]
66. Buckley, L.B.; Jetz, W. Insularity and the determinants of lizard population density. *Ecol. Lett.* **2007**, *10*, 481–489. [CrossRef] [PubMed]
67. Novosolov, M.; Rodda, G.H.; Feldman, A.; Kadison, A.E.; Dor, R.; Meiri, S. Power in numbers. Drivers of high population density in insular lizards. *Glob. Ecol. Biogeogr.* **2016**, *25*, 87–95.
68. Fick, S.E.; Hijmans, R.J. WorldClim 2: New 1-km spatial resolution climate surfaces for global land areas. *Int. J. Climatol.* **2017**, *37*, 4302–4315. [CrossRef]
69. Quinn, P.F.; Beven, K.J.; Lamb, R. The ln(a/tan beta) index: How to calculate it and how to use it within the TOPMODEL framework. *Hydrol. Process.* **1995**, *9*, 161–182. [CrossRef]
70. R Development Core Team. *R: A Language and Environment for Statistical Computing*; Core Team R: Vienna, Austria, 2004.
71. Burnham, K.P.; Anderson, D.R. *Introduction, Model Selection and Multimodel Inference*; Springer: New York, NY, USA, 2002.
72. Grömping, U. Relative importance for linear regression in R: The package relaimpo. *J. Stat. Softw.* **2006**, *17*, 1. [CrossRef]
73. Dibner, R.R.; Doak, D.F.; Murphy, M. Discrepancies in occupancy and abundance approaches to identifying and protecting habitat for an at-risk species. *Ecol. Evol.* **2017**, *7*, 5692–5702. [CrossRef] [PubMed]
74. He, F.; Gaston, K.J. Occupancy–Abundance relationships and sampling scales. *Ecography* **2000**, *23*, 503–511. [CrossRef]
75. Santos, T.; Díaz, J.A.; Pérez-Tris, J.; Carbonell, R.; Tellería, J.L. Habitat quality predicts the distribution of a lizard in fragmented woodlands better than habitat fragmentation. *Anim. Conserv.* **2008**, *11*, 46–56. [CrossRef]
76. Hanski, I. *Metapopulation Ecology*; Oxford University Press: New York, NY, USA, 1999.
77. Addicott, J.F.; Aho, J.M.; Antolin, M.F.; Padilla, D.K.; Richardson, J.S.; Soluk, D.A. Ecological neighborhoods: Scaling environmental patterns. *Oikos* **1987**, *49*, 340–346. [CrossRef]
78. Ryberg, W.A.; Hill, M.T.; Painter, C.W.; Fitzgerald, L.A. Landscape pattern determines neighborhood size and structure within a lizard population. *PLoS ONE* **2013**, *8*, e56856. [CrossRef]
79. Munguia-Vega, A.; Rodriquez-Estrella, R.; Shaw, W.W.; Culver, M. Localized extinction of an arboreal desert lizard caused by habitat fragmentation. *Biol. Conserv.* **2013**, *157*, 11–20. [CrossRef]
80. Almeida-Gomes, M.; Duarte Rocha, C.F. Diversity and distribution of lizards in fragmented Atlantic forest landscape in Southeastern Brazil. *J. Herpetol.* **2014**, *48*, 423–429. [CrossRef]
81. Crooks, K.R.; Sanjayan, M.A. Connectivity conservation: Maintaining connections for nature. In *Connectivity Conservation*; Crooks, K.R., Sanjayan, M., Eds.; Cambridge University Press: Cambridge, UK, 2006; pp. 1–20.
82. Boitani, L.; Falcucci, A.; Maiorano, L.; Rondinini, C. Ecological networks as conceptual frameworks or operational tools in conservation. *Conserv. Biol.* **2007**, *21*, 1414–1422. [CrossRef]
83. Minor, E.S.; Urban, D.L. A graph-theory framework for evaluating landscape connectivity and conservation planning. *Conserv. Biol.* **2008**, *22*, 297–307. [CrossRef]
84. Bodin, O.; Saura, S. Ranking individual habitat patches as connectivity providers: Integrating network analysis and patch removal experiments. *Ecol. Modell.* **2010**, *221*, 2393–2405. [CrossRef]
85. Beninde, J.; Feldmeier, S.; Werner, M.; Peroverde, D.; Schulte, U.; Hochkirch, A.; Veith, M. Cityscape genetics: Structural vs. functional connectivity of an urban lizard population. *Mol. Biol.* **2016**, *25*, 4984–5000.
86. Storfer, A.; Murphy, M.A.; Evans, J.S.; Goldberg, C.S.; Robinson, S.; Spear, S.F.; Dezzani, R.; Delmelle, E.; Vierling, L.; Waits, L.P. Putting the landscape in landscape genetics. *Heredity* **2007**, *98*, 128–142. [CrossRef]

 *sustainability*

Article

Analysis of the Adaptative Strategy of *Cirsium vulgare* (Savi) Ten. in the Colonization of New Territories

Jhony Fernando Cruz Román [1], Ricardo Enrique Hernández-Lambraño [1,2], David Rodríguez de la Cruz [1,2] and José Ángel Sánchez Agudo [1,2,*]

[1] Departamento de Botánica y Fisiología Vegetal, Área de Botánica, Universidad de Salamanca, Campus Miguel de Unamuno s/n, E-37007 Salamanca, Spain; jhonycruz@usal.es (J.F.C.R.); ricardohl123@usal.es (R.E.H.-L.); droc@usal.es (D.R.d.l.C.)

[2] Grupo de Investigación en Biodiversidad, Diversidad Humana y Biología de la Conservación, Universidad de Salamanca, Campus Miguel de Unamuno s/n, E-37007 Salamanca, Spain

[*] Correspondence: jasagudo@usal.es

Abstract: The current situation of global environmental degradation as a result of anthropogenic activities makes it necessary to open new research lines focused on the causes and effects of the main alterations caused in the ecosystems. One of the most relevant is how the niche dynamics of invasive species change between different geographical areas, since its understanding is key to the early detection and control of future invasions. In this regard, we analyzed the distribution pattern of *Cirsium vulgare* (Savi) Ten., a plant of the Asteraceae family originally from the Eurasian region that currently invades wide areas of the world. We estimated its niche shifts between continents using a combination of principal components analysis (PCA) and Ecological Niche Modelling (ENM) on an extensive set of data on global presences of its native and invaded ranges from Global Biodiversity Information Facility (GBIF). A set of bioclimatic variables and the Human Footprint (HFP) with a resolution of 10 km were selected for this purpose. Our results showed that the species has a marked global trend to expand toward warmer climates with less seasonality, although in some regions its invasiveness appears to be less than in others. The models had a good statistical performance and high coherence in relation to the known distribution of the species and allowed us to establish the relative weight of the contribution of each variable used, with the annual temperature and seasonality being the determining factors in the establishment of the species. Likewise, the use of non-climatic variable HFP has provided relevant information to explain the colonizing behavior of the species. The combination of this methodology with an adequate selection of predictor variables represents a very useful tool when focusing efforts and resources for the management of invasive species.

Keywords: ecological niche dynamics; MaxEnt; reciprocal niche models; biological invasions

 check for updates

Citation: Román, J.F.C.; Hernández-Lambraño, R.E.; Rodríguez de la Cruz, D.; Sánchez Agudo, J.Á. Analysis of the Adaptative Strategy of *Cirsium vulgare* (Savi) Ten. in the Colonization of New Territories. *Sustainability* **2021**, *13*, 2384. https://doi.org/10.3390/su13042384

Academic Editor: Carmelo M. Musarella

Received: 15 January 2021
Accepted: 18 February 2021
Published: 23 February 2021

Publisher's Note: MDPI stays neutral with regard to jurisdictional claims in published maps and institutional affiliations.

1. Introduction

Globalization, with its intense commercial activity of nations around the world, implies a massive flow of transportation of goods, people, and others from one territory to another. All the processes carried out in these movements involve strong environmental consequences such as the emission of pollutants, global warming, changes in land use, destruction of habitat, and landscape fragmentation [1–3]. No less important is the spread of some species with high colonizing capacity that give rise to biological invasions, which is considered to be one of the phenomena responsible for the planet's loss of biodiversity [4,5]. Although a large number of species can be transported voluntarily or involuntarily from their natural habitats to other territories [6], the majority do not survive because they do not adapt to new ecological conditions. Some of them become naturalized, coexisting in harmony with the native species, but a few are able to overcome adaptive barriers and become invasive, displacing native species and causing serious damage to the balance of the ecosystem [7–9]. Despite having more information available on biological invasions,

81

we ignore the majority of the processes that drive the dispersion of invasive species and the changes between ranks (native and invaded) [10]. To anticipate the behavior of the translocated species, it is essential to identify these key factors to prioritize areas for the detection and control of early invasions. Once established, these invaders can cause serious environmental and socioeconomic disruption because of their eradication and control costs [11–14]. In that sense, prevention can be a practical, effective, and profitable management strategy [15]. New methodologies for managing future ecological problems have been developed on the basis of advances in Geographic Information Systems (GIS) [16]. Ecological niche models (ENMs) are one of the newest tools for spatial ecological analysis, consisting of the correlation of species presence/absences or just presences with a background and predefined environmental variables (usually geographical, environmental, or topographical). The approach of these models is based on Hutchison's duality hypothesis, which means that these models are developed in two spaces: The geographical (two-dimensional) and the environmental–ecological (multidimensional) [17–19].

Hutchinson [20] differentiated two types of niches: (1) Fundamental, which consists of all the optimal biotic and abiotic conditions where a species can achieve its development and subsistence; and (2) realized or observed, which is restricted to the environments where the species is effectively found due to certain biotic interactions. To carry out the modelling process, it is necessary to use an algorithm that contains all the statistical processes to conduct the spatial information. Due to its efficiency and practicality, "MaxEnt" [21], an algorithm based on the principle of Maximum Entropy, is one of the most commonly used [10,22,23]. One of the main advantages of this software is that it is able to work only with presence data as it uses a self-generated background to match them with the environmental variables [10,21,24]. Another noteworthy feature of this algorithm is that it can be used to build models with robust results with a few points of the species' presence [21,24].

The main approach to investigating climatic niches in space and time has been to analyze climatic conditions across a species' distribution ranges over time [25]. One of the assumptions underlying the ENM is the principle of ecological niche conservation, which suggests that most species tend to conserve their ecological requirements in native and invaded ranges [26,27]. However, this approach has been refuted by some authors, including the authors of [21–23,28,29], who have concluded that, due to the complexity of the dynamics of biological invasions, species have a greater capacity to expand beyond their native environmental envelope than previously thought. Such niche shifts can be triggered by biotic factors, such as the absence of competitors and/or pathogens, evolutionary changes due to genetic drift, or other natural selection in the invaded range [30,31].

The relative weight of factors implicit in the dynamics of niche shift can be evaluated with metrics based primarily on observing the displacement of a niche centroid generated from environmental data from the combined ranges subjected to a principal component analysis (PCA) [29]. The most commonly used metrics for measuring centroid displacement are: (1) Expansion, which measures the proportion of the displacement in the species ecological centroid to new environmental features in the invaded range due to rapid adaptative evolution and/or biotic interactions; and (2) unfilling, which measures the proportion of ecological niche that is suitable for the development of the species. The unfilling metric provides a first approach to determine possible areas of future invasion of a species [29,32].

The present work delves into this hypothesis of niche shift across territories of the world using the species Spear thistle (*Cirsium vulgare* (Savi) Ten.). *C. vulgare*, which is considered one of the most prolific invasive species in the world. It is native to Eurasia and is naturalized in several countries, spreading to all continents except Antarctica [33–35]. This species represents a serious threat in protected areas and biodiversity sanctuaries worldwide. Different types of control methods have been implemented, ranging from mechanical removal and application of chemical agents to biological control [36]. However, no direct strategy has been shown to be effective with *C. vulgare*. There is a need to develop alternatives based eminently on the abovementioned prevention measures to make it possible to anticipate their invasion, and these alternatives must be supported by science-based research such as the current study.

We tested the application of the methodologies described above using georeferenced presence data from the Global Biodiversity Information Facility (GBIF) platform in conjunction with an extensive literature review of the species' distribution. Raster's layers were downloaded from Chelsa repositories and resampled to 10×10 km. Parametrization was done according to previous works at regional different levels dealing with biological invasions [15,22,23,37].

The objective of this work is to test the application of specific ecoinformatics methodologies to explain and predict the adaptability to new environments of certain invasive species that are currently distributed on a global scale.

2. Materials and Methods

2.1. The Target Species: Cirsium vulgare (Savi) Ten. (Asteraceae)

Spear thistle (*Cirsium vulgare* (Savi) Ten.) is an herbaceous biennial or perennial plant, occasionally annual, thorny, erect, up to 2 m high. Its basal leaves form a rosette, and the plant produces 150 to 250 flowers with purple corollas. Its fruit is an achene [38]. Its native distribution is located in Eurasia and it is naturalized widely in many countries of the world [33,38]. *C. vulgare* has become an invasive plant that is difficult to eradicate mainly due to the high production and resistance of its seeds, its viable life form, and its sequential germination pattern [39]. The impact of the species on the native flora is quite notable, since it can form dense shrub populations that occupy all of the existing space, displacing other species. In particular, *C. vulgare* displaces smaller species by monopolizing the light capture and more efficiently extracting underground resources through its dense roots [38,40]. It is believed that this species appeared in North America in the colonial period and spread throughout the continent [35]. In Oceania (Tasmania), there are records from the 1830s, and it is possible that it was transported there from South Africa [41]. In temperate zones of South America, there is a permanent risk of invasion by this specie due to the accidental transport of contaminated agricultural products such as seeds or fodder or its deliberate introduction as an ornamental plant [41]. *C. vulgare* usually grows in sunny areas in well-irrigated soils with high nutrient concentrations [31,42].

2.2. Occurrence Datasets

In the current study, 387,510 occurrence data from The Global Biodiversity Information Facility (www.gbif.org (accessed on: 3 April 2020)) were downloaded, processed, and filtered to eliminate incorrect, duplicate, or badly georeferenced citations, as well as to homogenize their resolutions. Thus, 10×10 km grids were generated in the geographical space to avoid bias [10]. Then, an exhaustive bibliographic review of the species was carried out, and the geographical ranges were established. We obtained 24,632 grids in the native range in Eurasia. For the invaded ranges, we obtained 2967 grids in North America, 176 in South America, 284 in Africa, and 7356 in Oceania (Figure 1).

The compiled database is the result of an extensive search of the occurrence of the species in the study area. Nevertheless, we acknowledge that this database may not represent the full range of environmental conditions in which the species can be found (e.g., other introduced areas) as in other studies elsewhere [10].

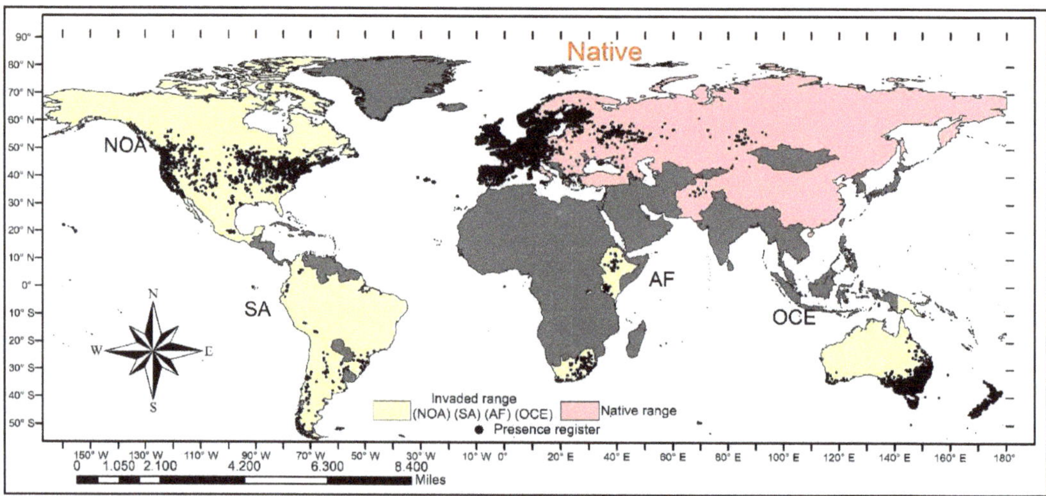

Figure 1. The map shows all the zones for the native and invaded ranges which were defined based on the filtered presences (black spots) and the bibliographic analysis of *C. vulgare*. The native range, represented with the light red color, covers almost all Eurasia. The invaded range, represented by the color yellow, includes: NOA—North America; SA—South America, excluding some countries (Venezuela, Uruguay, Paraguay, Surinam, Guyana, and French Guyana); AF—Africa, in areas belonging some countries (South Africa, Ethiopia, Rwanda, and Kenya); and OCE—Oceania, the whole continent. Zones that do not fall into either range are represented in grey.

2.3. Environmental Variables

As potential predictors to characterize the species' ecological niche, we used a set of variables related to climate and human influence (Table 1). Climatic variables (seasonal and annual patterns of temperature and precipitations) were available from climatology databases at high resolutions for the Earth's land surface areas (CHELSA; http://chelsa-climate.org/ (accessed on: 5 April 2020)), which provides improved climatic estimates in landscapes with complex topography at 30 arc-seconds spatial resolutions (~1 km). Solar radiation was available from the WorldClim database (https://www.worldclim.org (accessed on: 5 April 2020)) at 30 arc-seconds spatial resolutions. Because of its effect on invasive species distribution, we included a variable related to human footprint on the landscape (https://wcshumanfootprint.org (accessed on: 5 April 2020)). This variable measures the cumulative impact of direct pressures on nature from human activities. It includes 8 inputs: The extent of built environments, crop land, pastureland, human population density, nighttime lights, railways, roads, and navigable waterways.

Table 1. Selected explanatory variables, the first column of the table displays the abbreviation that was used in this study to represent the selected variables; in the second, the full name of each variable; and the third the type of variable, that is, whether it is climatic or non-climatic.

Code	Name	Type
Bio 1	Annual mean Temperature	
Bio 4	Temperature Seasonality	
Bio 12	Annual Precipitation	Climatic
Bio 15	Precipitation Seasonality	
Rad_sd	Standard deviance of Radiation	
HFP	Human Footprint	Non-Climatic

Finally, in order to avoid the cross-correlation within the selected environmental variables, a multicollinearity test was conducted using Pearson's correlation coefficient [10] in R software [43]. Variables with cross-correlation coefficient values of $r > \pm 0.75$ were excluded. The final explanatory variables selected were: Annual mean temperature (Bio 1), temperature seasonality (Bio 4), annual precipitation (Bio 12), precipitation seasonality (Bio 15), standard deviance of radiation (Rad_sd), and Human Footprint (HFP). Variables were resampled to 10 km to optimize processing time using an interpolation bilinear resampling technique. All spatial information processing was handled using the Spatial Analyst Tool from ArcGIS 10.5 [44].

2.4. Niche Shift Measurement

The niche shift of *C. vulgare* was measured on the basis of the environmental species envelope represented by explanatory variables, which contain climatic and non-climatic factors (Table 1). The process consists of calibrating a PCA with the areas effectively occupied by the plant in its native and invaded ranges and the environmental conditions within the whole study area [21,22,32].

The process was carried out below the R-program software with the library "ecospat" [45]. First, we extracted the environmental conditions of the native range and the areas of the invaded range. Then, the PCA was calibrated, and the first 2 axes were taken into account for the analysis. Second, in order to avoid spatial bias, we divided the environmental space into 100 × 100 cells and transformed the data into densities [10]. Third, based on Schoener's D metric (0 = totally different to 1 = complete overlap), we measured the proportion of native niche that does not overlap with the invaded niche, i.e., "unfilled," and the proportion of invaded niche that does not overlap with the native niche or "expansion" [32]. In addition, in order to avoid bias, we used the 90th percentile of all environmental space and we compared it with the whole environmental extent. Fourth, the distribution of density and median environmental space in both ranges was calculated to determine the overall trend of ecological niche shift [21,22,32]. Fifth, we applied a niche equivalence test, that is, the correspondence between an observed and expected D, by randomly reassigning the occurrences of the native and invaded ranges. In this test, the null hypothesis (the niches are not identical) is rejected if $p < 0.05$. Also, we applied the niche similarity test, which compares the observed and expected D by randomly reassigning the occurrences in a single range. The value of $p > 0.05$ implies that the niches are not more similar than expected by chance [22].

2.5. Reciprocal Ecological Niche Models: Calibration and Evaluation

To explore niche conservatism across ranges of *C. vulgare*, we generated Ecological Niche Models that were compared to each other in a reciprocal way [10,46] using the MaxEnt program [21]. The MaxEnt model is a maximum entropy-based machine learning program that estimates the habitats suitability for a species based on the environmental constraints [21]. To generate reciprocal models, we first made distribution models of potential suitable habitats with the same occurrence points and environmental variables that were used in our PCA. Then, we projected native models into introduced ranges and visually compared them with models calibrated with data occurrence in the introduced range. We then repeated this step but projected the introduced models into the native range and compared them in the same way.

In fitting these models, we set up 15 replicates using 80% of the data for calibration and the other 20% for evaluation, the selection of feature classes (autofeature), a regularization multiplier value of 1, a maximum of 500 iterations, and 10,000 background points. To select the background points, we generated a Kernel Density map to draw background points at random in MaxEnt. This limits the background points to areas that we assume were surveyed for the species, which provides MaxEnt with a background file with the same bias as the presence locations [47]. We measured variable importance by comparing the jackknife of training gain values when models were made with individual variables.

To avoid projections into environments outside which the models were trained upon, we used the 'fade-by-clamping' option in MaxEnt, which removes heavily clamped pixels from the final predictions [21]. Predictive performance of each model was assessed using 15-fold cross-validation and the area under a receiver operating characteristic curve (AUC), which measures a model's ability to discriminate presence from background records (0.5 = random, 1 = perfect).

3. Results

3.1. Environmental Niche Analysis

For the target regions of this study, North America (NOA), South America (SA), Africa (AF), and Oceania (OCE) (Figure 1), a clear niche shift was observed for *C. vulgare* between the native and the invaded range. The overlap of "D" niches (Table 2) between niches was much greater in NOA and OCE than in SA and AF. Metrics of niche displacement for NOA revealed that there is a large proportion of common environments occupied between ranges and that the species has expanded ("expansion") into warmer environments (Bio 1) and with less temperature variation (Bio 4). On the other hand, there is a very low proportion of environments with optimal conditions still not occupied ("unfilling," Figure 2a,e). The metrics also indicate that the presences are quite associated with anthropogenic activities (HFP), something similar occurring in OCE but in this case less expansion was observed (Figure 2d,h). In the case of SA (Figure 2b,f) and AF (Figure 2c,g), the climatic trend of the expansion was similar to that of NOA and OCE, but it was also observed that the species tends toward environments with greater seasonality of rainfall (Bio 15). The metric did not show significant variations when considering the whole extension of the environments or only marginal environments, that is, the 90th percentile, except for AF, where it was observed that the unfilled environments had a higher proportion considering the totality of the available environments. Similarly, it was found that, in all the analyses, the environments with less radiation had a generalized trend (Figure 2).

Table 2. Principal component analysis (PCA) metrics for *C. vulgare*. The metrics for the niche overlap (Schoener's D) the intersection, i.e., considering the percentile (90%) to eliminate marginal environments between the native and invaded ranges (marginal environments), and considering the full extent of available environments (whole environmental extent).

Range	Schoener's D Metric	Marginal Environments			Whole Environmental Extent		
	D. Overlap	Stability	Expansion	Unfilling	Stability	Expansion	Unfilling
Native vs. NOA	0.5486	0.8668	0.1331	0.0002	0.8642	0.1357	0.0002
Native vs. SA	0.3417	0.5411	0.4588	0.0557	0.4197	0.5802	0.0567
Native vs. AF	0.2522	0.2434	0.7565	0.3613	0.2291	0.7708	0.8097
Native vs. OCE	0.4761	0.9440	0.0559	0.1491	0.9410	0.0589	0.2713

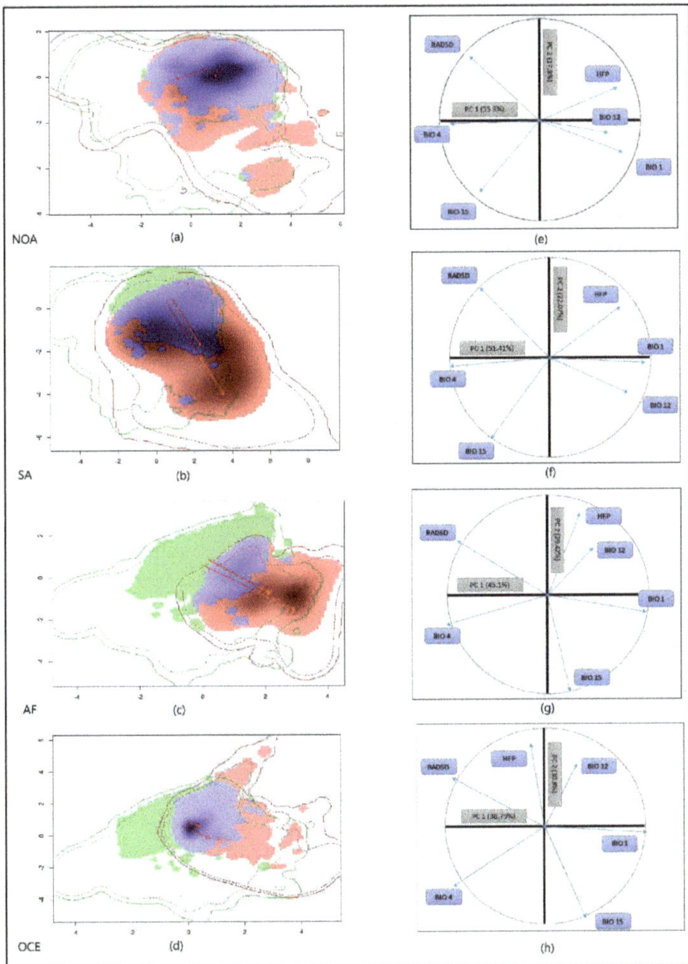

Figure 2. The niche overlap for each invaded range zones with the native range with 10 km of grid resolution is shown in (**a**) NOR—North America, (**b**) SA—South America, (**c**) AF—Africa, and (**d**) OCE—Oceania. Also, continuous and discontinuous red lines indicate 100% and 90% of the available background environments for *C. vulgare*, respectively. The solid red areas indicate the expansion, that is, areas that are actually occupied by the species (only for native range). The solid green color shows the areas that satisfy the requirements of the species but have not been occupied by it, that is, the unfilling (only invaded range). The solid blue color belongs to the stability, which is not more than the proportion of shared niche between the native and invaded ranges. The continuous red arrow shows the environmental distance between the median of the distribution density for each range, and the discontinuous red arrow shows the environmental distance between the median of the environmental space in each range. The contribution of climate variables in the first two axes of the PCA and the correlation of variables for each of the zones are shown, respectively, in (**e–h**).

According to the equivalence test carried out between both ranges, the niches occupied in SA were similar to those occupied in its native range, while OCE was the region with the most differences with respect to the niche of origin. In addition, the similarity test showed that except for the NOA region, the results were repeated. The results indicated that, in a certain proportion of the regions, niches were not more similar than expected by chance, i.e., in NOA, the niches were more similar to the niche of the native region than would be expected by chance (Figure 3).

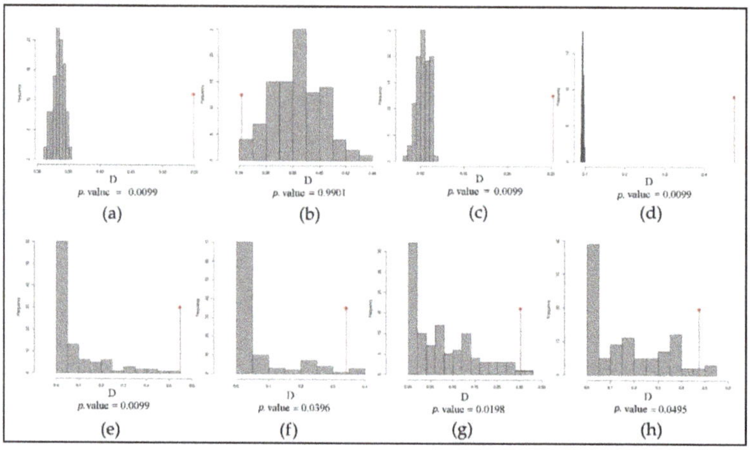

Figure 3. Equivalence and Similarity tests to compare the niches between the native and invaded ranges. The first row shows the equivalence values, i.e., the frequencies observed for the niche overlap index (D) in relation to the expected D for $p = 0.05$ (**a**–**d**). The niche similarity is shown in the second row (**e**–**h**). The first column compares native range with North America (**a**,**e**), the second column compares native range with South America (**b**,**f**), the third column compares native range with Africa (**c**,**g**), and the fourth column compares native range with Oceania (**d**,**h**).

3.2. Reciprocal Ecological Niche Models

The models generated for all regions showed a proper fitting of the models compared to random model, with good AUC values (Table 3), and a relatively low rate of omission, indicating that the presence itself of the species was correlated with the most suitable environments for it. The most important variable for *C. vulgare* in its native range was HFP, preceded by solar radiation (Rad_sd) (Table 4).

Table 3. Model accuracy results using the area under a receiver operating characteristic curve (AUC).

AUC Values									
Training					Test				
Native	NOA	SA	AF	OCE	Native	NOA	SA	AF	OCE
0.67	0.89	0.98	0.98	0.81	0.67	0.88	0.97	0.98	0.81

Table 4. Importance of the variables using the average $+/-$ the standard deviation of the contribution of the same variables as a product of the three replicates carried out for each model in the native and invaded ranges.

Variable	Native Range	Invaded Range			
		NOA	SA	AF	OCE
Bio 1	0.79 ± 8.37	15.3 ± 1.4	25.9 ± 4.4	31.4 ± 1.1	1.6 ± 4.4
Bio 4	4.4 ± 6.8	4.8 ± 5.0	43.8 ± 4.6	37 ± 2.3	47.8 ± 9.8
Bio 12	14.7 ± 6.4	12.9 ± 1.14	2.5 ± 6.8	5.3 ± 1.0	3.6 ± 1.4
Bio 15	2.2 ± 3.9	2.7 ± 2.6	3.6 ± 3.1	2.9 ± 1.9	13.8 ± 1.2
Rad_sd	29.7 ± 4.8	22.1 ± 2.1	5.4 ± 4.1	4.7 ± 2.9	25 ± 1.8
HFP	48.0 ± 2.4	41.9 ± 4.4	19.1 ± 1.2	18.4 ± 1.1	8.1 ± 1.2

Our results for the native region show a high adjustment of the suitability zones with the known presence of the species, which was expected given the greater historical extent of the colonization of this species (Figure 4). The prediction of the native model projected toward NOA (Figure 5a) was able to predict quite accurately the areas where *C. vulgare* is currently recorded. Projection calibrated in the invaded range (Figure 5b) showed that there are zones in the center of the subcontinent, with several areas of high habitat suitability extending westward. Variable analysis determined that HFP and Rad_sd were the most representative variables for this model (Table 4). In the case of SA (Figure 6a), the native model fairly predicted areas where the species occurs (to the south of Brazil, Chile and Argentina) and also predicted suitable habitats in Uruguay, but it did not hit areas with presence records in northern Argentina, southern Bolivia and Peru, and the mountainous areas of Ecuador and Colombia. The model of the invaded SA (Figure 6b) range predicted highly viable habitats along Chile and in the Andes Mountains in Ecuador and Colombia. Also, annual mean temperature (Bio 1) and seasonal temperature (Bio 4) (Table 4) were the explanatory variables with the highest contribution when calibrating the invaded model in SA. In the case of AF, the prediction of the native model (Figure 7a) was particularly accurate in North Africa in the Mediterranean region and in South Africa, where areas with records of *C. vulgare* were correctly predicted by our model, although it failed in the eastern areas of the African continent. On the other hand, the model calibrated in the invaded range of AF (Figure 7b) predicted areas of high habitat potential in Ethiopia and Kenya. The most important variables for AF were Bio 1 and Bio 4 (Table 4). For OCE, the prediction of the native model (Figure 8a) matches quite well with the records of the species, while the model of the invaded range for this region predicted areas of suitable habitat for *C. vulgare* across Papua New Guinea in northern Oceania. The model calibrated on the invaded range indicates that the species occupies a large part of Australia and that there are suitable environments for further expansion. Another suitable area is New Zealand, where the species is widely distributed (Figure 8b). The most important variables were Bio 4 and Rad_sd (Table 4).

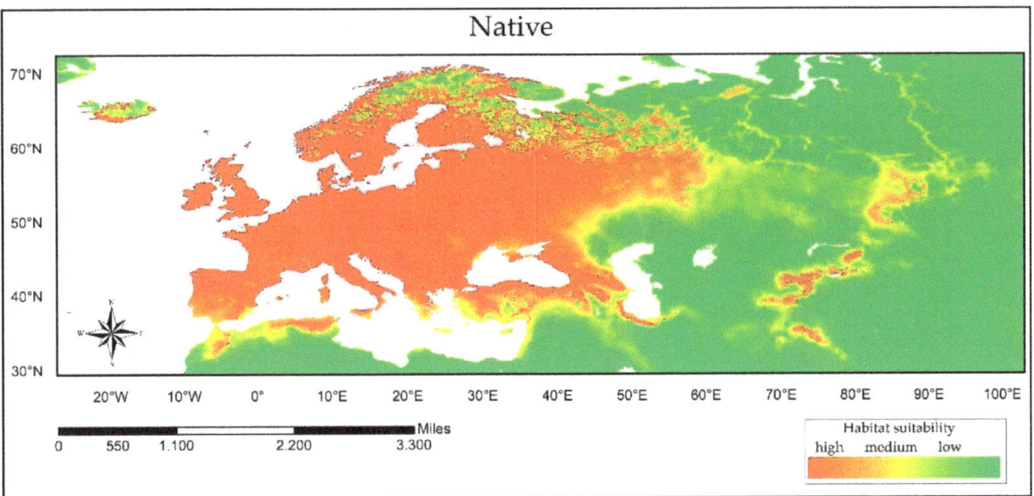

Figure 4. Native range model. Red indicates higher habitat suitability for *C. vulgare*. Yellow indicates medium suitability while light green indicates low or no suitability as it becomes lighter.

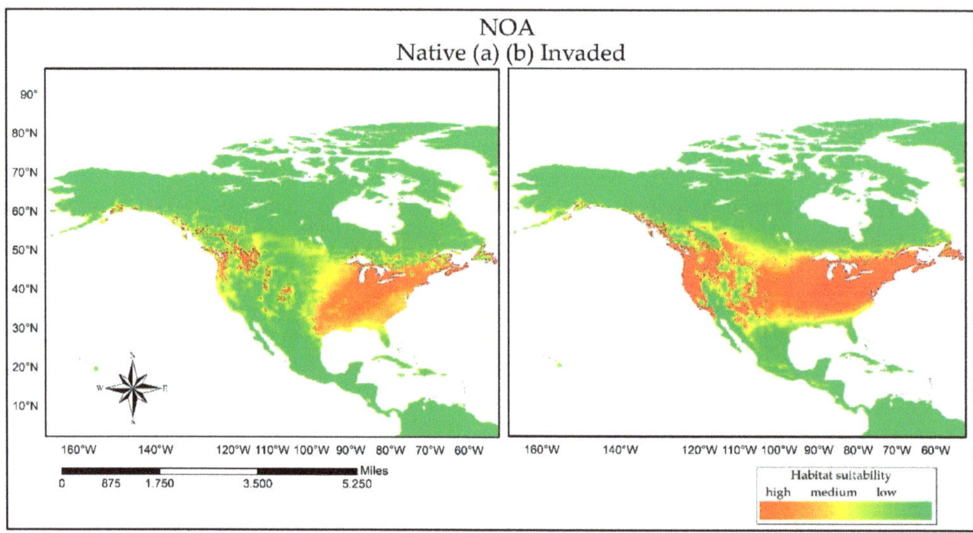

Figure 5. (**a**) Native model projected in NOA. (**b**) Model of the invaded range calibrated in NOA. Red indicates higher habitat suitability for *C. vulgare*. Yellow indicates medium suitability while light green indicates low or no suitability as it becomes lighter.

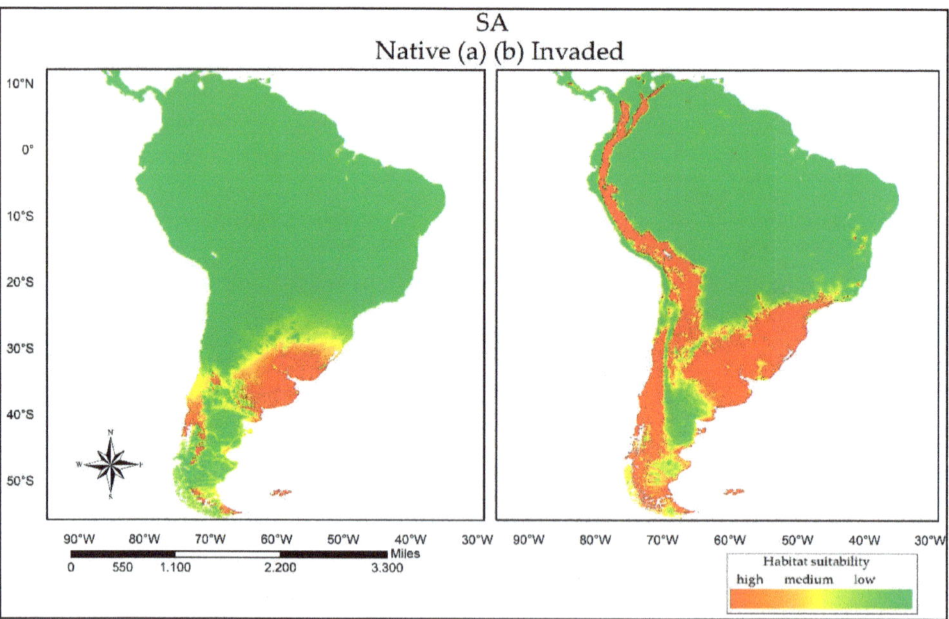

Figure 6. (**a**) Native model projected in SA. (**b**) Model of the invaded range calibrated in SA. Red indicates higher habitat suitability for *C. vulgare*. Yellow indicates medium suitability while light green indicates low or no suitability as it becomes lighter.

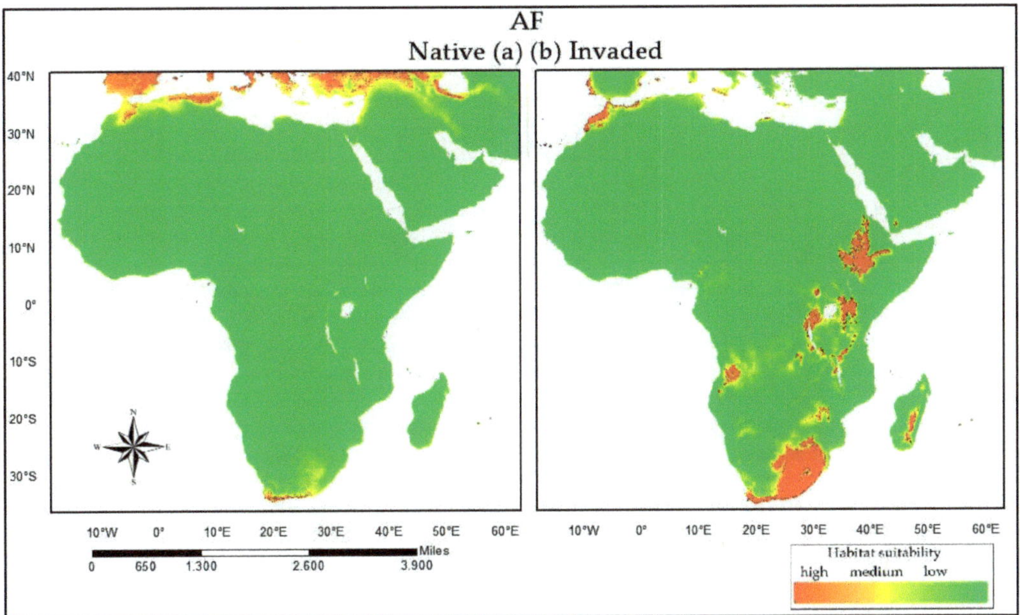

Figure 7. (**a**) Native model projected in AF. (**b**) Model of the invaded range calibrated in AF. Red indicates higher habitat suitability for *C. vulgare*. Yellow indicates medium suitability while light green indicates low or no suitability as it becomes lighter.

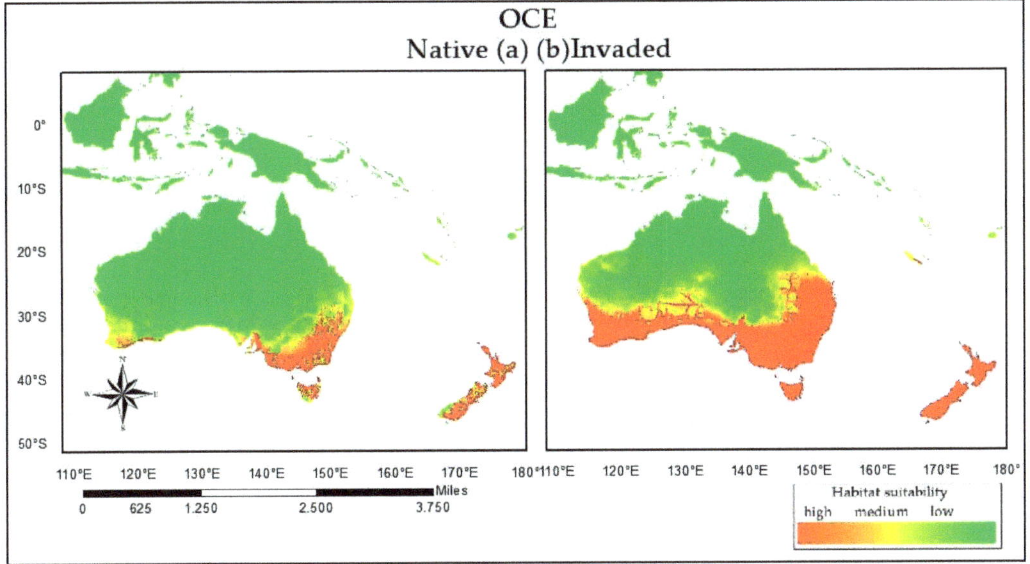

Figure 8. (**a**) Native model projected in OCE. (**b**) Model of the invaded range calibrated in OCE. Red indicates higher habitat suitability for *C. vulgare*. Yellow indicates medium suitability while light green indicates low or no suitability as it becomes lighter.

4. Discussion

The study and understanding of biological invasions are critical elements since they can provide crucial information of the impact of invasive species and its areas of potential invasion in order to develop effective strategies and prevention measures. Studies at the regional level [23,48] constructed with adequately selected predictor variables have allowed a better characterization of the ecological niche shifts at different scales and at different stages of invasion [29,32]. However, defining the reasons why that happens is complex due to the different processes involved in the invasion of a species and, of course, it is essential the application of adequate methodologies to confront elements of the dynamics of ecological niches. A displacement of the niche centroid may imply that the species has found new suitable conditions within the invaded range in a non-analogous climate (climate absent in its native distribution), which would imply a shift in the realized niche but within the tolerance range of the species' fundamental niche through a preadaptation mechanism, i.e., at some point, the species could have developed in the climate in question [49,50]. Another possibility is that the species underwent changes at the genetic level beyond its fundamental niche and adapted to new environmental conditions. It has been determined that *C. vulgare* has adapted and proliferated to the conditions of the dune ecosystems in Chile, presenting therophytic characteristics [51]. These new capacities may have been developed in the whole invaded range, although it is possible that this may be a preexisting characteristic of the species, coinciding with what happens with other therophytic species that have been described in some areas of the Mediterranean where *C. vulgare* has its native range. On the other hand, the establishment of species in new environments is not such a simple matter apart from phenotypic plasticity and genetic adaptation, as biological interactions play a crucial role in the processes of colonization of new territories [50]. A recent study showed that the species of the genus *Cirsium* (L.) Mill., since they generally share their environmental preferences and are often found in the same geographical area, are capable of sharing genetic material by means of cross-pollination. This would generate a greater genetic diversity and would imply possible adaptive capacities to new environments [52]. Some weed control studies [36,53,54] for *C. vulgare* have shown that this species has many natural competitors, mainly herbivorous insects that feed on the plant, reducing its capacity to produce seeds. These natural enemies have been used to control the proliferation of the plant mainly in North America, South Africa, and Oceania. However, the production capacity and resistance of the seeds of this species make it a very difficult adversary. *C. vulgare* is a weed with a fairly efficient competition capacity and with high resistance to chemical control systems, especially in South Australia where these cases are common with other species not necessarily of thistles [55].

When we analyzed the niche overlap metrics between regions, we observed that the use of marginal environments produced a variation in the values obtained, something previously verified in other invasive species [56]. In our case, the same case was fulfilled in almost all the regions. For example, in NOA, the values were practically the same, although the values in the expansion and unfilling were quite low (Table 2, Figure 2). This may be due to the similarity between niches that exist between this region and its native region (Figure 4). In the case of OCE, the expansion values showed a slight variation when using marginal environments, but the region also presented low values in the expansion. There was a great variation in the unfilling ratio when using these environments, which would mean possible future invasion areas. Besides, the SA region did not present such significant variability in the use of marginal environments, but the expansion values were much more representative compared to NOA and OCE. In addition, there is an important proportion of the native niche that the species has conserved but there is also a proportion of unfilling that has not been occupied. This could be because the species has not yet been introduced into these areas. Alternatively, it could be attributed to the existence of some geographical barrier, as it is a continent with large geographical features, or to biological interactions with other organisms—predatory pressure or competition for resources—that have not allowed it to settle [12,37]. In the case of AF, the metrics showed there was a rather low proportion of overlap of niches in relation to the native range. In fact, it was the lowest compared to the other three regions analyzed, but, interestingly, the expansion metrics were quite high and rather constant when including marginal environments. On the other hand, the unfilling metric varied drastically since was moderate without marginal environments, and the metric reached an important value when including these environments. It is interesting that the direction of displacement of the species in all regions was generally toward warmer environments with less variation in temperature.

According to AUC data, the performance of our models is statistically acceptable [10,37]. The analysis of variables importance indicates the very important role of annual and seasonal temperature in the distribution of the study species (Table 4). The predictions of the models (Figures 5–8) show some differences with regard to the current distribution of the species (Figure 1). In this sense, it is important to note that, for this sort of species, it is necessary to consider variables beyond the bioclimatic ones [10,29]. In our case, the inclusion of the Human Footprint (HFP) provided us with insight into how anthropogenic activities are specially linked to the distribution of *C. vulgare*. This has also been observed for example in studies carried out in Africa [57] and in New Zealand [36], where the invasive species studied mainly proliferate in wastelands, roadsides, and areas of pasture cultivation. It is clear that there is a correlation between human intervention and the proliferation and success of invasive species, and some studies have highlighted the vulnerability of invaded ecosystems, emphasizing landscape structure as a determining factor in the success of the invasion [48,58]. In our case, when the weight of the Human Footprint (HFP) was analyzed (Figure 2h), we observed a relationship between the proliferation of our species in areas where humans modified the natural landscape, one of the most outstanding in the fields of agricultural production [59].

This contribution aims to show that the combination of tools and methods for predicting and inferring possible future invasions with an adequate selection of climatic and non-climatic variables, taking into account a global vision of the distribution of species at a regional level, can be a starting point for understanding the general trends of biological invasions. Likewise, through the comparison of the different types of ecological needs of the target species, it may be possible to achieve basic information for developing effective control strategies fitted to each territory and to invest resources in a more profitable way to protect native biodiversity.

Author Contributions: Conceptualization, J.Á.S.A. and J.F.C.R.; methodology, R.E.H.-L., J.F.C.R. and J.Á.S.A.; software, R.E.H.-L. and J.F.C.R.; validation, R.E.H.-L. and J.F.C.R.; formal analysis, J.F.C.R., R.E.H.-L. and J.Á.S.A.; investigation, J.F.C.R.; writing—review and editing, J.F.C.R., J.Á.S.A. and D.R.d.l.C.; supervision, J.Á.S.A. and D.R.d.l.C. All authors have read and agreed to the published version of the manuscript.

Funding: This research received no external funding.

Institutional Review Board Statement: Not applicable.

Informed Consent Statement: Not applicable.

Data Availability Statement: Not applicable.

Acknowledgments: We wish to thank both the scientific and editorial reviewers for their invaluable help in correcting this article.

References

1. Sala, O.E.; Chapin, F.S., III; Armesto, J.J.; Berlow, E.; Bloomfield, J.; Dirzo, R.; Huber-Sanwald, E.; Huenneke, L.F.; Jackson, R.B.; Kinzig, A.; et al. Global Biodiversity Scenarios for the Year 2100. *Science* **2000**, *287*, 1770–1774. [CrossRef] [PubMed]
2. Haddad, N.M.; Brudvig, L.A.; Clobert, J.; Davies, K.F.; Gonzalez, A.; Holt, R.D.; Lovejoy, T.E.; Sexton, J.O.; Austin, M.P.; Collins, C.D.; et al. Habitat Fragmentation and Its Lasting Impact on Earth's Ecosystems. *Sci. Adv.* **2015**, *1*, e1500052. [CrossRef]
3. Rybicki, J.; Hanski, I. Species-Area Relationships and Extinctions Caused by Habitat Loss and Fragmentation. *Ecol. Lett.* **2013**, *16*, 27–38. [CrossRef] [PubMed]

4.   Pérez-García, J.N. Causas de la pérdida global de biodiversidad. *Rev. Asoc. Colomb. Cienc. Biol.* **2020**, 183–198. [CrossRef]

5.   Butchart, S.H.M.; Walpole, M.; Collen, B.; van Strien, A.; Scharlemann, J.P.W.; Almond, R.E.A.; Baillie, J.E.M.; Bomhard, B.; Brown, C.; Bruno, J.; et al. Global Biodiversity: Indicators of Recent Declines. *Science* **2010**, *328*, 1164–1168. [CrossRef] [PubMed]

6.   Seebens, H.; Blackburn, T.M.; Dyer, E.E.; Genovesi, P.; Hulme, P.E.; Jeschke, J.M.; Pagad, S.; Pyšek, P.; Winter, M.; Arianoutsou, M.; et al. No Saturation in the Accumulation of Alien Species Worldwide. *Nat. Commun.* **2017**, *8*, 14435. [CrossRef] [PubMed]

7.   Estrada, A.; Morales-Castilla, I.; Caplat, P.; Early, R. Usefulness of Species Traits in Predicting Range Shifts. *Trends Ecol. Evol.* **2016**, *31*, 190–203. [CrossRef] [PubMed]

8.   Mathakutha, R.; Steyn, C.; le Roux, P.C.; Blom, I.J.; Chown, S.L.; Daru, B.H.; Ripley, B.S.; Louw, A.; Greve, M. Invasive Species Differ in Key Functional Traits from Native and Non-Invasive Alien Plant Species. *J. Veg. Sci.* **2019**, *30*, 994–1006. [CrossRef]

9.   Alharbi, W.; Petrovskii, S. Patterns of Invasive Species Spread in a Landscape with a Complex Geometry. *Ecol. Complex.* **2018**, *33*, 93–105. [CrossRef]

10. Hernández-Lambraño, R.E.; González-Moreno, P.; Sánchez-Agudo, J.Á. Towards the Top: Niche Expansion of Taraxacum Officinale and Ulex Europaeus in Mountain Regions of South America. *Austral Ecol.* **2017**, *42*, 577–589. [CrossRef]

11. Born-Schmidt, G.; De Alba, F.; Servole, J.; Koleff, P. *Espaldon, maria victoria Principales Retos Que Enfrenta Mexico Ante Las Especies Exóticas Invasoras*; Comisión Nacional para el Conocimiento y Uso de la Biodiversidad (Conabio), el Programa de las Naciones Unidas para el Desarrollo (PNUD) y el Centro de Estudios Sociales y de Opinión Pública (CESOP): Mexico City, Mexico, 2017.

12. Zilio, M.I. *El Impacto Económico de las Invasiones Biológicas en Argentina: Cuánto Cuesta no Proteger la Biodiversidad*; Asociación Argentina de Economía Política (AAEP): Buenos Aires, Argentina, 2019.

13. Hoffmann, B.; Broadhurst, L. The Economic Cost of Managing Invasive Species in Australia. *NeoBiota* **2016**, *31*, 1–18. [CrossRef]

14. Jackson, T. Addressing the Economic Costs of Invasive Alien Species: Some Methodological and Empirical Issues. *Int. J. Sustain. Soc.* **2015**, *7*, 221–240. [CrossRef]

15. Broennimann, O.; Guisan, A. Predicting Current and Future Biological Invasions: Both Native and Invaded Ranges Matter. *Biol. Lett.* **2008**, *4*, 585–589. [CrossRef] [PubMed]

16. Medeiros, C.M. Aplicación de Modelos de Nicho Ecológico y Sistemas de Información Geográfica para la Conservación de la Biodiversidad. Ph.D. Thesis, Universidad de Salamanca, Salamanca, Spain, 2018. Available online: http://purl.org/dc/dcmitype/Text (accessed on 3 April 2020).

17. Soberón, J.; Osorio-Olvera, L.; Peterson, T. Diferenias conceptuales entre modelación de nichos y modelación de áreas de distribución. *Rev. Mex. Biodivers.* **2017**, *88*, 437–441. [CrossRef]

18. Mota Vargas, C.; Encarnación Luévano, A.; Ortega Andrade, H.M.; Prieto Torres, D.A.; Peña Peniche, A.; Rojas Soto, O.R. *Una Breve Introducción a los Modelos de Nicho Ecológico*; Universidad Autónoma del Estado de Hidalgo: Pachuca, Mexico, 2020; ISBN 978-607-482-598-5.

19. Colwell, R.K.; Rangel, T.F. Hutchinson's Duality: The Once and Future Niche. *Proc. Natl. Acad. Sci. USA* **2009**, *106* (Suppl. 2), 19651–19658. [CrossRef] [PubMed]

20. Hutchinson, G.E. Concluding Remarks. *Cold Spring Harb. Symp. Quant. Biol.* **1957**, *22*, 415–427. [CrossRef]

21. Phillips, S.J.; Anderson, R.P.; Schapire, R.E. Maximum Entropy Modeling of Species Geographic Distributions. *Ecol. Model.* **2006**, *190*, 231–259. [CrossRef]

22. Goncalves, E.; Herrera, I.; Duarte, M.; Bustamante, R.; Lampo, M.; Squez, G.; Sharma, G.; Garcia-Rangel, S. Global Invasion of Lantana Camara: Has the Climatic Niche Been Conserved across Continents? *PLoS ONE* **2014**, *9*. [CrossRef] [PubMed]

23. Battini, N.; Farias, N.; Giachetti, C.; Schwindt, E.; Bortolus, A. Staying Ahead of Invaders: Using Species Distribution Modeling to Predict Alien Species' Potential Niche Shifts. *Mar. Ecol. Prog. Ser.* **2019**, *612*, 127–140. [CrossRef]

24. Elith, J.; Graham, C.; Anderson, R.; Dudík, M.; Ferrier, S.; Guisan, A.; Hijmans, R.; Huettmann, F.; Leathwick, J.; Lehmann, A. Novel Methods Improve Prediction of Species' Distributions from Occurrence Data. *Ecography* **2006**, *29*, 129–151. [CrossRef]

25. Pearman, P.B.; Guisan, A.; Broennimann, O.; Randin, C.F. Niche Dynamics in Space and Time. *Trends Ecol. Evol.* **2008**, *23*, 149–158. [CrossRef] [PubMed]

26. Peterson, A.T. Predicting the Geography of Species' Invasions via Ecological Niche Modeling. *Q. Rev. Biol.* **2003**, *78*, 419–433. [CrossRef]

27. Carlos-Júnior, L.A.; Barbosa, N.P.U.; Moulton, T.P.; Creed, J.C. Ecological Niche Model Used to Examine the Distribution of an Invasive, Non-Indigenous Coral. *Mar. Environ. Res.* **2015**, *103*, 115–124. [CrossRef] [PubMed]

28. Parravicini, V.; Azzurro, E.; Kulbicki, M.; Belmaker, J. Niche Shift Can Impair the Ability to Predict Invasion Risk in the Marine Realm: An Illustration Using Mediterranean Fish Invaders. *Ecol. Lett.* **2015**, *18*, 246–253. [CrossRef] [PubMed]

29. Guisan, A.; Petitpierre, B.; Broennimann, O.; Daehler, C.; Kueffer, C. Unifying Niche Shift Studies: Insights from Biological Invasions. *Trends Ecol. Evol.* **2014**, *29*, 260–269. [CrossRef] [PubMed]

30. Hierro, J.L.; Maron, J.L.; Callaway, R.M. A Biogeographical Approach to Plant Invasions: The Importance of Studying Exotics in Their Introduced and Native Range. *J. Ecol.* **2005**, *93*, 5–15. [CrossRef]

31. Richardson, D.M.; Allsopp, N.; D'Antonio, C.M.; Milton, S.J.; Rejmánek, M. Plant Invasions—The Role of Mutualisms. *Biol. Rev.* **2000**, *75*, 65–93. [CrossRef] [PubMed]

32. Petitpierre, B.; Kueffer, C.; Broennimann, O.; Randin, C.; Daehler, C.; Guisan, A. Climatic Niche Shifts Are Rare Among Terrestrial Plant Invaders. *Science* **2012**, *335*, 1344–1348. [CrossRef] [PubMed]

33. Moore, R.J.; Frankton, C. *The Thistles of Canada*; Research Branch, Canada Department of Agriculture: Ottawa, ON, Canada, 1974.

34. Parsons, W.T.; Cuthbertson, E.G. *Noxious Weeds of Australia*; Inkata Press: Melbourne, VIC, Australia, 1992; ISBN 978-0-909605-81-0.

35. Mitich, L.W. Bull Thistle, Cirsium vulgare. *Weed Technol.* **1998**, *12*, 761–763. [CrossRef]

36. Cripps, M.; Navukula, J.; Casonato, S.; van Koten, C. Impact of the Gall Fly, Urophora Stylata, on the Pasture Weed, Cirsium Vulgare, in New Zealand. *BioControl* **2020**, *65*, 501–513. [CrossRef]

37. Broennimann, O.; Fitzpatrick, M.C.; Pearman, P.B.; Petitpierre, B.; Pellissier, L.; Yoccoz, N.G.; Thuiller, W.; Fortin, M.-J.; Randin, C.; Zimmermann, N.E.; et al. Measuring Ecological Niche Overlap from Occurrence and Spatial Environmental Data. *Glob. Ecol. Biogeogr.* **2012**, *21*, 481–497. [CrossRef]

38. Holm, L.; Doll, J.; Holm, E.; Pancho, J.V.; Herberger, J.P. *World Weeds: Natural Histories and Distribution*; John Wiley & Sons: Hoboken, NJ, USA, 1997.

39. Herrera, I.; Goncalves, E.; Pauchard, A.; Bustamante, R.O. *Manual de Plantas Invasoras de Sudamérica*; IEB Chile, Instituto de Ecología y Biodiversidad: Región de O'Higgins, Chile, 2016.

40. Petryna, L.; Moora, M.; Nuñes, C.O.; Cantero, J.J.; Zobel, M. Are Invaders Disturbance-Limited? Conservation of Mountain Grasslands in Central Argentina. *Appl. Veg. Sci.* **2002**, *5*, 195–202. [CrossRef]

41. Parsons, W.T.; Cuthbertson, E.G. *Noxious Weeds of Australia*, 2nd ed.; CSIRO Publishing: Collingwood, VIC, Australia, 2001; ISBN 978-0-643-06514-7.

42. Klinkhamer, P.G.; De Jong, T.J. Cirsium Vulgare (Savi) Ten. *J. Ecol.* **1993**, 177–191. [CrossRef]

43. R: El Proyecto R Para Computación Estadística. Available online: https://www.r-project.org/index.html (accessed on 4 February 2021).

44. About ArcGIS | Mapping & Analytics Software and Services. Available online: https://www.esri.com/en-us/arcgis/about-arcgis/overview (accessed on 31 January 2021).

45. Cola, V.D.; Broennimann, O.; Petitpierre, B.; Breiner, F.T.; D'Amen, M.; Randin, C.; Engler, R.; Pottier, J.; Pio, D.; Dubuis, A.; et al. Ecospat: An R Package to Support Spatial Analyses and Modeling of Species Niches and Distributions. *Ecography* **2017**, *40*, 774–787. [CrossRef]

46. Broennimann, O.; Treier, U.A.; Müller-Schärer, H.; Thuiller, W.; Peterson, A.T.; Guisan, A. Evidence of Climatic Niche Shift during Biological Invasion. *Ecol. Lett* **2007**, *10*, 701–709. [CrossRef] [PubMed]

47. Elith, J.; Phillips, S.J.; Hastie, T.; Dudík, M.; Chee, Y.E.; Yates, C.J. A Statistical Explanation of MaxEnt for Ecologists. *Divers. Distrib.* **2011**, *17*, 43–57. [CrossRef]

48. González-Moreno, P.; Delgado, J.; Vilà, M. Una Visión a Escala de Paisaje de Las Invasiones Biológicas. *Ecosistemas* **2015**, *24*, 84–92. [CrossRef]

49. Webber, B.; Le Maitre, D.; Kriticos, D. Comment on "Climatic Niche Shifts Are Rare Among Terrestrial Plant Invaders". *Science* **2012**, *338*, 193. [CrossRef] [PubMed]

50. Jack of All Trades, Master of Some? On the Role of Phenotypic Plasticity in Plant Invasions-Richards-2006-Ecology Letters-Wiley Online Library. Available online: https://onlinelibrary.wiley.com/doi/10.1111/j.1461-0248.2006.00950.x (accessed on 15 December 2020).

51. Martín, J.S.; Ramírez, C.; Martín, C.S. La flora de las dunas chilenas y sus adaptaciones morfológicas. *Bosque* **1992**, *13*, 29–39. [CrossRef]

52. Sheidai, M.; Zanganeh, S.; Haji-Ramezanali, R.; Nouroozi, M.; Noormohammadi, Z.; Ghsemzadeh-Baraki, S. Genetic Diversity and Population Structure in Four Cirsium (Asteraceae) Species. *Biologia* **2013**, *68*. [CrossRef]

53. Suwa, T.; Louda, S.M.; Leland Russell, F. No Interaction between Competition and Herbivory in Limiting Introduced Cirsium Vulgare Rosette Growth and Reproduction. *Oecologia* **2010**, *162*, 91–102. [CrossRef] [PubMed]

54. Moyo, C.; Harrington, K.C.; Kemp, P.D. Effectiveness of Spraying Herbicides in the Centre Compared to All over Rosettes of Cirsium Vulgare and Jacobaea Vulgaris. In Proceedings of the 19th Australasian Weeds Conference, Hobart, Australia, 1–4 September 2014; pp. 235–238.

55. Chauhan, B.S.; Jha, P. Glyphosate Resistance in Sonchus Oleraceus and Alternative Herbicide Options for Its Control in Southeast Australia. *Sustainability* **2020**, *12*, 8311. [CrossRef]

56. Paulo De Marco, J.; Diniz-Filho, J.A.F.; Bini, L.M. Spatial Analysis Improves Species Distribution Modelling during Range Expansion. *Biol. Lett.* **2008**, *4*, 577–580. [CrossRef] [PubMed]

57. Horo, J.T.; Tessema, T. Abundance and Distribution of Invasive Alien Plant Species in Illu Ababora Zone of Oromia National Regional State, Ethiopia. *J. Agric. Sci. Food Technol.* **2015**, *1*, 94–100. [CrossRef]

58. Cabra-Rivas, I.; Saldaña, A.; Castro-Díez, P.; Gallien, L. A Multi-Scale Approach to Identify Invasion Drivers and Invaders' Future Dynamics. *Biol. Invasions* **2016**, *18*. [CrossRef]

59. Zhang, X.; Wei, H.; Zhao, Z.; Liu, J.; Zhang, Q.; Zhang, X.; Gu, W. The Global Potential Distribution of Invasive Plants: Anredera Cordifolia under Climate Change and Human Activity Based on Random Forest Models. *Sustainability* **2020**, *12*, 1491. [CrossRef]

Article

Natural Protected Areas as Providers of Ecological Connectivity in the Landscape: The Case of the Iberian Lynx

Iván Barbero-Bermejo [1,2,*], Gabriela Crespo-Luengo [1,2], Ricardo Enrique Hernández-Lambraño [1,2], David Rodríguez de la Cruz [1,2] and José Ángel Sánchez-Agudo [1,2,*]

1 Grupo de Investigación en Biodiversidad, Diversidad humana y Biología de la Conservación, Campus Miguel de Unamuno s/n, Universidad de Salamanca, E-37007 Salamanca, Spain; gabrielacl94@usal.es (G.C.-L.); ricardohl123@usal.es (R.E.H.-L.); droc@usal.es (D.R.d.l.C.)
2 Departamento de Botánica y Fisiología Vegetal, Área de Botánica, Campus Miguel de Unamuno s/n, Universidad de Salamanca, E-37007 Salamanca, Spain
* Correspondence: ivanbarbero@usal.es (I.B.-B.); jasagudo@usal.es (J.A.S.-A.)

Abstract: The design of conservation plans for the improvement of habitats of threatened species constitutes one of the most plausible possibilities of intervention in the structure and composition of the landscape of a large territory. In this work we focus on the Iberian lynx in order to establish potential ecological corridors using ecoinformatic tools from the GIS environment to improve connectivity between the existing natural spaces within the scope of its historical distribution. We processed 669 records of the presence of the lynx and six predictor variables linked to the habitat of the species. With this, corridors have been generated between natural areas. The determination of possible bottlenecks or dangerous areas (e.g., hitches on highways) allows for focusing efforts on their conservation. This type of approach seeks to improve efficiency in the design of measures aimed at expanding the territory's capacity to host its populations, improving both its viability and that of all the other species that are linked to it. The proposals for action on the specific areas defined by the models elaborated in this work would imply interventions on the land uses and existing vegetation types in order to improve connectivity throughout the territory and increase the resilience of its ecosystems.

Keywords: landscape; GIS; Corridor Designer; MaxEnt; species distribution models; ecological corridors

Citation: Barbero-Bermejo, I.; Crespo-Luengo, G.; Hernández-Lambraño, R.E.; Rodríguez de la Cruz, D.; Sánchez-Agudo, J.Á. Natural Protected Areas as Providers of Ecological Connectivity in the Landscape: The Case of the Iberian Lynx. *Sustainability* **2021**, *13*, 41. https://doi.org/10.3390/su13010041

Received: 2 December 2020
Accepted: 21 December 2020
Published: 23 December 2020

Publisher's Note: MDPI stays neutral with regard to jurisdictional claims in published maps and institutional affiliations.

1. Introduction

Human activities shape and transform territories on a global scale through the modification of the forms and properties of the surface and subsoil; being, therefore, a dominant factor in the evolution of the landscape in this current period [1,2]. As a consequence of this transformation, severe alterations in ecosystem dynamics are generated, resulting in population and species reductions and extinctions. In this context, the fragmentation of habitats and populations appears as one of the main factors responsible for the serious and rapid loss of biodiversity [3,4].

Habitat fragmentation results in a loss of connectivity in the landscape. The preservation and restoration of connectivity has become one of the main conservation objectives [5]. Understanding the ecological processes which depend on connectivity and taking effective planning measures requires an understanding of how landscape features can affect it. Taylor et al. [6] define connectivity as the level of facilitation or resistance to the movement of organisms between patches of habitat with resources. This resistance is determined by a landscape matrix which can present different levels of alterations and is able to modulate connectivity between patches or provide resources to species. In this way, a well-preserved matrix with a low degree of alteration can act as a buffer zone for habitat patches and mitigate the isolation of these. In addition, the matrix between these patches is characterized

97

by topographic, anthropic, ecological, and other variables that will constitute an important factor for dispersion [7–9]. For the design of species conservation plans and strategies, it is fundamental to consider the potential and existing links that allow the flow of organisms between populations [10,11]. Increased consciousness about the ecological consequences of habitat loss and fragmentation [12,13] has resulted in a growing trend to incorporate preventive criteria in sectoral planning, completed mainly through environmental impact evaluation procedures and land use plans and programs [14].

The connectivity of habitat patches is important for the movement of genes, individuals, populations, and species on multiple time scales. Connectivity in the juvenile dispersal phase of many animals affects the success and recolonization of patches that may be unoccupied [15]. The capacity to migrate also has a great influence on the ability of species to adapt in response to climate change [16].

An example of a species that is seriously threatened by habitat fragmentation and loss of habitat is the Iberian lynx (*Lynx pardinus* Temminck, 1827), an endemic feline of the Iberian Peninsula [17] whose populations are severely fragmented as a main consequence of landscape homogenisation in the Iberian southwest. This is due to agricultural intensification during the 20th century [18]. The recent efforts made in conservation have improved its state of conservation. Populations have grown from only 100 specimens at the beginning of the 21st century to more than 600 at present, according to the latest census [19]. However, ensuring connectivity between these areas and those occupied by current populations is essential for their long-term persistence.

Recent studies show the interest in population connectivity studies to deal with the survival of this species [20,21]. These analyses have shown to be methodologies with great potential in the field of biodiversity conservation.

In this aspect, it is important to develop habitat quality models, as well as permeability models and ecological connectivity analyses [22]. For this purpose, Geographic Information Systems (GIS) constitute a set of powerful tools that allow the provision of input data for Species Distribution Models (SDMs), the use of which is increasingly common in the planning of conservation actions [23,24]. The MaxEnt method [25] is considered one of the most powerful and robust in this field [26,27].

Different approaches have been applied to connectivity studies, such as those based on movement simulations [28] or network connectivity analyses, such as least cost routes and the application of graph theory [29–31]. The latter have received increasing attention in recent years. In this study, we rely on the graph theory [32], which allows us to consider the landscape as a set of nodes (patches) and connecting elements (links) that can be interpreted either as physical corridors or as the dispersion potential of organisms [33]. This approach has been widely applied in order to maximize the efficiency of flow in networks and circuits and it is very useful in the field of computing and information technology. More recently, it has been used in the context of landscape and population ecology [34]. A graphical approach to the landscape can provide useful and relevant information about the dynamic processes taking place in the landscape, allowing important patches to be identified for connectivity. This is very useful for species conservation in heterogeneous landscapes [31] because the more detailed the ecological information incorporated in the model, the higher the correspondence with models of actual landscape dynamics [29,31]. Landscape indices/metrics describe the spatial structure of a landscape at a given time. In addition, these indices provide important data about the configuration of the landscape, allowing for comparisons between different compositions of the landscape [35].

Connectivity metrics combine topological and ecological characteristics of landscape elements. In particular, the probability of connectivity index (PC index) [36] is based on a probabilistic connection model that allows continuous modulation of the connection force [33]. The PC index can be defined as the probability that two points (presence records) that are randomly situated within the landscape are located in accessible habitat areas in a set formed by habitat patches and the links (connections) between them [36].

The main objective of this study is to provide a methodological approach to landscape assessment in order to maintain ecological connectivity, facilitating adequate landscape and land use management. We present this approach through the characteristics and habitat needs of a species of great interest in the Iberian Peninsula: the Iberian lynx. In this way, we aim to connect natural protected areas with good habitat and protection characteristics through a landscape matrix of very different qualities. To find out how the landscape influences the conservation of this species, we have set ourselves the following objectives: (a) to find out which factors most determine the distribution of the Iberian lynx in its historical range, (b) to generate a map of the potential distribution of the species in order to determine the suitability of the geographical area for its presence, (c) to evaluate different protected natural areas integrated into the historical distribution area of the species according to different factors: suitability of habitat and human influence, and (d) to evaluate the capacity for connection between these spaces and the generation of potential ecological corridors integrated into the landscape matrix that could be used by the species for its dispersion and conquest of future new territories.

2. Materials and Methods

2.1. Study Area

The study area covers approximately 176,254 km^2 located mainly in the central-west and south-west of the Iberian Peninsula (Figure 1) where there is historical evidence of the presence of Iberian lynx. The landscape is very heterogeneous and combines mountainous zones belonging to the Central System, Montes de Toledo, Sierra Morena, and a large part of the Sub-Baetic Systems with different river basins (Guadalquivir, Guadiana and Tajo), especially the depression formed by the river Guadalquivir.

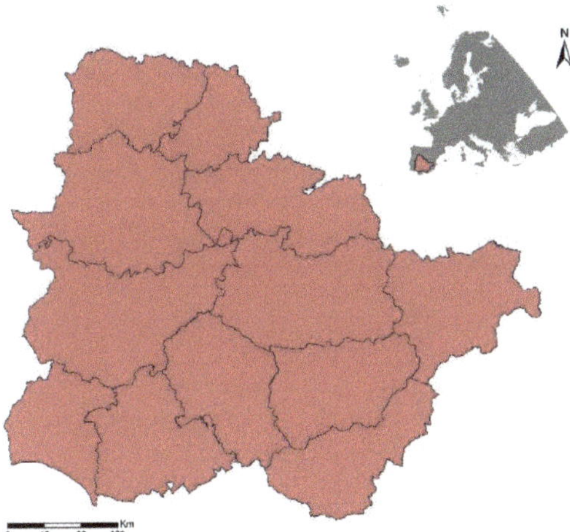

Figure 1. Studied area: south-west Spain, 12 provinces in 4 administrative regions (Andalucía, Extremadura, Castilla-La Mancha, Castilla y León).

2.2. Study Species

The Iberian lynx is a slender cat, approximately 1 m long and 8–15 kg in weight [37]. It is an emblematic Iberian species considered to be one of the most seriously threatened cat species in the world and is currently listed as "endangered" according to the International Union for Conservation of Nature (IUCN) categories [38,39]. The Iberian lynx is considered a strict species in terms of habitat requirement, being found in areas between 400 and 1300 m above sea level and closely linked to the Mediterranean mountain range, especially the extensive and dense scrub [40]. It has an elusive nature, avoiding open areas with human influence in all stages of its biological cycle [41]. In their dispersal phase, lynx are able to use remnants of suitable habitat, formed mainly by Mediterranean scrub, as "steps" to travel through a fragmented landscape [42] and through a poorer quality matrix formed by woodlands or forest plantations of pines or eucalyptus [43].

The reduction in its range has been linked to several key factors: on the one hand, agricultural and forestry transformations; on the other hand, the decline of its main prey, the rabbit (*Oryctolagus cuniculus* Linnaeus, 1758) due to myxomatosis and viral hemorrhagic disease [40,41,44]. Finally, the increase in non-natural mortality of this species caused mainly by illegal hunting and road traffic accidents is a serious obstacle to its recovery [40,42,45].

2.3. Habitat Suitability Modelling

2.3.1. Species Data

The Iberian lynx presence (n = 900) was obtained from the Global Biodiversity Information Facility (GBIF: https://www.gbif.org) [46] through human observation. Most of these records are integrated in the dataset of the National Biodiversity Inventory (2007) of the Ministry of Environment, Rural and Marine Affairs, collected between 1980 and 2007. The data were processed to eliminate duplicates and incorrect records, collecting a total of 669 presence records at 10 × 10 km resolution.

2.3.2. Environmental Data

Twenty-two variables were first collected relating to climatic, topographical, and land use characteristics. The trophic variable (rabbit density) was not used due to the lack of a thematic cartography in this study.

The bioclimatic variables were obtained from WorldClim-Global Climate Data version 2.0: http://worldclim.org/version2 for the period 1970–2000 at a resolution of 1 × 1 km. The slope was derived from a Digital Elevation Model (DEM) at 200 × 200 m resolution from the IGN—National Geographic Institute: https://www.ign.es, using the ArcGis Slope function [47]. The vegetation cover variable was obtained from the Spain Forest Map, project carried out during the years 1997 to 2006 and available at Ministerio para la Transición Ecológica (MITECO Spanish acronym, https://www.miteco.gob.es). We extracted the vegetation cover using raster calculator (ESRI, 2016).

Six of those 22 variables were finally selected following bibliographic sources on the biology and ecology of the Iberian lynx [41,45]. They are related to climatic, topographical, and land use aspects (Table 1).

Table 1. Brief description and source of variables used in the Iberian lynx distribution model.

Code	Description	Source
Slope	Gradient of the land and inclination	As of DEM [48]
FMS	Forest Map of Spain (vegetation cover)	[49]
Bio1	Average annual temperature	[50]
Bio4	Seasonal temperature (standard deviation * 100)	[50]
Bio12	Annual precipitation	[50]
Bio15	Seasonal precipitation (variation coefficient)	[50]

All environmental variables were standardized at geographic coordinates (Datum WGS-1984) and resampled at a spatial resolution of 10 × 10 km using bilinear interpolation sampling [47]. Previously, all continuous variables were checked for collinearity effects, through the "ggpairs" function of the "GGally" package [51] in R version 3.1.3 [52]. Those variables with ($|r| \geq 0.75$) were removed from the final set of predictor variables (Table S1).

2.3.3. Modelling Process with MaxEnt

Habitat suitability of Iberian lynx was modelled using MaxEnt version 3.4.1, a machine-learning process that uses presence-only data [25]. MaxEnt gives insight about what predictors are important and estimates the relative suitability of one place vs. another, as well as the probability of occurrence [53,54]. This approach has been demonstrated to perform well in a diverse set of modelling scenarios and is widely used in a great number of studies in ecology, biogeography and conservation [55–58].

Ten replicates were performed by Bootstrap procedure in "cloglog" format using 80% of the presence records as model training and the remaining 20% to evaluate the models. The background was of 10,000 random points from study area. Replicas used herein were performed using the MaxEnt default parameters [25]. Variable importance was calculated to assess the relevance of each predictor in the models and response curves were calculated to interrogate the relationship between the response (i.e., Iberian lynx presence) and each explanatory variable

The models were evaluated using the area under the curve of the receiver operating characteristic (AUC). AUC measures the ability of a model to discriminate between sites with occurrence, versus those where it is absent (herein, background points). The AUC ranges from 0 to 1 (0.5 = random, 1 = perfect).

Finally, the model was projected into the study area to generate a habitat suitability map of Iberian lynx. The criterion of maximizing the sum of sensitivity and specificity (maximum test sensitivity plus specificity) was used as a threshold to define the suitability of the cells [59,60].

2.4. The Design of Ecological Corridors

2.4.1. Selection of Core Areas

The Natural Protected Areas (NPAs) in the study area have been used as core areas. In particular, national parks and natural parks with protection category that have associated regulations and conservation measures for species and habitats were included.

To improve the analysis, we decided to merge those geographically adjacent NPAs and consider them as a unique unit (Figure S1). This is the case of Sierra Norte Sevilla, Sierra Hornachuelos, and Sierra de Aracena y Picos de Aroche (unit 2); Valle de Alcudia and Sierra Madrona, Sierra de Andújar, and Sierra de Cardeña y Montoro, (unit 4) and Quilamas and the Batuecas-Sierra Francia (unit 17). The selected NPAs (25) are shown in Table 2, where the numerical coding adopted is indicated in order to facilitate the interpretation and discussion of the results.

2.4.2. Resistance Raster

The next step in the connectivity analysis is the generation of a raster showing the resistance or suitability of the terrain—landscape matrix—for the movement of the species [30, 61]. A specific ArcGis extension called Corridor Designer (http://corridordesign.org) was used for this purpose [62], which allows for the design of ecological corridors based on a resistance raster.

This raster requires a previous weighting by the researcher of different selected variables, in this case: altimetric information representing the relief in the area (DEM) [48], topographic slope (Slope), a set of global data covering human impact (population density, land use, infrastructure, etc.) represented by the Human Influence Index (HII) was downloaded from the SEDAC—Socioeconomic Data and Applications Center database: https://sedac.ciesin.columbia.edu [63] and land cover information obtained from the National Forest Inventory through the Forest Map [49].

2.4.3. Corridor Design

The resistance map is used by the Corridor Designer as a basis for calculation of the corridors with a range of 0 to 100, where 0 represents the most restrictive value for the species and 100 the optimum. The resolution of this habitat model was 1 × 1 km. The lowest cost corridors are generated through a cost–distance analysis. Therefore, they are distance based corridors that minimize the cumulative cost of each start cell location through the above mentioned resistance raster. In this way the cost of making a move through each cell is defined.

Table 2. Natural Protected Areas (NPAs) selected as core areas to be connected in the corridor network. An alphanumeric coding is shown for a better understanding of the maps, the Protection Figure (Figure NPA), its area in km², the average suitability, the Human Influence Index (HII) and information about whether they currently host (Yes or No) an Iberian lynx population. CP: Current Population. NP: National Park; NatP: Natural Park; RegP: Regional Park; ZRI: Zone of Regional Interest; SCZ: Special Conservation Zone.

Cod.	NameNPAs	Protection Level	Surface Area (km²)	Suitability	HII	CP
1	Doñana	NP/NatP	1034.76	65.50	35.30	Y
2a	Sierra de Aracena	NatP	1868.01	46.10	30.35	N
2b	Sierra Norte de Sevilla	NatP	1774.83	43.40	30.24	N
2c	Sierra de Hornachuelos	NatP	600.31	46.20	25.19	N
3	Sierras Subbéticas	NatP	320.56	46.80	40.43	N
4a	Sierra de Cardeña	NatP	384.49	66.70	26.51	Y
4b	Sierra de Andújar	NatP	747.75	81.60	28.70	Y
4c	Valle de Alcudia	NatP	1488.20	72.10	21.79	Y
5	Sierra Mágina	NatP	199.61	58.00	28.63	N
6a	Sierra de Cazorla	NatP	2100.66	63.40	33.07	N
6b	Sierra de Castril	NatP	126.95	86.30	36.33	N
7	Sierra de Baza	NatP	535.98	55.40	31.66	N
8	Sierra de Huétor	NatP	121.29	53.60	38.57	N
9	Sierras de Tejeda	NatP	213.22	52.10	28.48	N
10	Sierra Nevada	NP/NatP	1195.71	44.50	33.80	N
11	Sierra Grande de Hornachos	ZRI	121.90	54.20	37.95	N
12	Sierra de San Pedro	ZRI	1151.77	45.40	29.98	N
13	Embalse Orellana	ZRI	425.95	31.40	37.71	Y
14	Cabañeros	NP	409.07	67.56	18.60	Y
15	Monfragüe	NP	180.07	59.30	22.76	N
16	Sierra de Gredos	RegP	863.82	50.50	27.11	N
17a	Batuecas-Sierra Francia	NatP	315.45	75.70	28.73	N
17b	Quilamas	SCZ	106.50	67.97	27.74	N
18	El Rebollar	SCZ	496.36	62.00	27.41	N
19	Despeñaperros	NatP	76.49	81.80	27.54	Y

Additionally, Corridor Designer considers biological information of the species through the following parameters: the threshold value, the minimum breeding area and the minimum host area of a population. The threshold value is a suitability value that discriminates between habitats where reproduction is possible and those where it is not [62]. It is related to the characteristics of the species and its habitat requirements. Majka et al. [62] recommend a standard threshold value of 60, but in our case, we selected a more restrictive value of 80 due to the narrow ecological requirements of the species. The minimum breeding area corresponds to a surface value large enough for the species to reproduce. The minimum breeding area for a population is defined as an area large enough to support reproduction for 10 years or more. The minimum breeding area and the minimum hosting area of a population adopt values of 500 and 2500 hectares, respectively (based on the connectivity study carried out by Puerto Marchena and Muñoz Reinoso [20]). At this stage of the process, this information is required to calculate suitable patches within the areas to be connected and use them as starting and finishing points, prioritizing the population patches.

Depending on the surface of the landscape considered, nested polygons (corridors) with different surfaces (polygon-corridor 1%, 2%, 3%, …) are generated as possible alternatives for transit between the nodes to be connected and can be considered as buffer areas. These are low-cost corridors that widen in quality areas and narrow where the habitat is unsuitable [64].

2.5. Corridor Evaluation

Once the ecological corridors are defined, we attempted to evaluate and identify those landscape elements that could be most vulnerable in order to prioritize their conservation or restoration. The identification of these elements or critical areas is summarized as follows: (1) analysis of bottlenecks along the corridors using the Bottleneck Analysis tool, (2) characterization of the core areas according to the index of human influence and the predicted average suitability, and (3) determination of the contribution of the landscape elements to the overall connectivity of the network using the Conefor sensinode 2.6 software.

Bottleneck analysis was performed with the Bottleneck Analysis tool of the Corridor Designer Evaluation extension [62]. This tool allows the location of bottlenecks along corridors, which are narrows that can make the corridor less effective. This is important for those species sensitive to edge effect, infrastructure, etc. For this purpose, we tested with a threshold value of 1500 m.

In addition, the core areas and corridors were classified according to habitat suitability (Maxent's model) and Human Influence Index (HII). On the other hand, we considered the habitat patches generated with Corridor Designer. The existence of quality habitat patches along the corridor may facilitate the transit of the species. This is fundamental in those areas where the corridor crosses areas of low-quality habitat [42], and also for those long corridors functioning as stepping stones. This could serve to facilitate measures and proposals for new reserves or protected areas in a strategic way.

Prioritization of Landscape Elements for Conservation and Restoration

The importance of landscape elements to the connectivity of protected areas for the Iberian lynx is quantified using a connectivity metrics approach, in our case, the Probability of Connectivity (PC) [36]. We used Conefor sensinode 2.6 software [65] available and updated at www.conefor.org.

Two input files are required: one relating to the nodes and the other to the links. In the first (nodes), the different patches to be connected are collected and each one of them is assigned an attribute, which is usually the area. We used the natural break method or Jenks optimization method [66] to classify in three intervals, both the size and the average suitability of the core areas. This method reduces the variation of each class, while maximizing the variance between classes. These areas were valued from 1 to 3 depending on the interval in which they have been located. Finally, we summed these values, obtaining values between 2 and 6. Thus, we obtained an attribute that integrates the surface and the average suitability (obtained through the species distribution model). The link file includes the references of the nodes that are connected and the length between them. This length is usually Euclidean, but we used the minimum cost distances obtained from the resistance map.

The PC approach is often used in connectivity analyses [67–69]. We focus on the negative effects for dispersion and re-colonization resulting from the loss or degradation of landscape elements. The PC index can be defined as the probability that two points (animals) that are randomly situated within the landscape are located in accessible habitat areas in a set formed by habitat patches and the links (connections) between them. The landscape elements are classified according to their contribution to the overall availability and connectivity of the habitat. This contribution is calculated by the percentage of variation of the PC index obtained after the individual elimination of each element [36].

For this index it is possible to differentiate three separate fractions that quantify the ways in which elements contribute to the overall connectivity and habitat availability of the landscape: dPCintra(k) measures intrapatch connectivity, while dPCconnector(k) and dPCflux(k) measure connectivity between patches in relation to a given landscape element k. Links only contribute to connectivity through the dPCconnector(k) fraction. Therefore, a patch will be more or less important (dPCk) depending on the intrinsic characteristics and its topological position within the network [33,70]

Finally, the calculation method of dPC k and dPC k connectors was based on the value of the mean, the median or maximum dispersion distance. We tested with a value of 60 km so that the dispersion probability of an individual travelling 60 km in his dispersive phase will be 0.5.

Another interesting metric is the Betweenness Centrality metric (BC), which refers to the degree to which the optimal patch paths pass through a particular node k. It provides a more complete view of how patches can be providers of connectivity [33,70].

We also carried out an analysis with Conefor sensinode to determine the elements of the proposed corridor network that contribute most to the availability and connectivity of the habitat. This was analyzed with the dPCconnector fraction. We selected the priority elements with the highest values of overall importance, and represented the results in ArcGis 10.5.

3. Results

3.1. Evaluation of the Suitability Model

The final SDM showed AUC values of 0.779 ± 0.014 (average ± standard deviation for 10 replicates of the models). An AUC > 0.7 constitutes a result with good discriminatory power [71], indicating that the model is better than random, and that the suitable environmental areas predicted by MaxEnt are highly correlated with presence records. The method for selecting the suitability threshold is a very important step in the final processing of SDMs and depends fundamentally on the objective of the study. Liu et al. [60] consider that if the purpose is the search for new populations of a species, it will be of greater importance to minimize errors of commission (false positives), while if the objective is the establishment of a conservation area, it is possible to tolerate greater errors of commission. Thus, in our work the selection of suitable areas has been carried out according to the threshold established through the criterion: maximum test sensitivity plus specificity, with a value of 0.416 (41.6%) and two suitability maps have been drawn up (Figure 2), one in continuous and the other in binary (suitable–not suitable).

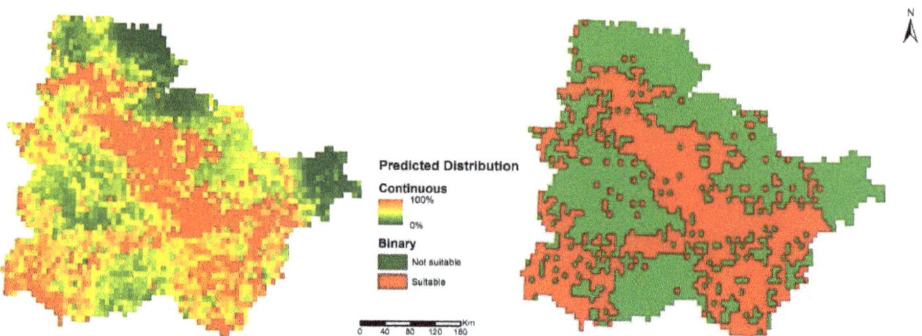

Figure 2. Suitability maps for the Iberian lynx based on the average of the 10 replicates carried out. Left: suitability map in continuous. Right: suitability map in binary.

The variables that best explain the distribution of the lynx are, in this order: slope, vegetation cover, and the seasonal precipitation (Bio15), followed by the seasonal temperature (Bio4) and the annual precipitation (Bio12), and with a lower contribution, the average annual temperature (Bio1). The environmental variables with the greatest gain when used in isolation are slope, vegetation cover, and Bio15 (Figure 3) because they provide the most useful information on their own.

As can be shown on the map of MaxEnt (Figure 2), two large areas with high suitability values can be seen: one that crosses the study area from SE to NW and another, more isolated, in the south of the study area. On the other hand, there is a large area of low suitability that mainly affects the west central part of the area under analysis and is related to the large presence of large crop cultivation areas on the banks of the Guadiana river and Tierra de Barros [72]. The case of Doñana is similar: connectivity with the western Sierra Morena is limited as it is located in a hostile environment, with extensive areas of cultivation and human infrastructure that obstruct the dispersal of the lynx towards the Sierra Morena [42].

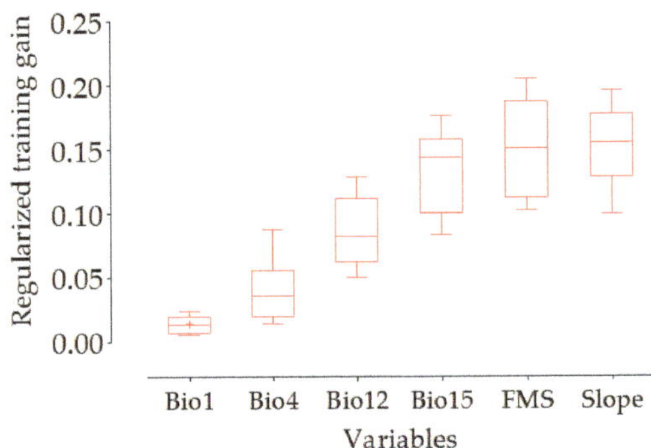

Figure 3. Boxplots represent model gain when variables are used in isolation.

There are areas of high suitability along the Central System, from the Sierra de Gata to the Sierra de Gredos and the Tiétar Valley in the northwest of the study area. The lack of information in the Portuguese territory limits the analysis of the suitability in this geographical area, although there is a glimpse of an area that apparently would present high suitability along the Portuguese border, from the west central corner of the study area to the Sierra de Gata and the Rebollar in the province of Salamanca, passing through Serra Malcata (Portugal). Similarly, there is an area of high suitability along the Portuguese border in the southwest corner of the area under analysis, which we suppose could extend to areas such as the Vale do Guadiana (current lynx population) and the Algarve, both in Portugal. The human influence rate along this entire border strip is low, mainly in the northwest of the territory studied, so it can be said that in absence of information about rabbit density, it is an area that holds good conditions to host the lynx. In fact, it was considered by some authors [17,73–75] to be an area of great importance and the last strongholds of the lynx population in the Iberian Peninsula.

3.2. Suitability of Natural Protected Areas

The 25 NPAs selected (Table 2) have an average surface area of 674 km^2. The least extensive NPA is the Despeñaperros Natural Park with 76.5 km^2, while the most extensive is Sierra de Cazorla, Segura, and Las Villas with 2100 km^2.

The average suitability of the NPAs (between 0 and 100%) is 59% according to the suitability map obtained with the MaxEnt prediction. The most suitable NPAs are Sierra de Castril with 86%, Despeñaperros with 82%, and Sierra de Andújar with 82%. On the other hand, there is only one NPA that is less suitable than the threshold value, Embalse de Orellana and Sierra de Pela, with only 31%, whose inclusion in this study was due only to its strategic geographical location. The core areas have been represented on a map for connectivity analysis and were classified according to their suitability (Figure 4 and Table 2).

The Human Influence Index (HII) [63] (WCS and CIESIN, 2005) has been evaluated independently for each NPA (Table 2). The core areas present values of around 30%. The natural park of the Sierras Subbéticas presents the highest index of human activity, with values of 40.4%. On the other hand, Cabañeros National Park presents the lowest index, with 18.6%.

3.3. Corridor Design

Figure 5 shows the design of a network with 29 corridors—represented as nested polygons—among 19 NPAs considered for the Iberian lynx. The first two polygons (0.1%

and 1% polygons) are represented as a proposed corridor, while the third (2% polygon) is represented as a buffer zone.

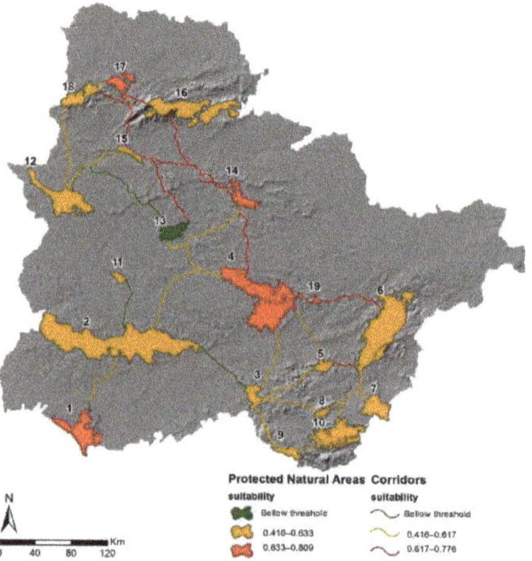

Figure 4. Classification of corridors and core-areas (Natural Protected Areas: NPAs) according to their average suitability.

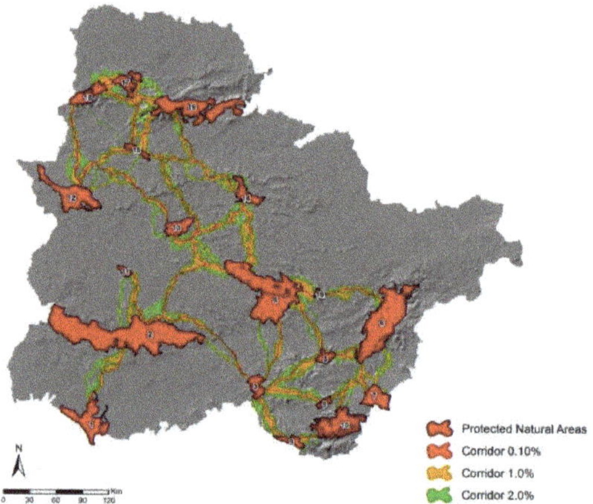

Figure 5. Corridor network designed through Corridor Designer that integrates 19 core areas.

The average length (Table S2) of the 29 corridors is 79.2 km. Corridor 13–12 is the longest at 168.5 km and Corridor 17–18 is the shortest at 11.5 km. The average width of the 29 corridors is 3.89 km (calculated on the 1% polygon). The corridor with the smallest average width is Corridor 2–3 with an average width of 1.2 km. Corridors 3–10 and 6–19 have the greatest narrowing, with widths of 0.164 km. The average minimum width of the 29 corridors is around 1.7 km.

3.4. Corridors Evaluation

Corridor 2–3 has important bottlenecks, with 73.1% narrowing below the established threshold (1500 m). On average, 16.4% of the kilometers that run through the proposed corridors are in areas with widths of less than 1500 m (Table S2).

The corridors were classified according to the habitat suitability of the territory predicted by MaxEnt (Figure 4). The average suitability of all the corridors is 55.3%, with those with the highest suitability being 4–14, 17–16, and 6–5 with values above 70%. On the other hand, the three corridors with suitability levels below 40% are 13–12, 2–3, and 2–11.

Landscape elements were classified in three categories by using the natural breaks of Jenks. Only four nodes (NPAs)—Sierra Morena West (2), Sierra Morena East (4), Sierras Subbeticas (3) and Sierra de Cazorla (6) with values above 3.5%, highlighting Sierra Morena East with 9.7% and eight connectors (4–19, 6–7, 16–17, 1–2 and 3–9) with dPC values above 5%—were assigned to the superior category (Figure 6).

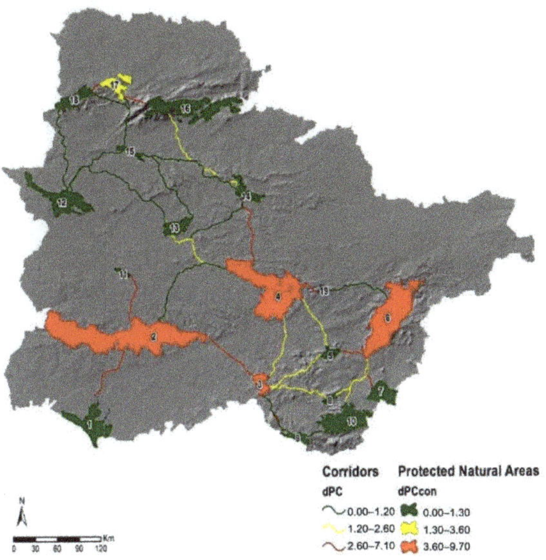

Figure 6. The dPCconnector fraction values for the nodes (NPAs) and links (corridors) used in the connectivity analysis. The landscape elements with the highest dPC are those that contribute most to overall connectivity.

4. Discussion

Our study is based on the use of existing protected natural areas included in the landscape matrix to generate a network of corridors that will allow the identification of natural areas, connectors, and zones where conservation and restoration work can be most interesting in terms of the conservation of the Iberian lynx and its potential dispersal. The latter is of great interest, since it will allow, if necessary, for greater ease on the part of administrations to carry out restoration and conservation measures. Our approach is complementary to others, particularly the interesting study carried out by Blazquez-Cabrera et al. [21], in which they analyzed the importance and priority of restoring ecological corridors established between current Iberian lynx populations in the Iberian Peninsula.

With this type of study, we intend to combine the characteristics—fundamentally abiotic—of the landscape with the biological aspects of the species to promote their conservation.

4.1. Connectivity Studies in Environmental Assessments. The Practical Approach

Connectivity studies are essential in environmental impact assessments and strategic environmental assessments. They make it possible to discriminate between different alternatives in the design of large infrastructures that are promoted in the landscape, mainly communication routes (highways, roads, railways, etc.) [76]. In addition, they allow the characterization of possible wildlife passage areas to reduce barrier effects, as well as the identification of "black points" of wildlife mortality due to traffic accidents [77,78] and of natural habitat restoration areas that can be used as "steps" by wildlife. The graphic approach allows for the simplification of a landscape mosaic and its complex network of functional connections into elements that form nodes and links. The nodes represent habitat sites surrounded by hostile habitats and the links symbolize the dispersal capacity of a species between two nodes. This method allows us to represent the landscape as a network of interconnected patches and to perform complex computational analyses on the connectivity of the landscape [29,34,79]. Usually, landscape connectivity studies have a strong local character; however, nowadays these analyses are increasingly applied to large territories and regions [11]. These regional connectivity studies provide great value as support tools for the future management of the landscape matrix. There are interesting examples of large-scale connectivity studies such as the initiative known as Yellowstone to Yukon (Y2Y) launched in 1998 [80], or the analysis of the connectivity of forest ecosystems integrated into the Natura 2000 Network in Spain, promoted by de la Fuente et al. [68].

The results obtained from this proposal allow the identification of key landscape elements according to their importance, both for their intrinsic characteristics (human influence index, average suitability) and for their contribution to connectivity, either by analyzing bottlenecks or by quantifying their contribution to overall connectivity.

4.2. Quality of Natural Protected Areas

As already mentioned, the NPAs included in this study were selected for their adequate ecological integrity, their strategic geographical location, and their protection category, which allows for the implementation of conservation measures. This is important, as some studies affirm that protected areas have lower rates of habitat loss [81,82]. In our case, some of the NPAs that currently have a lynx population are the most protected, such as Doñana, Sierra de Andújar, or Valle de Alcudia. In this sense, it would be expected that the variable of human influence (HII) was linked to the protection level, and therefore to the distribution of lynx, as argued by Palomares et al. [41]. HII values are between 18.6% (Cabañeros National Park) and 40.4% (Sierras Subbéticas Natural Park). In general, the most suitable areas for lynx have low HII (less than 30%), while the most anthropized NPAs (high HII) were the Sierras Subbéticas, Sierra Grande de Hornachos, Embalse de Orellana, and—surprisingly—Doñana, due to the occupation of human infrastructures and agricultural plots.

4.3. Connectivity between NPAs. The Important Role of Sierra Morena

Although further research is still needed on the dispersal behavior of the Iberian lynx, some studies suggest distances of approximately 30–40 km in its wandering [41,42], but there are individuals who may well exceed these values. The average length of our corridors is 79.2 km, although they are very different in size. Each of these corridors should not be interpreted as a continuous unit of territory with optimal conditions for the species, but as a route connecting patches of suitable habitat (functioning as stepping-stones). As suggested by Ferreras [42] lynx may use these remnants of suitable habitat as "steps" to travel through a fragmented landscape. Furthermore, in terms of the distance between the fragments, the fragmentation of a territory seems to be directly related to the dispersal capacity of its species. Thus, in Doñana, moderate fragmentation effects seem to incite dispersers to risk longer journeys, exploring larger areas [42]. This has also been previously verified in experimental studies with wood mice and capercaillies [83,84] where the rate of movement between patches increased with habitat fragmentation.

According to the graphical approach, in our study area, node four (Sierra Morena Oriental) is the one that contributes most to the global connectivity of the network. In this case, the loss of this NPA would imply the division of the network into two large parts that would result in a great loss of connectivity. Without this node, NPAs such as two, three and six, as well as the adjacent ones, would be totally disconnected from those located north of the Sierra Morena Oriental. The Sierras Subbéticas (3) are a clear example of how a core area can function as a stepping stone to connectivity, since its loss would generate important disconnections between the NPAs of the Baetic systems (9, 8, 10), and those of the western and eastern Sierra Morena (two and four). Connectors such as 1–2 and 2–3 contribute greatly to network connectivity. The first one does so mainly because its elimination would determine the disconnection of an important NPA such as Doñana National Park. Similarly, Corridor 2–3 is a vital connector, establishing the connection between the Western Sierra Morena and the Baetic and sub-Baetic systems. This would be in line with the results obtained by Blazquez-Cabrera et al. [21] in which they identify two main points of action where restoration efforts should be focused. This is the axis formed by the populations of Doñana, Sierra Norte, and Matachel and the axis formed by the populations of Andújar, Guarrizas, and Sierra Morena Oriental, which coincides to a large extent with the results we have obtained on priority of areas and connectors.

As mentioned above, the Sierra Morena oriental patch (4) is a strategic node that is vital for maintaining connectivity in the corridor network. This node acts as a link between two large components in the study area and without it, global connectivity would be severely reduced and divided. It has the highest values of PC index for the three fractions (intra, flow, and connector) and also of BC (PC) centrality, showing its great importance both for the maintenance of connectivity and for the availability of its habitat and its possible function as a source of dispersal flow. If we consider the other characteristics used for the categorization of NPAs used in this study, the three natural parks forming this node are characterized by low values of the human influence index (less than 30%) and high habitat suitability (around 70%). All of this shows the fundamental role of this node in maintaining lynx populations and its function as a source of individuals and genetic structure for other areas in the process of recolonization.

4.4. The Importance of Corridor Width

Moreover, narrowing along the corridors can limit the efficiency of the course of the species, generating bottlenecks. It is important to consider the number of bottlenecks that a corridor has in its path, as well as the proximity between them. If bottlenecks are not too long and are spaced far enough apart, they may not be considered barriers to movement.

In general, the corridors generated in our study area have bottlenecks in low proportions along the course. Only the corridor linking the Sierra de Hornachuelos (2) with the Sierras Subbéticas (3) stands out, as it presents important narrowing in a large proportion (more than 70%). This could be due to the poor quality of the landscape matrix for the Iberian lynx, with high levels of human influence due to the proximity of large population centers, such as the city of Cordoba, as well as different highways and land dedicated to dry farming.

5. Conclusions

The use of tools in the field of eco-informatics, such as GIS in combination with SDMs and connectivity analyses, is a strategy for planning and managing biodiversity at broad territorial scales. This study presents a methodological approach aimed at facilitating decision-making in the management of the biotic and anthropic components of a landscape, in order to favor the conservation of its biodiversity. Our contribution, focused on the Iberian lynx and oriented to the theoretical design of corridors that connect different natural areas, aims to contribute to providing biological coherence to future anthropic intervention plans in the southwest of the peninsula, a territory of great natural wealth that is subject to numerous threats derived from the current development model.

Our results show that there are large areas and natural protected areas that maintain good characteristics to be able to host the Iberian lynx, both in terms of environmental suitability and human influence. Many of these areas do not host current populations of the species, although they did in the past, such is the case in El Rebollar, Sierra de Francia or Monfragüe; places where reintroduction and habitat management strategies can be carried out to facilitate their presence. Similarly, the analysis of connectivity in the study area provides relevant information on the contribution of connectors and NPAs to the general connectivity of the landscape, detecting critical points for the flow of individuals between spaces. Of all those discussed here, block number four Sierra Morena Oriental (4a. Sierra de Cardeña y Montoro, 4b. Sierra de Andújar and 4c. Valle de Alcudia and Sierra Madrona) stands out in all the aspects treated in this study. Currently these places host the largest populations of Iberian lynx in Spain, which only confirms the importance of this area and the need for its maintenance and conservation.

Supplementary Materials: The following are available online at https://www.mdpi.com/2071-1050/13/1/41/s1, Figure S1: Merged natural protected areas. Eight natural areas that are to be considered as three core areas, Table S1: Collinearity effects between variables used in the predictive model, Table S2: List of the proposed corridors that make up the network.

Author Contributions: Conceptualization, J.Á.S.-A. and I.B.-B.; methodology, R.E.H.-L., I.B.-B., G.C.-L. and J.Á.S.-A.; software, I.B.-B.; validation, R.E.H.-L. and I.B.-B.; formal analysis, I.B.-B. and G.C.-L.; investigation, I.B.-B.; writing—review and editing, I.B.-B., J.Á.S.-A. and D.R.d.l.C.; supervision, J.Á.S.-A. All authors have read and agreed to the published version of the manuscript.

Funding: This research received no external funding.

Institutional Review Board Statement: Not applicable.

Informed Consent Statement: Not applicable.

Acknowledgments: I.B-B. has been funded by the Salamanca City of Culture and Knowledge Foundation, through the "VIII Centenary Programme to retain young talent for the initiation of research at the University of Salamanca, financed by the Salamanca City Council. Finally, we want to thank Sarah Young for revising the English of this manuscript.

Conflicts of Interest: The authors declare no conflict of interest.

References

1. Zalasiewicz, J.; Williams, M.; Smith, A.; Barry, T.L.; Coe, A.L.; Bown, P.R.; Brenchley, P.; Cantrill, D.; Gale, A.; Gibbard, P.; et al. Are we now living in the Anthropocene? *Gsa Today* **2008**, *18*, 4. [CrossRef]
2. Zalasiewicz, J.; Williams, M.; Steffen, W.; Crutzen, P. The new world of the Anthropocene. *Environ. Sci. Technol.* **2010**, *44*, 2228–2231. [CrossRef]
3. Lovejoy, T.E.; Bierregaard, R.O., Jr.; Rylands, A.B.; Malcolm, J.R.; Quintela, C.E.; Harper, L.H.; Brown, K.S., Jr.; Powell, A.H.; Powell, G.V.N.; Schubart, H.O.R. *Edge and Other Effects of Isolation on Amazon Forest Fragments*; Sinauer Associates: Sunderland, MA, USA, 1986.
4. Crooks, K.R.; Sanjayan, M. *Connectivity Conservation: Maintaining Connections for Nature*; Conservation Biology; Cambridge University Press: Cambridge, UK, 2006; Volume 14.
5. Beier, P.; Penrod, K.L.; Luke, C.; Spencer, W.D.; Cabañero, C. South Coast Missing Linkages: Restoring connectivity to wildlands in the largest metropolitan area in the USA. In *Connectivity Conservation: Maintaining Connections for Nature*; Crooks, K.R., Sanjayan, M., Eds.; Conservation Biology; Cambridge University Press: Cambridge, UK, 2006; pp. 555–586. ISBN 9780521673815.
6. Taylor, P.D.; Fahrig, L.; Henein, K.; Merriam, G. Connectivity is a vital element of landscape structure. *Oikos* **1993**, *68*, 571–573. [CrossRef]
7. Wiens, J.A. Metapopulation dynamics and landscape ecology. In *Metapopulation Biology*; Academic Press: New York, NY, USA, 1997; pp. 43–62.
8. Tischendorf, L.; Fahrig, L. On the usage and measurement of landscape connectivity. *Oikos* **2000**, *90*, 7–19. [CrossRef]
9. Moilanen, A.; Hanski, I. On the use of connectivity measures in spatial ecology. *Oikos* **2001**, *95*, 147–151. [CrossRef]
10. Nelson, M.; Allen, J.; Ailing, A.; Dempster, W.F.; Silverstone, S. Earth applications of closed ecological systems: Relevance to the development of sustainability in our global biosphere. *Adv. Space Res.* **2003**, *31*, 1649–1655. [CrossRef]
11. Beier, P.; Spencer, W.; Baldwin, R.F.; McRae, B.H. Toward best practices for developing regional connectivity maps. *Conserv. Biol.* **2011**, *25*, 879–892. [CrossRef]
12. Schmitz, S. La géographie humaine et ses revues «internationales»: Globalisation ou fragmentation? *Ann. Géo* **2003**, *632*, 402–411. [CrossRef]
13. Múgica, M.; De Lucio, J.V.; Martínez, C.; Sastre, P.; Atauri-Mezquida, J.A.; Montes, C. *Integración Territorial de Espacios Naturales Protegidos y Conectividad Ecológica en Paisajes Mediterráneos*; RENPA, Junta de Andalucía: Sevilla, Spain, 2002.
14. Gurrutxaga, M.; Lozano, P.J.; del Barrio, G. GIS-based approach for incorporating the connectivity of ecological networks into regional planning. *J. Nat. Conserv.* **2010**, *18*, 318–326. [CrossRef]
15. Clergeau, P.; Burel, F. The role of spatio-temporal patch connectivity at the landscape level: An example in a bird distribution. *Landsc. Urban Plan.* **1997**, *38*, 37–43. [CrossRef]
16. Flather, C.H.; Bevers, M. Patchy reaction-diffusion and population abundance: The relative importance of habitat amount and arrangement. *Am. Nat.* **2002**, *159*, 40–56. [CrossRef] [PubMed]

17. Rodriguez, A.; Delibes, M. Current range and status of the Iberian lynx *Felis pardina* Temminck, 1824 in Spain. *Biol. Conserv.* **1992**, *61*, 189–196. [CrossRef]
18. Rodríguez, A.; Delibes, M. Internal structure and patterns of contraction in the geographic range of the Iberian lynx. *Ecography* **2002**, *25*, 314–328. [CrossRef]
19. Life + IBERLINCE. LIFE+ Iberlince project (LIFE10NAT/ES/570): Recuperación de la Distribución Histórica del Lince Ibérico (Lynx pardinus) en España y Portugal. Available online: www.iberlince.eu (accessed on 10 November 2019).
20. Puerto, A.; Muñoz, J.C. Red de conectores ecológicos para el lince ibérico en la provincia de Huelva. In *Tecnologías de la Información Geográfica: La Información Geográfica al servicio de los ciudadanos, Proceedings of the Congreso Nacional de Tecnologías de la Información Geográfica*; Secretariado de Publicaciones de la Universidad de Sevilla: Seville, Spain, 2010; pp. 1028–1038.
21. Blazquez-Cabrera, S.; Ciudad, C.; Gastón, A.; Simón, M.Á.; Saura, S. Identification of strategic corridors for restoring landscape connectivity: Application to the Iberian lynx. *Anim. Conserv.* **2019**, *22*, 210–219. [CrossRef]
22. Illanas, S.; Gastón, A.; Blázquez-Cabrera, S.; Simón, M.A.; Saura, S. Selección del hábitat y permeabilidad del territorio para el lince ibérico (*Lynx pardinus*) en Andalucía: Influencia del estado de comportamiento y de la resolución cartográfica. *Cuad. Soc. Española Cienc. For.* **2017**, *43*, 193–208. [CrossRef]
23. Guisan, A.; Zimmermann, N.E. Predictive habitat distribution models in ecology. *Ecol. Model.* **2000**, *135*, 147–186. [CrossRef]
24. Elith, J.; Burgman, M.A. Predictions and their validation: Rare plants in the Central Highlands, Victoria, Australia. In *Predictions Species Occurences: Issues of Accuracy and Scale*; Scott, J.M., Heglund, P.J., Morrison, M.L., Raphael, M.G., Wall, W.A., Samson, F.B., Eds.; Island Press: Covelo, CA, USA, 2002; pp. 303–314.
25. Phillips, S.J.; Anderson, R.P.; Schapire, R.E. Maximum entropy modeling of species geographic distributions. *Ecol. Model.* **2006**, *190*, 231–259. [CrossRef]
26. Wisz, M.S.; Hijmans, R.J.; Li, J.; Peterson, A.T.; Graham, C.H.; Guisan, A. Effects of sample size on the performance of species distribution models. *Divers. Distrib.* **2008**, *14*, 763–773. [CrossRef]
27. Kramer-Schadt, S.; Niedballa, J.; Pilgrim, J.D.; Schröder, B.; Lindenborn, J.; Reinfelder, V.; Stillfried, M.; Heckmann, I.; Scharf, A.K.; Augeri, D.M. The importance of correcting for sampling bias in MaxEnt species distribution models. *Divers. Distrib.* **2013**, *19*, 14. [CrossRef]
28. Schumaker, N.H. *A Users Guide to the PATCH Model*; U.S. Environmental Protection Agency, Environmental Research Laboratory: Corvallis, OR, USA, 1998.
29. Urban, D.; Keitt, T. Landscape connectivity: A graph-theoretic perspective. *Ecology* **2001**, *82*, 1205–1218. [CrossRef]
30. Adriaensen, F.; Chardon, J.P.; De Blust, G.; Swinnen, E.; Villalba, S.; Gulinck, H.; Matthysen, E. The application of 'least-cost' modelling as a functional landscape model. *Landsc. Urban Plan.* **2003**, *64*, 233–247. [CrossRef]
31. Minor, E.S.; Urban, D.L. Graph theory as a proxy for spatially explicit population models in conservation planning. *Ecol. Appl.* **2007**, *17*, 1771–1782. [CrossRef] [PubMed]
32. Harary, F. Some historical and intuitive aspects of graph theory. *SIAM Rev.* **1960**, *2*, 123–131. [CrossRef]
33. Saura, S.; Rubio, L. A common currency for the different ways in which patches and links can contribute to habitat availability and connectivity in the landscape. *Ecography* **2010**, *33*, 523–537. [CrossRef]
34. Bunn, A.; Urban, D.; Keitt, T. Landscape connectivity: A conservation application of graph theory. *J. Environ. Manag.* **2000**, *59*, 265–278. [CrossRef]
35. Gustafson, E.J. Quantifying Landscape Spatial Pattern: What Is the State of the Art? *Ecosystems* **1998**, *1*, 143–156. [CrossRef]
36. Saura, S.; Pascual-Hortal, L. A new habitat availability index to integrate connectivity in landscape conservation planning: Comparison with existing indices and application to a case study. *Landsc. Urban Plan.* **2007**, *83*, 91–103. [CrossRef]
37. Rodríguez, A. Lince ibérico—*Lynx pardinus* (Temminck, 1827). In *Enciclopedia Virtual de los Vertebrados Españoles*; Museo Nacional de Ciencias Naturales: Madrid, Spain, 2004.
38. The IUCN Red List of Threatened Species. Available online: https://www.iucnredlist.org/ (accessed on 10 November 2020).
39. Rodríguez, A.; Calzada, J. *Lynx pardinus* (Iberian Lynx), 2015. The IUCN Red List of Threatened Species. Available online: https://www.iucnredlist.org/species/12520/174111773 (accessed on 10 November 2020).
40. Rodríguez, A.; Delibes, M. *El Lince Ibérico (Lynx pardina) en España: Distribución y Problemas de Conservación*; Instituto Nacional para la Conservación de la Naturaleza: Madrid, Spain, 1990.
41. Palomares, F.; Delibes, M.; Ferreras, P.; Fedriani, J.M.; Calzada, J.; Revilla, E. Iberian lynx in a fragmented landscape: Predispersal, dispersal, and postdispersal habitats. *Conserv. Biol.* **2000**, *14*, 809–818. [CrossRef]
42. Ferreras, P. Landscape structure and asymmetrical inter-patch connectivity in a metapopulation of the endangered Iberian lynx. *Biol. Conserv.* **2001**, *100*, 125–136. [CrossRef]
43. Palomares, F. Vegetation structure and prey abundance requirements of the Iberian lynx: Implications for the design of reserves and corridors. *J. Appl. Ecol.* **2001**, *38*, 9–18. [CrossRef]
44. Delibes, M.; Rodriguez, A.; Ferreras, P. *Action Plan for the Conservation of the Iberian Lynx (Lynx pardinus) in Europe*; Council of Europe: Strasbourg, France, 2000.
45. Ferreras, P.; Beltrán, J.F.; Aldama, J.J.; Delibes, M. Spatial organization and land tenure system of the endangered Iberian lynx (*Lynx pardinus*). *J. Zool.* **1997**, *243*, 163–189. [CrossRef]
46. GBIF: The Global Biodiversity Information Facility. Available online: https://www.gbif.org/es/ (accessed on 10 November 2019).
47. ESRI. *ArcMap 10.5*; ESRI: Redlands, CA, USA, 2016.

48. IGN Modelo Digital del Terreno con Paso de Malla de 200 m (MDT200). Datum ETRS89, Huso 29 y 30. Available online: http://centrodedescargas.cnig.es/CentroDescargas/index.jsp (accessed on 25 August 2018).

49. MAPAMA. Mapa Forestal de España 1:50,000 (MFE). Available online: https://www.mapama.gob.es/es/biodiversidad/servicios/banco-datos-naturaleza/informacion-disponible/mfe50.aspx (accessed on 16 July 2018).

50. Fick, S.; Hijmans, R.J. WorldClim 2: New 1-km spatial resolution climate surfaces for global land areas. *Int. J. Climatol.* **2017**, *37*, 4302–4315. [CrossRef]

51. Schloerke, B.; Crowley, J.; Cook, D.; Briatte, F.; Marbach, M.; Thoen, E.; Elberg, A.; Larmarange, J. GGally: Extension to 'ggplot2''. R Package, Version 1.4.0. Available online: https://CRAN.R-project.org/package=GGally\T1\textquoteright (accessed on 10 December 2018).

52. R Core Team. *R: A Language and Environment for Statistical Computing*; R Foundation for Statistical Computing: Vienna, Austria, 2013.

53. Elith, J.; Kearney, M.; Phillips, S. The art of modelling range-shifting species. *Methods Ecol. Evol.* **2010**, *1*, 330–342. [CrossRef]

54. Elith, J.; Graham, C.H.; Anderson, R.P.; Dudík, M.; Ferrier, S.; Guisan, A.; Hijmans, R.J.; Huettmann, F.; Leathwick, J.R.; Lehmann, A.; et al. Novel methods improve prediction of species' distributions from occurrence data. *Ecography* **2006**, *29*, 129–151. [CrossRef]

55. Elith, J.; Leathwick, J.R. Species distribution models: Ecological explanation and prediction across space and time. *Annu. Rev. Ecol. Evol. Syst.* **2009**, *40*, 677–697. [CrossRef]

56. Franklin, J. *Mapping Species Distributions: Spatial Inference and Prediction*; Cambridge University Press: Cambridge, UK, 2010; ISBN 9780521700023.

57. Hernández-Lambraño, R.; de la Cruz, D.R.; Sánchez-Agudo, J. Spatial oak decline models to inform conservation planning in the Central-Western Iberian Peninsula. *For. Ecol. Manag.* **2019**, *441*, 115–126. [CrossRef]

58. Crespo-Luengo, G.; Hernández-Lambraño, R.E.; Barbero-Bermejo, I.; Sánchez-Agudo, J.Á. Analysis of Spatio-Temporal Patterns of Red Kite *Milvus milvus* Electrocution. *Ardeola* **2020**, *67*, 247–268. [CrossRef]

59. Cantor, S.B.; Sun, C.C.; Tortolero-Luna, G.; Richards-Kortum, R.; Follen, M. A comparison of C/B ratios from studies using receiver operating characteristic curve analysis. *J. Clin. Epidemiol.* **1999**, *52*, 885–892. [CrossRef]

60. Liu, C.; Berry, P.M.; Dawson, T.P.; Pearson, R.G. Selecting thresholds of occurrence in the prediction of species distributions. *Ecography* **2005**, *28*, 385–393. [CrossRef]

61. Beier, P.; Majka, D.R.; Spencer, W.D. Forks in the road: Choices in procedures for designing wildland linkages. *Conserv. Biol.* **2008**, *22*, 836–851. [CrossRef]

62. Majka, D.; Jenness, J.; Beier, P. ArcGIS Tools for Designing and Evaluating Corridors, 2007. CorridorDesign. Available online: http://corridordesign.org (accessed on 21 December 2020).

63. WCS; CIESIN. Last of the Wild Project, Version 2, (LWP-2): Global Human Footprint Dataset (Geographic), 2005. Palisades. Available online: http://dx.doi.org/10.7927/H4M61H5F (accessed on 10 November 2019).

64. Beier, P.; Majka, D.R.; Newell, S.L. Uncertainty analysis of least-cost modeling for designing wildlife linkages. *Ecol. Appl.* **2009**, *19*, 2067–2077. [CrossRef]

65. Saura, S.; Torné, J. Conefor Sensinode 2.2: A software package for quantifying the importance of habitat patches for landscape connectivity. *Environ. Model. Softw.* **2009**, *24*, 135–139. [CrossRef]

66. Jenks, G.F. The Data Model Concept in Statistical Mapping. *Int. Yearb. Cartogr.* **1967**, *7*, 186–190.

67. Clauzel, C.; Bannwarth, C.; Foltete, J.-C. Integrating regional-scale connectivity in habitat restoration: An application for amphibian conservation in eastern France. *J. Nat. Conserv.* **2015**, *23*, 98–107. [CrossRef]

68. de la Fuente, B.; Beck, P.S.A. Invasive Species May Disrupt Protected Area Networks: Insights from the Pine Wood Nematode Spread in Portugal. *Forests* **2018**, *9*, 282. [CrossRef]

69. Duflot, R.; Avon, C.; Roche, P.; Bergès, L. Combining habitat suitability models and spatial graphs for more effective landscape conservation planning: An applied methodological framework and a species case study. *J. Nat. Conserv.* **2018**, *46*, 38–47. [CrossRef]

70. Bodin, Ö.; Saura, S. Ranking individual habitat patches as connectivity providers: Integrating network analysis and patch removal experiments. *Ecol. Model.* **2010**, *221*, 2393–2405. [CrossRef]

71. Swets, J.A. Measuring the accuracy of diagnostic systems. *Science* **1988**, *240*, 1285–1293. [CrossRef]

72. Gragera, F. Nuevos datos sobre la distribución pasada y actual del Lince ibérico (*Felis pardina* T.) en la provincia de Badajoz. *Aegypius* **1993**, *11*, 77–79.

73. Oreja, J.A.G.; Vázquez, J.G.G. Situación del lince ibérico en Sierra de Gata. *Doñana Acta Vertebr.* **1996**, *23*, 91–98.

74. Oreja, J.A.G. Non-natural mortality of the Iberian lynx in the fragmented population of Sierra de Gata (W Spain). *Miscel Lània Zoològica* **1998**, *21*, 31–35.

75. Ordiz, A.; Llaneza, L. Situación del lince ibérico Lynx pardinus en la Sierra de Gata y aledaños. *Galemys: Boletín informativo de la Sociedad Española para la conservacion y estudio de los mamiferos* **2004**, *16*, 15–23.

76. Sastre, P.; de Lucio, J.V.; Martínez, C. Modelos de conectividad del paisaje a distintas escalas. Ejemplos de aplicación en la Comunidad de Madrid. *Ecosistemas* **2002**, *11*, 1–10.

77. Colino-Rabanal, V.J.; Lizana, M.; Peris, S.J. Factors influencing wolf Canis lupus roadkills in Northwest Spain. *Eur. J. Wildl. Res.* **2011**, *57*, 399–409. [CrossRef]

78. Malo, J.E.; Suárez, F.; Díez, A. Can we mitigate animal-vehicle accidents using predictive models? *J. Appl. Ecol.* **2004**, *41*, 701–710. [CrossRef]
79. Pascual-Hortal, L.; Saura, S. Comparison and development of new graph-based landscape connectivity indices: Towards the priorization of habitat patches and corridors for conservation. *Landsc. Ecol.* **2006**, *21*, 959–967. [CrossRef]
80. Chester, C.C. Yellowstone to Yukon: Transborder conservation across a vast international landscape. *Environ. Sci. Policy* **2015**, *49*, 75–84. [CrossRef]
81. Geldmann, J.; Barnes, M.; Coad, L.; Craigie, I.D.; Hockings, M.; Burgess, N.D. Effectiveness of terrestrial protected areas in reducing habitat loss and population declines. *Biol. Conserv.* **2013**, *161*, 230–238. [CrossRef]
82. Gray, C.L.; Hill, S.L.L.; Newbold, T.; Hudson, L.N.; Börger, L.; Contu, S.; Hoskins, A.J.; Ferrier, S.; Purvis, A.; Scharlemann, J.P.W. Local biodiversity is higher inside than outside terrestrial protected areas worldwide. *Nat. Commun.* **2016**, *7*, 12306. [CrossRef]
83. Andreassen, H.P.; Hertzberg, K.; Ims, R.A. Space-use responses to habitat fragmentation and connectivity in the root vole Microtus oeconomus. *Ecology* **1998**, *79*, 1223–1235. [CrossRef]
84. Whitcomb, S.D.; Servello, F.A.; O'Connell, A.F., Jr. Patch occupancy and dispersal of spruce grouse on the edge of its range in Maine. *Can. J. Zool.* **1996**, *74*, 1951–1955. [CrossRef]

sustainability

Article

Effect of Protected Areas on Human Populations in the Context of Colombian Armed Conflict, 2005–2018

Roberto Rodríguez-Díaz [1], Víctor Javier Colino-Rabanal [2,*], Alejandra Gutierrez-López [1] and María José Blanco-Villegas [1]

[1] Section of Physical Anthropology, Department of Animal Biology, Parasitology, Ecology, Edaphology and Agronomic Chemistry, Campus Miguel de Unamuno, University of Salamanca, 37071 Salamanca, Spain; roberrd@usal.es (R.R.-D.); alejaglopez@gmail.com (A.G.-L.); mache@usal.es (M.J.B.-V.)

[2] Section of Zoology, Department of Animal Biology, Parasitology, Ecology, Edaphology and Agronomic Chemistry, Campus Miguel de Unamuno, University of Salamanca, 37071 Salamanca, Spain

* Correspondence: vcolino@usal.es; Tel.: +34-676-646-770

Abstract: It is widely recognised that conservation policies in protected areas must also favour the development and viability of human populations. Although much research has focused on economic consequences, understanding the real impact of conservation on local populations requires a more holistic standpoint. Using quasi-experimental matching methods and a diachronic perspective, the biodemographic and socio-economic effects of Colombia's National Natural Parks (NNPs) were evaluated (all in a context of internal conflict and post-conflict). The analyses were made for the set of NNPs and then grouped into four natural regions (Andes, Caribbean, Amazon-Orinoquía and Pacific) and two conflict intensities. Differences were found mainly for NNPs with low-intensity conflict, but only for biodemographic variables, not for socio-economic ones. Starting from a situation of disadvantage, a relative improvement in the conditions of the NNP municipalities was observed throughout the 13-year period in relation to the control group. Results should be taken with caution due to the conflict situation, but the lack of correlation between biodemographic and socio-economic aspects highlights the need to include more complex approaches in protected area management policies.

Keywords: armed conflict; Colombia; conservation and development; fertility; human biodemography; infant mortality; local populations; population structure; protected areas; socio-economic effects

Citation: Rodríguez-Díaz, R.; Colino-Rabanal, V.J.; Gutierrez-López, A.; Blanco-Villegas, M.J. Effect of Protected Areas on Human Populations in the Context of Colombian Armed Conflict, 2005–2018. *Sustainability* **2021**, *13*, 146. https://dx.doi.org/doi:10.3390/su13010146

Received: 2 October 2020
Accepted: 22 December 2020
Published: 25 December 2020

Publisher's Note: MDPI stays neutral with regard to jurisdictional claims in published maps and institutional affiliations.

1. Introduction

Protected areas are an essential mechanism for conserving the biological and cultural diversity of a territory [1]. Initially, protected areas were conceived as spaces isolated from human populations [2–4]. However, since the 1980s [5], the idea began to spread that they should also be socially inclusive and contribute to the development of nations and the reduction of poverty [6]. Moreover, policies that fail to take into account the various relationships between conservation needs and the demands of poverty reduction are more likely to fail [7]. The issue has been the subject of considerable controversy, with much debate over the role that protected areas play in the livelihoods and development of communities within their area of influence [8–11].

Some authors argue that the establishment of protected areas can alter socio-economic dynamics, increasing poverty conditions and conflict over the use of territory [3,12,13]. Limitations are imposed on future land-use options, with potentially significant opportunity costs [14], which are borne by local populations [15]. Meanwhile, other authors claim that these areas contribute to improving the quality of life of local populations, through income from tourism [16], access to ecosystem services [17], diversification of lifestyles [18] and/or modernisation of infrastructure [19]. Thus, it seems that there is no

115

single answer. The relationships between poverty and conservation actions are dynamic and locally specific [20].

In addition to these socio-economic dynamics, demographic changes have also been observed. Attracted by the opportunities provided by protected areas, many people have settled in the vicinity of these zones [21]. Apart from attraction mechanisms, the increase in population could also be explained by frontier engulfment models or by incidental processes [22]. However, Joppa et al. [23] considered these results to be only an artifice. In their study of 45 countries, they found no clear pattern in population trends, with both increases and decreases around protected areas. In addition, sometimes the population may decrease, but at the same time there may be an increase in residential pressure [24].

Nevertheless, the impact of protected areas on human populations shows many dimensions, so more holistic approaches must be sought that go beyond purely monetary analyses [25]. In this sense, surprisingly few observational or experimental details are available on the consequences that these socio-economic and demographic effects—but also other possible effects of a different nature—have on the biology of human populations under the influence of protected areas. Biodemographic factors (demographic variables that make it possible to study human populations from a biological perspective) are also relevant as indicators of the well-being of populations but, above all, these factors offer the best vision of the dynamics involved in the medium- and long-term viability of populations. In this sense, it would be very interesting to analyse the relationship between changes in socio-economic conditions and biodemographic aspects to attempt to verify the dependence or independence between them. It could be the case that an improvement in the former does not directly translate into an improvement in the latter, so that the objectives of well-being and viability (in demographic and biological terms) are not being met. It could also be that the effects of natural areas could manifest themselves in the form of changes in these biodemographic aspects, but without a reflection in socio-economic conditions. To date, only a few authors, such as Naidoo et al. [26], have included biodemographic variables. Thus, in their study of 34 countries, they found that protected areas contribute to improved child growth rates, possibly through the positive impact of tourism on the living standards of local people. In another study of developing countries, no differences were found in infant mortality rates between areas near protected areas and national averages, although the results should be taken with caution given the poor spatial resolution of the baseline data [27].

To increase our knowledge about the biodemographic effects of protected areas on local populations, this study analyses the impact that Colombian National Natural Parks (NNP) have on the municipalities under their influence. Colombia is an ideal candidate for this type of study as it has an important natural heritage and a network of natural parks that is more than 50 years old and covers the country's main biomes. The Colombian National System of Protected Areas develops different strategies that seek to guarantee the quality of life of human populations through sustainable local development and the social participation of the communities that live in and around the protected areas [28]. In the Colombian context, the effects of protected areas on the livelihoods of the population have been analysed mainly via political perspectives, conflict, displacement and poverty [29–32]. Protected areas have thus played a relevant role throughout the armed conflict, which is demonstrated by an increase in the presence and violence of guerrillas in their vicinity [33], and which has made it difficult to achieve their conservation objectives [34].

Quasi-experimental matching was used to study the effects of NNPs on local populations [16,35–38]. Matching methods, by selecting for analysis a sample of control localities with similar characteristics but not included in an NNP, reduce possible biases caused by confounding baseline effects. For example, Andam et al. [16] used this design to demonstrate that, although the protected areas in Costa Rica and Thailand had poverty levels above the average in their respective countries, the creation of these areas was a slight improvement over similar areas not included within natural parks.

The analysis was carried out for all of Colombia's NNPs. In addition, given the great diversity among humans and ecosystems present in Colombia, the analyses were repeated for each of the country's four natural regions (Andes, Caribbean, Amazon-Orinoquía and Pacific). Moreover, as Colombia is a country where armed conflict has had a great impact, with unpredictable repercussions at both the population and social levels, a detailed analysis has been carried out by grouping the areas analysed according to the intensity of the conflict. The municipality was used as the unit of analysis because it is the fundamental territorial entity of the political-administrative division of the Colombian State. In addition, to introduce a diachronic perspective, uncommon in previous studies, a time interval of 13 years was considered based on data from the 2005 and 2018 Colombian population and housing censuses.

2. Materials and Methods

2.1. Study Area

2.1.1. Basic Data and Natural Regions

Colombia is located in the intertropical convergence zone in the northwest of South America. Its surface area is 1,141,748 km^2 [39]. It has a population of 45.5 million people, organised administratively in 32 departments, which are in turn divided into 1101 municipalities [40]. Its climate is predominantly tropical, with a temperature that varies according to altitude. Colombia is a very diverse country, both environmentally and ethnically. The genetic composition of the Colombian population is 40–60% European, 28–40% Amerindian and 10–25% African [41–43]. On the mainland, four major natural regions can be distinguished [40], each of which maintains a certain homogeneity in terms of relief, climate, biomes and human populations: the Andes (282,540 km^2), the Caribbean (151,118 km^2), the Amazon-Orinoquía (624,958 km^2) and the Pacific (83,170 km^2) (see Figure 1).

Figure 1. Map of Colombia with the National Natural Parks and the four major natural regions considered in the study.

The Andean Region is bound by three branches of the Andes: the Western, Central and Eastern Cordilleras, which reach an altitude of 5000 m and between which are interspersed the inter-Andean valleys of the main rivers, the Cauca and the Magdalena. This complexity of relief explains the diversity of climates, from tropical and temperate to high mountain. There are significant variations in rainfall (from 500 to 2000 mm per year). Altitudinal variations explain the diversity of ecosystems in the area. It is also the most populated region (75% of Colombia's total), where the main cities and most of the country's economic activity are concentrated. The Caribbean Region comprises the coastal plains located in the north and west of the country. It also includes the Sierra Nevada de Santa Marta and the Guajira peninsula. From the human point of view, it is characterised by marked ethnic and racial integration. Agriculture, fishing, and livestock, along with tourism activities, are its main sources of income. The Pacific Region is a mountainous territory characterised by one of the rainiest areas in the world (more than 4000 mm per year), with a dense tropical forest and very abundant rivers despite their short course. It is a sparsely populated area (3% of the country's total) and has the highest percentage of Afro-Colombians (in the Department of Chocó, 74% of the population). The Orinoco and Amazon Regions (joined here to agglutinate enough NNPs) are vast plains, the Orinoco basin of savannahs and the Amazon basin of tropical rainforest. The Orinoco is a natural and cultural region shared by Colombia and Venezuela with a livestock vocation, where 3% of the Colombian population lives. The Amazon is the least populated area (0.5%) and has a higher proportion of indigenous population (in the Department of Vaupés, exceeding 80%).

2.1.2. National Natural Parks (NNPs)

This geomorphological and environmental complexity explains why Colombia is one of the megadiverse countries, second only to Brazil. A total of 58,312 species have been recorded, of which approximately 15% are endemic, and 1302 (2.2%) are threatened [44]. The country ranks first in the world in terms of the number of orchids and birds; second in terms of plants, amphibians, butterflies and freshwater fish; third in terms of palm tree and reptile species; and fourth in terms of mammal biodiversity. In recent decades, human pressures on the country's natural areas and values have increased significantly [45,46]. To conserve and protect this rich natural heritage, the Colombian National Natural Park System (SINAP) was created, which is made up of 59 protected areas, representing a total of 17,541,489 hectares of the national surface area [47]. For Colombia, NNPs are areas that allow for ecological self-regulation and whose ecosystems in general have not been substantially altered by human exploitation or occupation, and where plant and animal species, geomorphological complexes and historical or cultural manifestations have national scientific, educational, aesthetic and recreational value and, for their perpetuation, are subject to an adequate management regime [28]. The conservation status and the human pressure on the environment, both in the protected areas and outside them, are irregular and diverse, as a consequence of the high environmental, economic, social, ecological and geographical variability of Colombian territory [45,48]. The dominant strategy in Colombia has been a proactive one, allocating the largest proportion of protected land to intact, hard-to-reach and species-rich areas such as the Amazon [49]. The first NNP was declared in 1960; 6 were declared in the 1960s, 24 in the 1970s, 12 in the 1980s, 4 in the 1990s and 13 since 2000. By natural region, 26 are located in the Andes, 12 in the Caribbean, 4 in the Pacific, 14 in the Amazon-Orinoquía region and 3 on islands. This study only considers continental NNPs with a declaration date up to 2005, since this is the year of the first census included in the data [40]. Island or marine NNPs were excluded (i.e., Malpelo, Islas Corales del Rosario and San Bernardo, Isla Gorgona and Old Providence McBean Lagoon) due to the absence of census data for these areas and the impossibility of finding control municipalities with similar characteristics. Protected areas smaller than 10,000 ha were not included (these are mostly flora and fauna sanctuaries). Considering these criteria, a total of 38 NNPs were finally included in the analyses. By natural region: 19 NNPs were in the Andes, 5 in the Caribbean, 3 in the Pacific, and 11 in Amazon-Orinoquía.

2.2. Context of Violence in Colombia

The effect of NNPs on human populations in Colombia cannot be dissociated from the historical and political context of recent decades. Since the 1960s, the country has been marked by an asymmetric and low-intensity armed conflict anchored in agrarian and land tenure conflicts dating back almost a century [50]. This conflict has involved numerous actors, including the Colombian government, guerrillas, paramilitaries, drug cartels, criminal gangs and organised armed groups. The intensity of the conflict has gone through different stages, the period 1988–2012 being the bloodiest due to the rapid growth of Revolutionary Armed Forces of Colombia FARC factions (the main guerrilla group) and paramilitary incursions. In recent years, the violence has been decreasing up to the signing of the 2016 Peace Accords between the Colombian Government and FARC. However, at present, there are still residual trouble spots. The intensity of the conflict has also not been homogeneous throughout Colombian territory. The most affected areas have varied over time, with the departments of Antioquia, Santander, Norte de Santander, Cauca, Valle del Cauca, César, Magdalena or Meta recording the highest number of victims. It is estimated that since the armed conflict began in 1958, there have been up to nine million victims.

Protected areas have played an important role during the armed conflict [51]. The guerrillas, especially in the stages when the Colombian armed forces and paramilitary groups were at their strongest, found a refuge in the NNPs where they could continue their activities. This seems to be explained not only by the fact that the protected areas are located in remote areas that are difficult to access, but also by the fact that, because other legal activities were restricted, these areas were less frequented and offered better conditions as a refuge for guerrilla groups. This presence explains the increase in violence in the vicinity of protected areas [52,53]. The effect of violence on the natural ecosystems of protected areas is complex [46,54]. The practices of armed groups in relation to land use have varied in space and time. In specific areas, the presence of guerrillas has led to what some authors have called 'gunpoint conservation', a phenomenon that has been observed in different conflict regions globally [55–57]. Armed groups mined and defended certain areas for alleged environmental reasons, but also because that was where they had established their base camps. This control may have given some protection to the environmental values of these areas, at least indirectly [55]. However, many of these areas outside government control have been subject to significant changes in land use, with negative consequences for conservation [53,54,58]. In the post-conflict phase, there has been a growing conflict over the exploitation of the territories previously controlled by the guerrillas, with an increase in deforestation [46,53,57,59].

2.3. Quasi-Experimental Matching Method

Colombian NNPs tend to concentrate on peripheral areas, with a high degree of isolation and specific geomorphological and environmental characteristics [49]. This is a pattern that is repeated practically all over the world [60,61]. These locations also condition the socio-economic possibilities and biodemographic characteristics of human populations. Without careful selection of the control municipalities, the possible differences found for municipalities in NNPs could be related not so much to their being close to the NNP, but to the very conditions of their location, regardless of whether they are in an NNP or not. Thus, to avoid these biases, a comparison must be made between municipalities located in an NNP (treatment) and municipalities with similar characteristics not affected by this protection regime (control). The matching method allows the selection of control populations with the most similar confounding baseline characteristics to ones near NNPs. This quasi-experimental matching method, by minimising the biases by controlling for confounders, ensures that the observed effects are mainly due to the NNP designation.

Table 1 shows the confounding baseline characteristics considered in this study and the indicators used for each one. These characteristics are involved in the socio-economic conditions and biodemographic dynamics of the municipalities in NNPs, but not directly linked to protection status. They are related to the level of isolation (distance from the departmental capital), the orography as an estimator of accessibility (slope) or the environmental conditions (altitude and rainfall). Because some biodemographic variables can vary between ethnic groups, ethnic composition (% Afro-descendants and % Indigenous) was also taken into consideration. Furthermore, due to the context of violence, variables related to conflict intensity (% of displaced persons and conflict rate) were also included. All of these variables were obtained for all Colombian municipalities.

Table 1. Indicators of the confounding baseline characteristics used to select a set of control localities as similar as possible to the National Natural Parks (NNP) municipalities through matching techniques.

Cofounder	Indicator	Description
Size	Area	Municipal area in km^2
	Population size	Number of inhabitants registered in the municipality
Population composition	% Afro-descendants	Proportion of people according to ethnicity (African descendants)
	% Indigenous	Proportion of people according to ethnicity (Indigenous)
	Rurality	Percentage of rural population
Orography	Slope	Average slope of the terrain
	Altitude	Average height above sea level
Climatology	Rainfall	Average amount of rain collected over a year
Location	Distance to the departmental capital	Distance from the municipality to the capital of its Department
Armed conflict	% Displaced	Proportion of residents who had to leave the municipality because of the armed conflict
	Armed conflict index	Presence of armed groups and number of events of the internal conflict (2000–2012). Grouped by intensity of the conflict from 1 (strongly affected municipalities and persistent conflict) to 7 (municipalities without conflict)

The Euclidean distances between all Colombian municipalities were calculated from the indicators in Table 1. By means of matching, a set of control municipalities most like the municipalities in NNP was selected. There is one control municipality per NNP municipality ($n = 181$). In the final selection, not only the Euclidean distances obtained from the indicators of the confounding characteristics were taken into account, but three other requirements also had to be met. To avoid any influence of the NNPs on the control municipalities, controls must be located at least 30 km away from the nearest NNP. Furthermore, as the analysis was also carried out by grouping the NNP municipalities by natural region (4 regions) and by conflict intensity (2 levels), an effort was made to ensure that for each municipality in NNP there would be a control municipality placed in the same natural region and in the same conflict intensity category. Thus, as a result of the matching, we obtained a set of 181 control municipalities with similar characteristics (in terms of confounder indicators) as the NNP populations and with a similar distribution between the four natural regions and the different conflict intensities.

2.4. Biodemographic and Socio-Economic Variables

The biodemographic and socio-economic variables used to characterise the influence of NNPs on human populations are described in Table 2. As special emphasis is placed on the impact of NNPs on biodemographic aspects, nine variables were selected to cover the main aspects of human population biology. Those biodemographic variables available from the 2005 and 2018 censuses [40] that best define human population dynamics were selected and grouped into three blocks: population structure, fertility, and infant mortality. The description of the population structure makes it possible to assess the population stability, and to this end, the ageing ratio (AGR), the mean mortality age (MMA) and the changes in the population size (intercensal growth [ICG]), were measured. From the fertility and infant mortality estimates, inferences can be made about the medium- and long-term viability of the population. Regarding fertility, the number of children born alive per woman (BAL), the relationship between weight and height of the newborns (WES) and the duration of pregnancy (DPR) were considered. Infant mortality was analysed along three stages: foetal (FOM), neonatal (NEM) and post-neonatal (POM).

Table 2. Biodemographic (related to population structure, fertility, and infant mortality) and socio-economic variables used to estimate the influence of NNPs on human populations. All data were obtained from population and housing census published by the Colombian National Administrative Department of Statistics (DANE) [40].

	Type	Variable	Abbrev	Description
Biodemographic variables	Population structure	Ageing Ratio	AGR	Ratio of the proportion of elderly people (65 years and over) and young people (under 15 years) multiplied by 100
		Mean Mortality Age	MMA	Average age of mortality excluding first-year mortality (calculated for the periods 2001–2005 and 2014–2018)
		Intercensal Growth	ICG	Difference between the population size in the current census and that from the previous one
	Fertility	Born Alive	BAL	Live births per woman (calculated for the periods 2001–2005 and 2014–2018)
		Weight/Size	WES	Birth weight divided by height in newborns (for the periods 2001–2005 and 2014–2018)
		Duration of pregnancy	DPR	Average length of gestation (for 2001–2005 and 2014–2018)
	Infant mortality	Foetal Mortality	FOM	Proportion of pregnancies not carried to term (for the periods 2001–2005 and 2014–2018)
		Neonatal Mortality	NEM	Proportion of stillbirths by pregnancy (for the periods 2001–2005 and 2014–2018)
		Post neonatal Mortality	POM	Proportion of deaths before age 1 by birth (for the periods 2001–2005 and 2014–2018)

Table 2. *Cont.*

Type	Variable	Abbrev	Description
Socio-economic variables	Illiteracy	ILL	Proportion of the population over 15 years of age that is illiterate
	Unsatisfied Basic Needs	UBN	Proportion of the population unable to meet their basic needs
	Insufficient income	INI	Proportion of households with insufficient income to cover basic expenses (2005 only)
	Per Capita Income	PCI	Per capita income by municipality (for 2018 only)

Socio-economic variables were included for a double purpose: first, to identify the possible effects of the NNPs on them, and second, especially, to compare their behaviour with that of the biodemographic variables. The aim was to check whether changes in socio-economic conditions directly lead to changes in the biology of the populations. Four socio-economic variables were included. One of them was linked to educational level (illiteracy [ILL]) and the others were related to the standard of living and access to material goods: unsatisfied basic needs (UBN), proportion of households with insufficient income (INI) in 2005 and per capita income (PCI) in 2018.

This set of variables is intended to address a hitherto little explored approach, namely the direct influence of the effect of protected areas on the biodemography of the local populations. To this end, the effect of the NNPs on the biodemographic variables and on the socio-economic variables is estimated, considering their possible interrelationship.

2.5. Sources of Information

The biodemographic and socio-economic variables were obtained from the General Population and Housing Census of 2005 and 2018, provided by the National Administrative Department of Statistics [40]. The census data were collected by DANE between May and November 2005 through face-to-face interviews. In 2018 the face-to-face interviews were conducted between April and October 2018 and, in addition, electronic interviews were conducted between January and April 2018. In these censuses, economic, social, housing, activity and personal indicators were collected.

Data on conflict intensity were obtained from the Conflict Analysis Resource Centre (CERAC) database [62–64]. This database establishes a classification that categorises Colombian municipalities according to the effect caused by the conflict between 2000 and 2012, according to which conflict intensity in each municipality is assessed based on the average number of armed conflict events during the study period with respect to the national average (3 events per municipality). A municipality is considered to have been strongly affected (high-intensity) by the conflict when the average number of events is greater than or equal to the national average and slightly affected (low-intensity) otherwise.

2.6. Statistical Analysis

The statistical differences between NNP and control municipalities in relation to the biodemographic and socio-economic variables were estimated by means of unpaired sample tests: Student's *t*-test in the case of normality and Wilcoxon in the case of non-normality [65,66].

3. Results

3.1. Differences between NNP and Control Municipalities for 2005 and 2018

Table 3 shows the average values of the 9 biodemographic and the 4 socio-economic variables in the NNP and control populations for the periods 2005 and 2018 for the country as a whole and for each of the subgroups considered (the four natural regions and the two levels of violence intensity). A total of 6 of the 13 variables considered (AGR, MMA, WES, FOM, NEM and POM) showed significant differences between NNP and control populations (highlighted in bold in Table 3). These differences were not constant over time either for the entire country or for each of the natural regions or each conflict intensity considered.

In 2005, for Colombia as a whole, there were significant differences between the NNP and control municipalities for two variables related to infant mortality (FOM and NEM). FOM (NNP = 22.619, Control = 18.817) and NEM (NNP = 12.773, Control = 10.911) were higher in the municipalities in NNP (Table 3 and Figure 2). For the rest of the variables no differences were found. However, these overall results contained notable variations between natural regions. In the Andes (Table 3 and Figure 3), there were significant differences for the same variables FOM (NNP = 19.484, Control = 15.712) and NEM (NNP = 11.530, Control = 9.516), but there were also differences in two biodemographic variables related to population structure, AGR and MMA. The Andean municipalities in NNP showed a lower AGR (NNP = 0.075, Control = 0.085) and a lower MMA (NNP = 58.908, Control = 60.826). The Caribbean (Table 3 and Figure 2) followed a similar pattern to the Andes. Thus, in terms of infant mortality the differences were also significant in FOM (NNP = 25.574, Control = 19.755) and POM (NNP = 11.932, Control = 7.302). For Amazon-Orinoquía and Pacific there were no differences between municipalities in NNP and the control. No significant differences in socio-economic variables were found for any of the cases.

In summary, in 2005, in general the data for NNP populations are worse than for non-NNP populations. There is a lower level of ageing, probably related to a lower age of mortality. Variables linked to infant and perinatal mortality also show worse results in NNP populations. All effects of NNPs are observed on the biodemographic variables, none on the socioeconomic ones.

In 2018 and for the set of NNPs (Table 3 and Figure 1), the differences in variables related to infant mortality remained, but for this period they were expressed through the POM (NNP = 6.107, Control = 5.165). POM was greater in the populations of NNP municipalities. Unlike 2005, there were also differences for two other biodemographic variables (Figure 2). One related to population structure, MMA, and another related to fertility, WES. Municipalities in NNP showed a lower MMA (NNP = 63.750, Control = 65.595) and a higher WES (NNP = 62.474, Control = 61.993). Differences in WES were found for the Andes (NNP = 62.383, Control = 61.841) and the Caribbean (NNP = 62.642, Control = 61.767). These two areas, with similar response patterns, maintained significantly lower values in AGR (NNP = 0.105 and Control = 0.117 in Andes; NNP = 0.064, Control = 0.076 in Caribbean) and MMA (NNP = 67.177 and Control = 69.097 in Andes, NNP = 61.218, Control = 65.415 in Caribbean). For the Amazon-Orinoquía region, the only variation identified was for POM (NNP = 7.137, Control = 3.966). For the Pacific, as for 2005, no variations were found between NNP and control areas (Table 3). No significant differences in socio-economic variables were found in 2018.

Table 3. Mean and standard deviation (in brackets) for each biodemographic and socio-economic variable in each period (2005 and 2018) and subdivision (4 natural regions and 2 conflict intensities). The orientation of the arrows located to the right of each variable show the changes over the 13-year period except for insufficient income (INI, available only for 2005) and per capita income (PCI, available only for 2018). The variable and the period for which the difference between NNP and control was significant are highlighted in bold.

			COLOMBIA 2005	COLOMBIA 2018		ANDES 2005	ANDES 2018		CARIBBEAN 2005	CARIBBEAN 2018		AMAZON-ORINOQ 2005	AMAZON-ORINOQ 2018		PACIFIC 2005	PACIFIC 2018		LOW-INTENSITY 2005	LOW-INTENSITY 2018		HIGH-INTENSITY 2005	HIGH-INTENSITY 2018	
		Variable																					
BIODEMOGRAPHIC	Population Structure	AGR NNP	0.066 (0.024)	0.088 (0.035)	⇑	**0.075 (0.023)**	**0.105 (0.032)**	⇑	**0.048 (0.010)**	**0.064 (0.015)**	⇑	0.045 (0.014)	0.057 (0.017)	⇑	0.049 (0.014)	0.055 (0.010)	⇑	**0.066 (0.025)**	**0.091 (0.036)**	⇑	0.056 (0.015)	0.076 (0.026)	⇑
		AGR Control	0.072 (0.029)	0.098 (0.042)	⇑	**0.085 (0.029)**	**0.117 (0.040)**	⇑	**0.055 (0.011)**	**0.076 (0.018)**	⇑	0.044 (0.010)	0.056 (0.013)	⇑	0.057 (0.014)	0.059 (0.010)	⇑	**0.075 (0.031)**	**0.103 (0.044)**	⇑	0.058 (0.015)	0.077 (0.025)	⇑
		MMA NNP	55.182 (8.537)	63.750 (7.849)	⇑	58.908 (6.947)	67.177 (5.597)	⇑	52.879 (6.665)	61.218 (6.694)	⇑	46.421 (6.152)	58.355 (6.245)	⇑	46.050 (7.011)	50.372 (7.963)	⇑	**55.606 (8.971)**	**64.096 (8.267)**	⇑	53.275 (5.977)	62.195 (5.450)	⇑
		MMA Control	56.932 (8.499)	65.595 (7.600)	⇑	60.826 (6.721)	69.097 (5.360)	⇑	56.162 (5.732)	65.415 (5.114)	⇑	45.898 (5.851)	58.312 (5.425)	⇑	49.036 (4.663)	52.308 (5.712)	⇑	**57.943 (8.364)**	**66.363 (7.673)**	⇑	52.906 (8.002)	62.894 (6.163)	⇑
		ICG NNP	0.031 (0.267)	0.002 (0.245)	⇓	−0.017 (0.195)	−0.043 (0.228)	⇓	0.042 (0.501)	0.120 (0.165)	⇓	0.182 (0.240)	0.070 (0.333)	⇓	0.132 (0.162)	0.020 (0.206)	⇓	0.018 (0.273)	−0.001 (0.259)	⇓	0.089 (0.234)	0.019 (0.176)	⇓
		ICG Control	−0.003 (0.269)	0.023 (0.251)	⇑	−0.047 (0.242)	−0.020 (0.213)	⇑	0.017 (0.338)	0.174 (0.152)	⇑	0.161 (0.201)	0.064 (0.328)	⇑	−0.058 (0.449)	−0.020 (0.388)	⇑	−0.014 (0.254)	0.010 (0.258)	⇑	0.064 (0.310)	0.075 (0.217)	⇑
	Fertility	BAL NNP	0.038 (0.0185)	0.030 (0.011)	⇓	0.036 (0.010)	0.027 (0.009)	⇓	0.036 (0.012)	0.038 (0.013)	⇑	0.055 (0.034)	0.034 (0.009)	⇓	0.017 (0.011)	0.027 (0.012)	⇑	0.036 (0.015)	0.028 (0.010)	⇓	0.043 (0.029)	0.036 (0.009)	⇓
		BAL Control	0.037 (0.0167)	0.027 (0.008)	⇓	0.036 (0.009)	0.024 (0.007)	⇓	0.037 (0.012)	0.037 (0.005)	⇒	0.054 (0.029)	0.034 (0.006)	⇓	0.014 (0.009)	0.021 (0.008)	⇑	0.036 (0.016)	0.026 (0.008)	⇓	0.043 (0.019)	0.035 (0.007)	⇓
		WES NNP	64.171 (1.811)	**62.474 (1.432)**	⇓	63.955 (1.988)	**62.383 (1.491)**	⇓	64.454 (1.073)	**62.642 (1.713)**	⇓	64.755 (1.400)	62.966 (0.976)	⇓	64.239 (1.964)	61.868 (0.732)	⇓	64.114 (1.869)	**62.505 (1.506)**	⇓	64.428 (1.529)	62.333 (1.040)	⇓
		WES Control	64.212 (1.879)	**61.993 (1.407)**	⇓	63.961 (1.963)	**61.841 (1.504)**	⇓	64.492 (1.945)	**61.767 (1.484)**	⇓	65.119 (1.507)	62.747 (0.935)	⇓	63.996 (1.226)	62.074 (0.740)	⇓	64.109 (1.978)	**61.915 (1.427)**	⇓	64.741 (1.319)	62.390 (1.287)	⇓
		DPR NNP	250.754 (15.534)	267.656 (1.926)	⇑	256.290 (11.755)	267.967 (1.700)	⇑	248.358 (12.748)	266.094 (1.613)	⇑	229.700 (16.522)	268.242 (2.218)	⇑	252.659 (4.336)	266.759 (2.140)	⇑	250.609 (16.060)	267.692 (1.942)	⇑	251.405 (13.144)	267.491 (1.881)	⇑
		DPR Control	251.669 (13.698)	267.554 (1.853)	⇑	257.858 (9.373)	267.785 (1.703)	⇑	245.210 (11.402)	265.950 (1.768)	⇑	232.400 (13.470)	268.946 (1.069)	⇑	252.289 (4.226)	266.020 (1.309)	⇑	251.097 (13.925)	267.626 (1.856)	⇑	254.079 (13.081)	267.330 (1.881)	⇑

124

Table 3. *Cont.*

		Variable	COLOMBIA 2005	COLOMBIA 2018	ANDES 2005	ANDES 2018	CARIBBEAN 2005	CARIBBEAN 2018	AMAZON-ORINOQ 2005	AMAZON-ORINOQ 2018	PACIFIC 2005	PACIFIC 2018	LOW-INTENSITY 2005	LOW-INTENSITY 2018	HIGH-INTENSITY 2005	HIGH-INTENSITY 2018
Infant Mortality	FOM	NNP	**22.619** **(15.698)**	10.743 (11.192) ⇓	**19.484** **(13.155)**	6.706 (5.456) ⇓	**25.574** **(9.487)**	18.465 (14.524) ⇓	22.225 (14.409)	13.656 (12.021) ⇓	45.213 (27.682)	20.054 (6.885) ⇓	22.314 (16.956)	10.557 (11.832) ⇓	**23.989** **(7.940)**	11.570 (7.869) ⇓
		Control	18.817 (10.837)	9.062 (9.519) ⇓	**15.712** **(7.436)**	6.036 (5.438) ⇓	**19.755** **(7.044)**	12.746 (5.567) ⇓	20.987 (9.662)	10.350 (6.976) ⇓	40.086 (19.251)	29.039 (23.575) ⇓	18.221 (11.129)	8.648 (9.894) ⇓	**19.616** **(6.484)**	9.727 (5.377) ⇓
	NEM	NNP	**12.773** **(8.582)**	7.376 (5.237) ⇓	**11.530** **(7.127)**	6.623 (5.547) ⇓	13.642 (4.634)	9.206 (2.797) ⇓	11.063 (6.679)	6.590 (2.799) ⇓	25.736 (16.823)	12.546 (7.146) ⇓	12.533 (9.248)	7.042 (4.518) ⇓	13.854 (4.464)	8.871 (7.605) ⇓
		Control	**10.911** **(6.225)**	7.264 (4.776) ⇓	**9.516** **(4.569)**	6.721 (5.116) ⇓	12.454 (4.922)	7.623 (3.614) ⇓	10.255 (4.686)	7.529 (4.004) ⇓	21.689 (12.146)	10.440 (4.721) ⇓	10.525 (6.353)	7.260 (5.104) ⇓	11.725 (3.737)	7.013 (2.763) ⇓
	POM	NNP	9.845 (8.585)	**6.107** **(6.366)** ⇓	7.954 (7.711)	5.081 (5.777) ⇓	**11.932** **(6.483)**	6.708 (3.687) ⇓	11.162 (8.307)	**7.137** **(5.647)** ⇓	19.478 (12.620)	12.258 (12.572) ⇓	9.781 (9.251)	**6.322** **(6.850)** ⇓	10.135 (4.621)	5.144 (3.388) ⇓
		Control	7.906 (6.165)	**5.165** **(9.064)** ⇓	6.196 (4.136)	3.316 (3.074) ⇓	**7.302** **(3.263)**	5.264 (1.790) ⇓	10.732 (6.392)	**3.966** **(3.780)** ⇓	18.397 (11.854)	16.011 (14.374) ⇓	7.700 (6.392)	**5.312** **(10.051)** ⇓	7.891 (3.809)	4.462 (1.881) ⇓
SOCIO-ECONOMIC	ILL	NNP	0.166 (0.098)	0.104 (0.061) ⇓	0.146 (0.085)	0.094 (0.055) ⇓	0.248 (0.136)	0.139 (0.079) ⇓	0.135 (0.034)	0.084 (0.026) ⇓	0.242 (0.101)	0.165 (0.070) ⇓	0.163 (0.098)	0.105 (0.063) ⇓	0.180 (0.102)	0.101 (0.050) ⇓
		Control	0.163 (0.087)	0.103 (0.054) ⇓	0.135 (0.068)	0.087 (0.039) ⇓	0.235 (0.095)	0.131 (0.057) ⇓	0.143 (0.050)	0.082 (0.024) ⇓	0.302 (0.065)	0.214 (0.052) ⇓	0.159 (0.085)	0.102 (0.054) ⇓	0.174 (0.093)	0.101 (0.053) ⇓
	UBN	NNP	35.791 (24.141)	26.016 (18.158) ⇓	25.718 (16.004)	18.813 (12.127) ⇓	44.429 (13.560)	36.019 (18.497) ⇓	54.342 (30.795)	31.968 (16.187) ⇓	66.876 (29.170)	56.534 (21.877) ⇓	35.420 (24.860)	25.518 (18.516) ⇓	37.462 (20.924)	28.254 (16.574) ⇓
		Control	35.661 (22.535)	25.213 (18.099) ⇓	24.665 (11.460)	16.534 (10.717) ⇓	51.676 (19.476)	37.586 (17.553) ⇓	47.542 (25.321)	31.678 (14.499) ⇓	74.420 (24.111)	59.907 (12.058) ⇓	34.343 (21.814)	24.184 (17.859) ⇓	37.342 (20.811)	26.606 (15.148) ⇓
	INI	NNP	0.827 (0.085)		0.826 (0.090)		0.815 (0.070)		0.809 (0.067)		0.896 (0.076)		0.841 (0.081)		0.763 (0.075)	
		Control	0.834 (0.099)		0.833 (0.105)		0.828 (0.075)		0.791 (0.080)		0.935 (0.038)		0.843 (0.096)		0.785 (0.097)	
	PCI	NNP		0.011 (0.008)		0.011 (0.004)		0.010 (0.006)		0.015 (0.018)		0.005 (0.003)		0.011 (0.009)		0.011 (0.004)
		Control		0.019 (0.041)		0.013 (0.007)		0.012 (0.014)		0.044 (0.090)		0.005 (0.004)		0.020 (0.045)		0.015 (0.011)

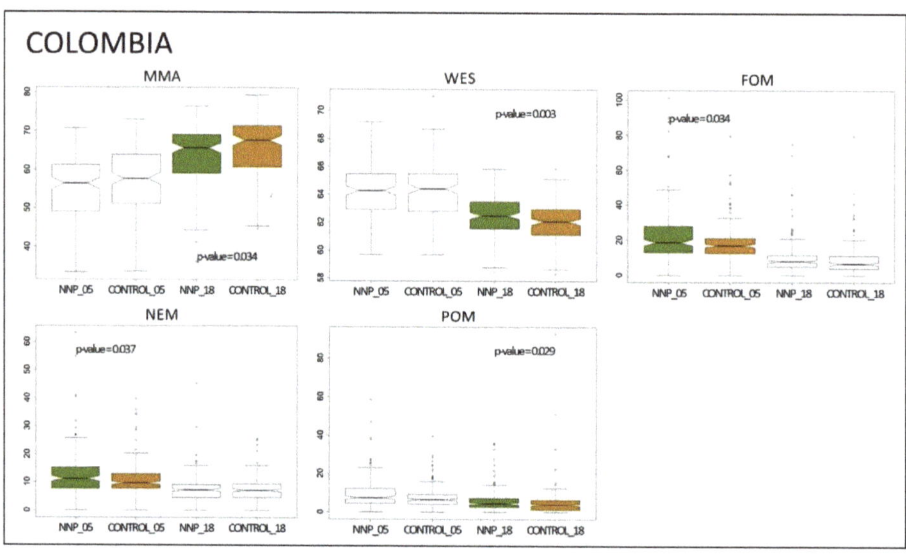

Figure 2. Boxplots with the variables for which significant differences were detected between NNP and control municipalities for all Colombian NNP considered as a whole: mean mortality age (MMA), weight/length (WES), foetal mortality (FOM), neonatal mortality (NEM) and post-neonatal mortality (POM). The boxplot is only shown in colour if the differences are significant during this period, otherwise the boxplot is shown in white.

In general terms, in 2018, two opposing changes can be observed. On the one hand, in the adult population in NNP, there is less aging, again probably linked to a lower mortality age. On the other hand, neonates show a greater weight at birth. Similar to 2005, no effect of NNP on socioeconomic variables is detected.

3.2. Variation between 2005 and 2018

In general, the populations of the studied municipalities experienced changes in the values of the biodemographic and socio-economic variables over the 13-year period between 2005 and 2018. Thus, the values of two biodemographic variables related to population structure, AGR and MMA, increased throughout this period both inside and outside NNPs (Table 3). There was also a population increase (but at a lower rate than in 2005), but ICG was the only variable that showed a different evolution in NNP and control municipalities, declining within NNPs and increasing outside them. As for fertility, BAL and WES decreased, while DPR increased. The variables FOM, NEM and POM experienced a notable decrease. Socio-economic indicators behaved in a similar way to biodemographic indicators. Throughout this period, a remarkable socio-economic development was consolidated, as evidenced by the reduction in ILL values and two poverty indicators (UBN and INI/PCI). All of these changes occurred for the entire population under study (p-value < 0.05) and for the populations of all four natural regions, whether they were in NNP or not.

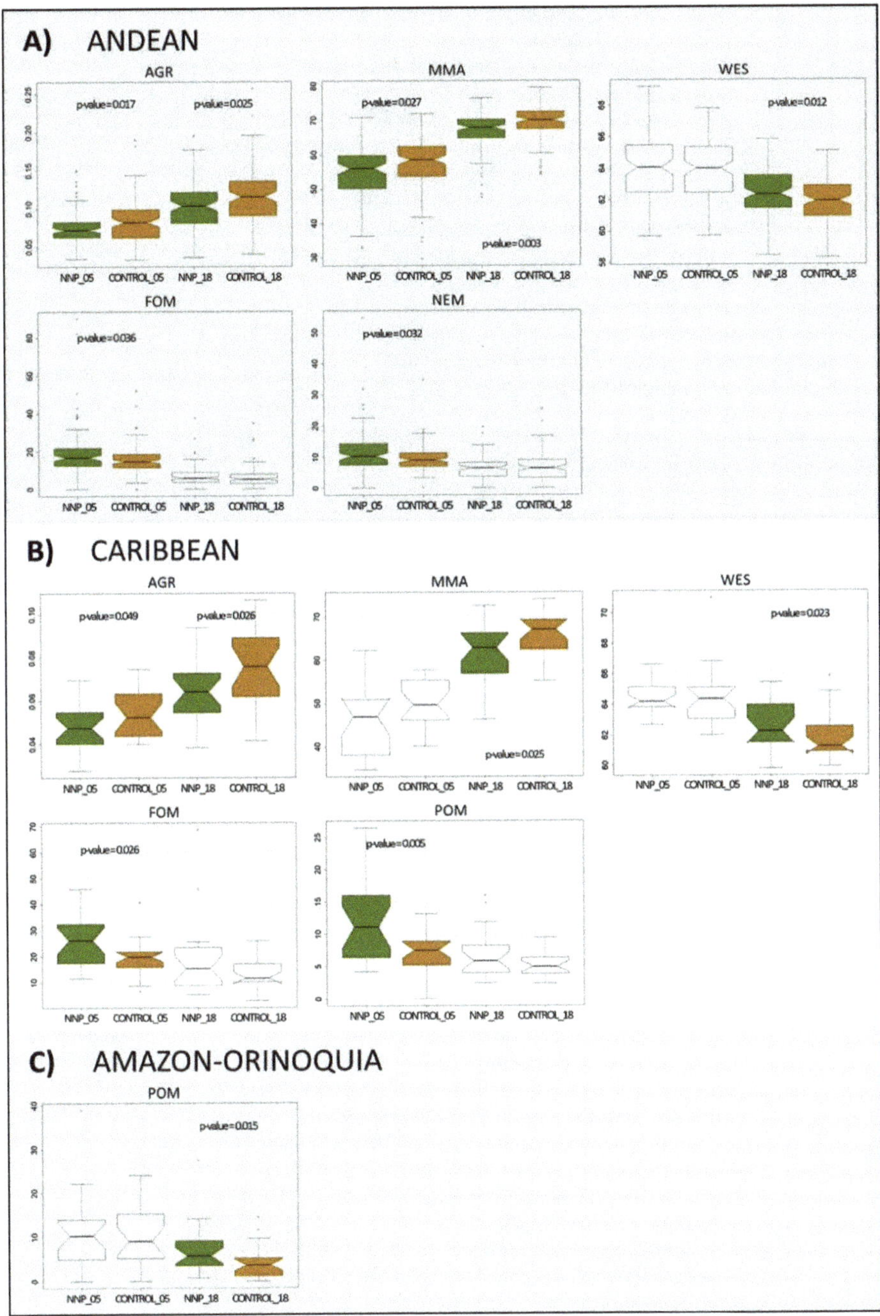

Figure 3. Boxplots of the variables with significant differences between NNP and control municipalities for 3 natural regions (for Pacific no differences were found): (**A**) Andes, (**B**) Caribbean, (**C**) Amazon-Orinoquía. Variables: ageing rate (AGR), mean mortality age (MMA), weight/height (WES), neonatal mortality (NEM), post-neonatal mortality (POM) and foetal mortality (FOM). The boxplot is only shown in colour if the differences are significant during this period, otherwise the boxplot is shown in white.

More interesting for the study of the impact of the NNP on local populations is the analysis of the relative changes in the NNP municipalities in relation to control over this 13-year period. Table 4 shows the direction of change for those variables where significant differences were found between NNP and control municipalities in at least one of the two censuses. In relative terms and for all of Colombia, the municipalities in NNP showed an improvement for two of the variables on infant mortality (NEM and FOM) and one on fertility (WES). WES in 2018 was comparably higher in NNPs. The rates of NEM and FOM were reduced proportionally more in the NNP populations, and by 2018 there was no longer a difference with populations outside NNP. However, the rates of the other variable for infant mortality, POM, worsened (was further reduced in control municipalities) although it is likely that the remarkable change in POM between 2005 and 2018 outside NNP in Amazon-Orinoquía (2005 = 10.732, 2018 = 3.966) explained the significant results for the whole country. The other variable for which some relative decline was identified for NNP populations was MMA, related to population structure. MMA values increased in the NNP municipalities, but to a lesser extent than in the controls.

Table 4. Change in the differences between the NNP and control municipalities between 2005 and 2018. + means that, in relation to the controls, a negative effect for the NNP disappeared or a positive one appeared; − means that a positive effect disappeared, or a negative effect appeared; = means the effect was maintained; ∅ means no effect observed.

		Variable	COLOMBIA	ANDES	CARIBBEAN	AMAZON-ORINOQUIA	PACIFIC	LOW INTENSITY	HIGH INTENSITY
Biodemographic	Population structure	AGR	∅	=	=	∅	∅	=	∅
		MMA	−	=	−	∅	∅	=	∅
		ICG	∅	∅	∅	∅	∅	∅	∅
	Fertility	BAL	∅	∅	∅	∅	∅	∅	∅
		WES	+	+	+	∅	∅	+	∅
		DPR	∅	∅	∅	∅	∅	∅	∅
	Infant mortality	FOM	+	+	+	∅	∅	∅	+
		NEM	+	+	∅	∅	∅	∅	∅
		POM	−	∅	+	−	∅	−	∅
Socio-economic		ILL	∅	∅	∅	∅	∅	∅	∅
		UBN	∅	∅	∅	∅	∅	∅	∅
		INI *							
		PCI *							

* The INI and PCI data are only available for one of the two censuses, so the evolution of these two variables could not be analysed.

Grouping by natural regions (Table 4 and Figure 2), the Andean NNPs maintained their relative position to the control group in terms of population structure (AGR and MMA) and showed an improvement in variables related to fertility (WES) and infant mortality (FOM and NEM). In the Caribbean NNPs, there were also relative improvements in WES and in two variables on infant mortality (FOM and POM). FOM and POM reduced at a greater rate in NNP, matching those of the control group. In terms of population structure, differences remained for AGR and there was a slight worsening in MMA. Amazon-Orinoquía NNPs regressed in POM. In the Pacific there was no change at all.

Regarding socio-economic variables, if in 2005 there were no differences between PPN and control areas, both groups followed similar changes over the 13 years, so no significant differences were found in 2018, either.

3.3. National Natural Parks and Level of Violence

Grouped by conflict intensity, Figure 4 shows those variables with significant differences between NNP and control localities. For Low-Intensity, in 2005 the main differences were those related to population structure: a higher AGR in NNP (NNP = 0.066, Control = 0.075) and a lower MMA (NNP = 55.606, Control = 57.943). In 2018 the differences between the two groups were repeated: AGR (NNP = 0.091, Control = 0.103) and MMA (NNP = 64.096, Control = 66.363). Among the fertility variables, only in WES (NNP = 62.505, Control = 61.915) were differences significant. In those related to infant mortality only POM (NNP = 6322, Control = 5312) differed significantly. None of the socio-economic variables showed any differences. For 2005, in the NNPs subject to high-intensity conflict (Figure 4) differences were only found in FOM (NNP = 23,989, Control = 19,616). No differences were identified in 2018.

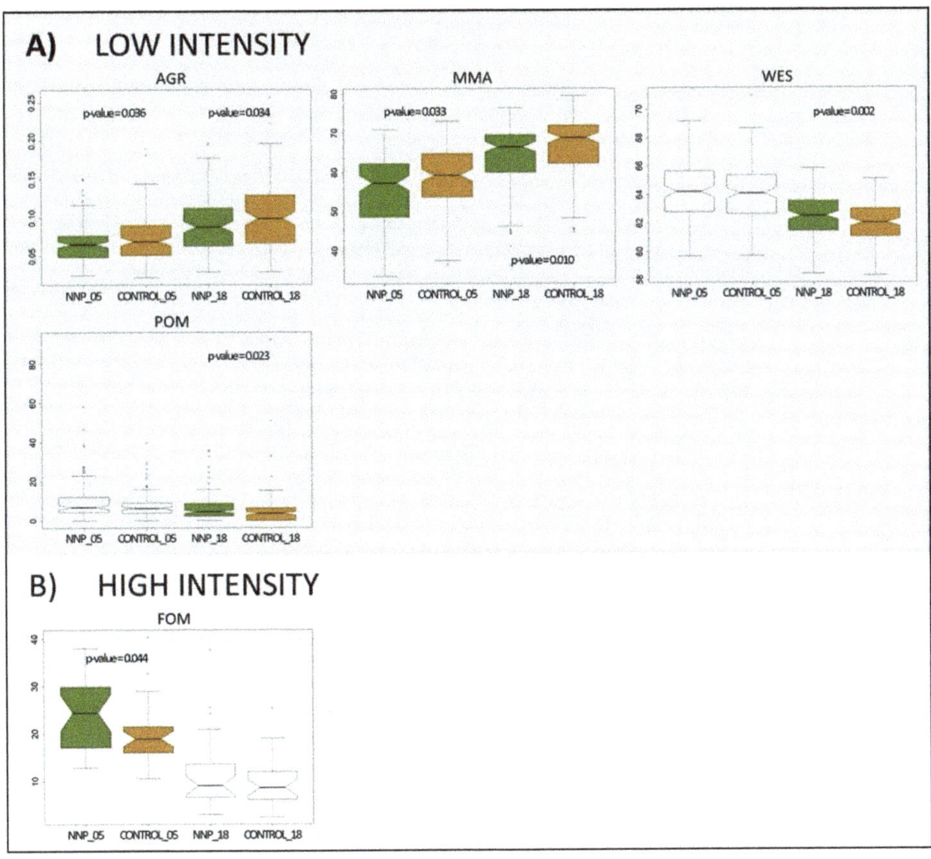

Figure 4. Boxplots of the variables with statistically significant differences between NNPs and control municipalities according to the incidence of the armed conflict: (**A**) low intensity; (**B**) high intensity. Variables: ageing ratio (AGR), mean mortality age (MMA), weight/height (WES), post-neonatal mortality (POM) and foetal mortality (FOM). The boxplot is only shown in colour if the differences are significant during this period, otherwise the boxplot is shown in white.

4. Discussion

The main goal in defining protected areas is to preserve environmental values and bio-diversity [8], but, at the same time, it is also a priority to guarantee adequate development of the local populations [5,6,28,51]. This human development is linked to a compendium of cultural, environmental, social and biological factors that are often difficult to study separately. Analysis of the effects of protected areas on local populations has thus frequently focused on the analysis of the socio-economic impacts, but other effects remain virtually unexplored. This study is one of the first to focus on the effects that NNPs may have on biodemographic aspects and the relationship between these and socio-economic changes.

4.1. Differences between Municipalities in NNP and Control

All of the differences between the NNP and control populations were observed in biodemographic variables. For at least one of the variables in each of the biodemographic blocks (population structure, fertility, infant mortality), differences were identified between the two groups. These variables can be determining factors in the stability and balance of human populations—as well as in their viability and continuity—and ultimately define their biological structure.

The availability of population and housing censuses for 2005 and 2018 allowed us to introduce a diachronic approach to the analysis. Over these 13 years and as a result of social and economic development, notable progress was observed in biodemographic and socio-economic terms for the entire population under study. However, this progress was differentially more rapid in the NNP populations than in the controls. Following the 2016 peace agreement, the impact of which should begin to be felt in the 2018 census, most of the differences that were noted for 2005 in the NNP municipalities in relation to the control group disappeared. By 2018, the location of a municipality within a protected area represented neither an advantage nor a disadvantage for the human populations (Tables 3 and 4). In general, there has been a shift from a situation in 2005 in which NNPs have a lower proportion of elderly people and a lower mean age at death, with no differences in fertility, and higher infant mortality (especially in foetuses and newborns), to the scenario in 2018 in which newborns begin to show greater weight than in the control group, differences in variables related to population structure are maintained and there is also a notable relative improvement in infant mortality rates (Table 4 and Figures 1–3). These biodemographic variables vary by both individual-level responses and meso-level factors [67].

Contrary to the conclusions of other studies [15,16,26,52,68–70], our results for Colombia's NNPs showed that socio-economic variables remained unchanged regardless of the period or natural region considered. With the data handled in the study, and in a context of armed conflict, it is difficult to determine the mechanisms underlying the biodemographic changes described and why these occur independently of the socio-economic aspects. In this respect, the biological dynamics of human populations are sensitive to different factors, including socio-economic ones, but also environmental, geographical or cultural [71]. Even fortuitous or specific events can also be decisive.

Some differences were found between natural regions. If there are socio-economic winners and losers depending on location [37,72], this is also probably true for the biodemographic variables. However, more research is needed on these differences between protected areas and on the causes of this spatial heterogeneity.

4.2. NNP Municipalities and Armed Conflict

Almost all of the effects associated with the presence of NNPs discussed in the previous section were observed in the municipalities where the intensity of the armed conflict was classified as low [64]. In areas with high-intensity conflict it is very difficult to ensure the linkage of any effect with the NNPs, because the violence could be masking outcomes.

The armed conflict affected the entire Colombian population, although not homogeneously [64]. Thus, protected areas, whose environmental values have been preserved in many cases due to their remote location or special isolation, were frequently used as a refuge by the guerrillas to carry out their operations [52,53]. This peripheral location has hampered the capacity and presence of the state in these areas, with a consequent possible effect on the welfare and health conditions of human populations. In Colombia, this capacity seems to have been limited during the conflict to the vicinity of the main populated areas, where violence was less intense and state services reached the population [46,58]. It has been estimated that if the protected areas had not also increased guerrilla activities, the poverty reduction effects would have been more than triple [38].

Attributing causes to explain the changes observed after the peace process is complicated and is not the subject of this study. Nevertheless, it should be noted that the process has not been homogeneous and that post-conflict scenarios in protected areas have been particularly complex. The Colombian administration may not have reached all of these new territories as effectively, and much of the activity there is still outside government control. Thus, with the abandonment of guerrilla positions in some of the more isolated protected areas, key groups in land management (such as large landowners, peasants, cartels, etc.) seem to have expanded their activities in these areas, favouring large-scale livestock farming and speculative land markets or coca crops, with the resulting increase in deforestation and threats to biodiversity [46,53,73]. It is foreseeable that these activities have also had their effects on local population dynamics [51]. In this sense, if the effects of NNPs on local populations in situations of armed conflict remain to be explored, we have even less knowledge of the dynamics of post-conflict scenarios. Therefore, it seems necessary to establish systems to monitor the functionality of the reserved areas [48] and, in particular, their effect on these dynamics.

5. Conclusions

In this study for Colombia, an effect associated with the presence of NNPs was verified on the biodemography of human populations. These effects were independent of socio-economic factors, which remained unchanged over the period analysed. In other words, no socio-economic changes were observed to which biodemographic alterations could be attributed. Nevertheless, these results should be treated with caution, because the presence of the armed conflict is, on the one hand, a source of population stress capable of displacing any other factor that may influence population dynamics, and on the other hand, a source of noise that makes it extremely difficult to interpret the differences observed. Despite these limitations, and in light of the results, it seems necessary to consider broader approaches that include both socio-economic and bio-demographic dynamics in the study of the effect of protected areas on human populations.

Moving forward, to operationalize the results of this line of research, it is not enough to know that NNP exert a quantifiable influence on human populations. It is also necessary to study in depth the underlying causes that explain the interrelations and dependencies between socio-economic and biodemographic aspects and NNP. These causes are out of the scope of this study and should be explored in future research. In the case of Colombia, where as a consequence of the internal conflict the positive economic effects that tourism provides to protected areas are still taking off, the differences between municipalities within and outside NNPs are probably explained by differential access to education or health systems, nutrition, or differences in access to the ecosystem services provided by nature. In any case, to guarantee future viability and improve the living standards of human populations under the influence of NNPs, it is required to define policies specifically designed to improve not only socio-economic indicators but also biodemographic ones such as those studied here.

Author Contributions: Conceptualization, R.R.-D., V.J.C.-R. and M.J.B.-V.; methodology, R.R.-D. and V.J.C.-R.; formal analysis, R.R.-D., A.G.-L. and V.J.C.-R.; investigation, R.R.-D. and V.J.C.-R.; resources, R.R.-D., A.G.-L. and V.J.C.-R.; data curation, R.R.-D. and A.G.-L.; writing—original draft preparation, V.J.C.-R., R.R.-D. and A.G.-L.; writing—review and editing, V.J.C.-R., R.R.-D. and M.J.B.-V.; visualization, R.R.-D., V.J.C.-R. and A.G.-L.; supervision, M.J.B.-V.; project administration, R.R.-D., V.J.C.-R. and M.J.B.-V.; funding acquisition, A.G.-L. All authors have read and agreed to the published version of the manuscript.

Funding: This research was partially funded by International Scholarship "Universidad de Salamanca-Banco Santander" for Latin America.

Institutional Review Board Statement: Not applicable.

Informed Consent Statement: Not applicable.

Data Availability Statement: No new data were created or analyzed in this study. Data sharing is not applicable to this article.

Acknowledgments: The authors wish to acknowledge the source of various datasets used in the analysis for this study: General Population and Housing Census provided by the Colombian National Administrative Department of Statistics (DANE).

Conflicts of Interest: The authors declare no conflict of interest. The funders had no role in the design of the study; in the collection, analyses, or interpretation of data; in the writing of the manuscript; or in the decision to publish the results.

References

1. IUCN; WCMC. *Guidelines for Protected Area Management Categories*; IUCN: Gland, Switzerland, 1994.
2. Scherl, L.M.; Wilson, A.; Wild, R.; Blockhus, J.; Franks, P.; McNeely, J.A.; McShane, T. *Can Protected Areas Contribute to Poverty Reduction? Opportunities and Limitations*; The World Conservation Union: Gland, Switzerland; Cambridge, UK, 2004.
3. Adams, W.M.; Hutton, J. People, parks, and poverty: Political ecology and biodiversity conservation. *Conserv. Soc.* **2007**, *5*, 147–183.
4. Lele, S.; Wilshusen, P.; Brockington, D.; Seidler, R.; Bawa, K. Beyond exclusion: Alternative approaches to biodiversity conservation in the developing tropics. *Curr. Opin. Environ. Sustain.* **2010**, *2*, 94–100. [CrossRef]
5. Unesco. *World National Parks Congress*; Unesco: Bali, Indonesia, 1982.
6. Millennium Ecosystem Assessment. *Ecosystems and Human Well-Being: Policy Responses*; Millennium Ecosystem Assessment: Washington, DC, USA, 2005.
7. Sanderson, S.E.; Redford, K.H. Contested relationships between biodiversity conservation and poverty alleviation. *Oryx* **2003**, *37*, 389–390. [CrossRef]
8. Adams, W.M.; Aveling, R.; Brockington, D.; Dickson, B.; Elliot, J.; Hutton, J.; Roe, D.; Vira, B.; Wolmer, W. Biodiversity conservation and the eradication of poverty. *Science* **2004**, *306*, 1146–1149. [CrossRef] [PubMed]
9. Roe, D. The origins and evolution of the conservation-poverty debate: A review of key literature, events, and policy processes. *Oryx* **2008**, *42*, 491–503. [CrossRef]
10. Brockington, D.; Wilkie, D. Protected areas and poverty. *Philos. Trans. R Soc. B* **2015**, *370*, 20140271. [CrossRef]
11. Naughton-Treves, L.; Alix-Garcia, J.; Chapman, C.A. Lessons about parks and poverty from a decade of forest loss and economic growth around Kibale National Park, Uganda. *Proc. Natl. Acad. Sci. USA* **2011**, *108*, 13919–13924. [CrossRef]
12. Brockington, D.; Igoe, J.; Schmidt-Soltau, K.A. Conservation, human rights, and poverty reduction. *Conserv. Biol.* **2006**, *20*, 250–252. [CrossRef] [PubMed]
13. West, P.; Igoe, J.; Brockington, D. Parks and peoples: The social impact of protected areas. *Annu. Rev. Anthropol.* **2006**, *35*, 251–277. [CrossRef]
14. Norton-Griffiths, M.; Southey, C. The opportunity costs of biodiversity conservation in Kenya. *Ecol. Econ.* **1995**, *12*, 125–139. [CrossRef]
15. Ferraro, P.J. The local costs of establishing protected areas in low-income nations: Ranomafana National Park, Madagascar. *Ecol. Econ.* **2002**, *43*, 261–275. [CrossRef]
16. Andam, K.S.; Ferraro, P.J.; Sims, K.R.; Healy, A.; Holland, M.B. Protected areas reduced poverty in Costa Rica and Thailand. *Proc. Natl. Acad. Sci. USA* **2010**, *107*, 9996–10001. [CrossRef]
17. Ferraro, P.J.; Hanauer, M.M. Quantifying causal mechanisms to determine how protected areas affect poverty through changes in ecosystem services and infrastructure. *Proc. Natl. Acad. Sci. USA* **2014**, *111*, 4332–4337. [CrossRef]
18. Baird, T.D.; Leslie, P.W. Conservation as disturbance: Upheaval and livelihood diversification near Tarangire National Park, northern Tanzania. *Glob. Environ. Change* **2013**, *23*, 1131–1141. [CrossRef]
19. Ezebilo, E.E.; Mattsson, L. Socio-economic benefits of protected areas as perceived by local people around Cross River National Park, Nigeria. *For. Policy Econ.* **2010**, *12*, 189–193. [CrossRef]
20. Upton, C.; Ladle, R.; Hulme, D.; Jiang, T.; Brockington, D.; Adams, W.M. Are poverty and protected area establishment linked at a national scale? *Oryx* **2008**, *42*, 19–25. [CrossRef]
21. Wittemyer, G.; Elsen, P.; Bean, W.T.; Burton, A.C.; Brashares, J.S. Accelerated human population growth at protected area edges. *Science* **2008**, *321*, 123–126. [CrossRef]
22. Scholte, P.; De Groot, W.T. From debate to insight: Three models of immigration to protected areas. *Conserv. Biol.* **2009**, *24*, 630–632. [CrossRef]
23. Joppa, L.N.; Loarie, S.R.; Pimm, S.L. On population growth near protected areas. *PLoS ONE* **2009**, *4*, e4279. [CrossRef]
24. Castro-Prieto, J.; Martinuzzi, S.; Radeloff, V.C.; Helmers, D.P.; Quiñones, M.; Gould, W.A. Declining human population but increasing residential development around protected areas in Puerto Rico. *Biol. Conserv.* **2017**, *209*, 473–481. [CrossRef]
25. Woodhouse, E.; Homewood, K.M.; Beauchamp, E.; Clements, T.; McCabe, J.T.; Wilkie, D.; Milner-Gulland, E.J. Guiding principles for evaluating the impacts of conservation interventions on human well-being. *Phil. Trans. R Soc. B* **2015**, *370*, 20150103. [CrossRef]
26. Naidoo, R.; Gerkey, D.; Hole, D.; Pfaff, A.; Ellis, A.M.; Golden, C.D.; Herrera, D.; Johnson, K.; Mulligan, M.; Ricketts, T.H.; et al. Evaluating the impacts of protected areas on human well-being across the developing world. *Sci. Adv.* **2019**, *5*, eaav3006. [CrossRef] [PubMed]
27. De Sherbinin, A. Is poverty more acute near parks? An assessment of infant mortality rates around protected areas in developing countries. *Oryx* **2008**, *42*, 26–35. [CrossRef]

28. PNNC (Parques Nacionales Naturales de Colombia). Sistema Nacional de Áreas Protegidas-SINAP. Available online: http://www.parquesnacionales.gov.co/portal/es/sistema-nacional-de-areas-protegidas-sinap/mapa-sinap/ (accessed on 15 January 2020).

29. Díaz, M. Conflicto de ocupación en áreas protegidas. Conservación versus derechos de comunidades. *Opin. Juríd.* **2008**, *7*, 53–69.

30. Durán, C.A. Gobernanza en los Parques Nacionales Naturales Colombianos: Reflexiones a partir del caso de la comunidad Orika y su participación en la conservación del Parque Nacional Natural Corales del Rosario y San Bernardo. *Rev. Estud. Soc.* **2009**, *32*, 60–73.

31. Rojas, Y. La historia de las áreas protegidas en Colombia, sus firmas de gobierno y las alternativas para la gobernanza. *Soc. Econ.* **2014**, *27*, 155–175.

32. De Pourcq, K.; Thomas, E.; Van Damme, P.; Léon-Sicard, T. Análisis de los conflictos entre comunidades locales y autoridades de conservación en Colombia: Causas y recomendaciones. *Gest. Ambiente* **2017**, *20*, 122–139. [CrossRef]

33. Canavire-Bacarreza, G.; Hanauer, M.M. Estimating the impacts of Bolivia's protected areas on poverty. *World Dev.* **2013**, *41*, 265–285. [CrossRef]

34. Ferguson, B.; Gardner, C.J.; Andriamarovololona, M.M.; Healy, T.; Muttenzer, F.; Smith, S.; Hockley, N.; Gingembre, M. Governing ancestral land in Madagascar: Have policy reforms contributed to social justice? In *Governance for Justice and Environmental Sustainability: Lessons across Natural Resource Sectors in Sub-Saharan Africa*; Sowman, M., Wynberg, R., Eds.; Routledge: London, UK, 2014; pp. 63–93.

35. Stuart, A.E. Matching methods for causal inference: A review and a look forward. *Stat. Sci.* **2010**, *25*, 1–21. [CrossRef]

36. Diamond, A.; Sekhon, J.S. Genetic matching for estimating causal effects: A general multivariate matching method for achieving balance in observational studies. *Rev. Econ. Stat.* **2013**, *95*, 932–945. [CrossRef]

37. Hanauer, M.M.; Canavire-Bacarreza, G. Implications of heterogeneous impacts of protected areas on deforestation and poverty. *Philos. Trans. R Soc. B* **2015**, *370*, 20140272. [CrossRef] [PubMed]

38. Hanauer, M.M.; Canavire-Bacarreza, G. *Civil Conflict Reduced the Impact of Colombia's Protected Areas*; Sonoma State University: Rohnert Park, CA, USA; Inter-American Development Bank: Washington, DC, USA, 2018; 31p.

39. Cancillería de Colombia. Símbolos Patrios y Otros Datos de Interés. Available online: https://www.cancilleria.gov.co/colombia/nuestro-pais/simbolos (accessed on 18 March 2020).

40. DANE (Departamento Administrativo Nacional de Estadística). Available online: http://www.dane.gov.co (accessed on 15 January 2020).

41. Ruiz-Linares, A.; Adhikari, K.; Acuña-Alonzo, V.; Quinto-Sanchez, M.; Jaramillo, C.; Arias, W.; Fuentes, M.; Pizarro, M.; Everardo, P.; de Avila, F.; et al. Admixture in Latin America: Geographic structure, phenotypic diversity and self-perception of ancestry based on 7,342 individuals. *PLoS Genet.* **2014**, *10*, e1004572. [CrossRef] [PubMed]

42. Ibarra, A.; Restrepo, T.; Rojas, W.; Castillo, A.; Amorim, A.; Martínez, B.; Burgos, G.; Ostos, H.; Álvarez, K.; Camacho, M.; et al. Evaluating the X chromosome-specific diversity of Colombian populations using insertion/deletion polymorphisms. *PloS Genet.* **2014**, *9*, e87202. [CrossRef] [PubMed]

43. Homburguer, J.R.; Moreno-Estrada, A.; Gignoux, C.R.; Nelson, D.; Sanchez, E.; Ortiz-Tello, P.; Pons-Estel, B.A.; Acevedo-Vasquez, E.; Miranda, P.; Langefeld, C.D.; et al. Genomic insights into the ancestry and demographic history of South America. *PloS Genet.* **2015**, *11*, e1005602. [CrossRef] [PubMed]

44. Sistema de Información sobre Biodiversidad de Colombia. Biodiversidad en cifras. Available online: https://cifras.biodiversidad.co/ (accessed on 25 February 2020).

45. Ayram, C.; Etter, A.; Díaz-Timoté, J.; Rodríguez Buriticá, S.; Ramírez, W.; Corzo, G. Spatiotemporal evaluation of the human footprint in Colombia: Four decades of anthropic impact in highly biodiverse ecosystems. *Ecol. Indic.* **2020**, *117*, 106630. [CrossRef]

46. Suarez, A.; Arias-Arévalo, P.A.; Martínez-Mera, E. Environmental sustainability in post-conflict countries: Insights for rural Colombia. *Environ. Dev. Sustain.* **2018**, *20*, 997–1015. [CrossRef]

47. PNGIBSE (Política Nacional para la Gestión Integral de la Biodiversidad y sus Servicios Ecosistémicos). Ministerio de Ambiente y Desarrollo Sostenible, Instituto Alexander Von Humboldt, Bogotá, Colombia. Available online: http://www.humboldt.org.co/images/documentos/pdf/documentos/pngibse-espaol-web.pdf (accessed on 12 March 2020).

48. Sierra, C.A.; Mahecha, M.; Poveda, G.; Álvarez-Dávila, E.; Gutierrez-Velez, V.H.; Reu, B.; Feilhauer, H.; Anáya, J.; Armenteras, D.; Benavides, A.; et al. Monitoring ecological change during rapid socio-economic and political transitions: Colombian ecosystems in the post-conflict era. *Environ. Sci. Policy* **2017**, *76*, 40–49. [CrossRef]

49. Forero-Medina, G.; Joppa, L. Representation of global and national conservation priorities by Colombia's Protected Area Network. *PLoS ONE* **2010**, *5*, e13210. [CrossRef]

50. Sánchez, F.; Palau, M.M. Conflict, decentralisation and local governance in Colombia, 1974–2004. In *Households in Conflict Network*; Working Paper 14; Universidad de los Andes: Bogotá, Colombia, 2006.

51. Aguilar, M.; Sierra, J.; Ramírez, W.; Ragas, O.; Calle, Z.; Vargas, W.; Murcia, C.; Aronson, J.; Barrera Castaño, J.I. Toward a post-conflict Colombia: Restoring to the future. *Restor. Ecol.* **2015**, *23*, 4–6. [CrossRef]

52. Canavire-Bacarreza, G.; Diaz-Gutierrez, J.E.; Hanauer, M.M. Unintended consequences of conservation: Estimating the impact of protected areas on violence in Colombia. *J. Environ. Econ. Manag.* **2018**, *89*, 46–70. [CrossRef]

53. Hoffmann, C.; García Márquez, J.R.; Krueger, T. A local perspective on drivers and measures to slow deforestation in the Andean-Amazonian foothills of Colombia. *Land Use Policy* **2018**, *77*, 379–391. [CrossRef]

54. Dávalos, L. The San Lucas mountain range in Colombia: How much conservation is owed to the violence? *Biodivers. Conserv.* **2001**, *10*, 69–78. [CrossRef]

55. Alvarez, M. Forests in the time of violence: Conservation implications of the Colombian war. *J. Sustain. For.* **2003**, *16*, 47–68. [CrossRef]

56. Enaruvbe, G.O.; Keculah, K.M.; Atedhor, G.O.; Osewole, A.O. Armed conflict and mining induced land-use transition in northern Nimba County, Liberia. *Glob. Ecol. Conserv.* **2019**, *17*, 2351–9894. [CrossRef]

57. Murillo-Sandoval, P.J.; Van Dexter, K.; Van Den Hoek, J.; Wrathall, D.; Kennedy, R. The end of gunpoint conservation: Forest disturbance after the Colombian peace agreement. *Environ. Res. Lett.* **2020**, *15*, 034033. [CrossRef]

58. Bonilla-Mejía, L.; Higuera-Mendieta, I. Protected areas under weak institutions: Evidence from Colombia. *World Dev.* **2019**, *122*, 585–596. [CrossRef]

59. Clerici, N.; Salazar, C.; Pardo-Díaz, C.; Jiggins, C.D.; Richardson, J.E.; Linares, M. Peace in Colombia is a critical moment for Neotropical connectivity and conservation: Save the northern Andes-Amazon biodiversity bridge. *Conserv. Lett.* **2018**, *12*, 1–7. [CrossRef]

60. Joppa, L.N.; Pfaff, A. High and far: Biases in the location of protected areas. *PLoS ONE* **2009**, *4*, e8273. [CrossRef]

61. Baldi, G.; Texeira, M.; Martin, O.A.; Grau, H.R.; Jobbágy, E.G. Opportunities drive the global distribution of protected areas. *PeerJ* **2017**, *5*, e2989. [CrossRef]

62. Restrepo, J.; Spagat, M.; Vargas, J. The dynamics of the Colombian civil conflict: A new data set. *Homo Oecon.* **2004**, *21*, 396–428.

63. Restrepo, J.; Spagat, M.; Vargas, J. The severity of the Colombian conflict: Cross-country datasets versus new micro data. *J. Peace Res.* **2006**, *43*, 99–115. [CrossRef]

64. CERAC (Centro de Recursos Para el Análisis de Conflictos). Typology of Colombian Municipalities According to the Internal Armed Conflict. 2014. Available online: http://www.cerac.org.co/es/l%C3%ADneas-de-investigaci%C3%B3n/analisis-conflicto/tipologia-por-municipios-del-conflicto-armado.html (accessed on 17 December 2020).

65. Bauer, D.F. Constructing confidence sets using rank statistics. *J. Am. Stat. Assoc.* **1972**, *67*, 687–690. [CrossRef]

66. Hollander, M.; Wolfe, D.A. *Nonparametric Statistical Methods*; John Wiley & Sons: New York, NY, USA, 1973; (or second edition 1999); pp. 27–33.

67. Castro Torres, A.F.; Urdinola, B.P. Armed conflict and fertility in Colombia, 2000–2010. *Popul. Res. Policy Rev.* **2019**, *38*, 173–213. [CrossRef]

68. Coad, L.; Campbell, A.; Miles, L.; Humphries, K. *The Costs and Benefits of Protected Areas for Local Livelihoods: A Review of the Current Literature*; United Nations Environment Programme: Nairobi, Kenya; World Conservation Monitoring Centre: Cambridge, UK, 2008.

69. Pfaff, A.; Robalino, J.; Lima, E.; Sandoval, C.; Herrera, L.D. Governance, location and avoided deforestation from protected areas: Greater restrictions can have lower impact due to differences in location. *World Dev.* **2014**, *55*, 7–20. [CrossRef]

70. Specht, M.J.; Santos, B.A.; Marshall, N.; Melo, F.P.; Leal, I.R.; Tabarelli, M.; Baldauf, C. Socio-economic differences among resident, users and neighbour populations of a protected area in the Brazilian dry forest. *J. Environ. Econ. Manag.* **2019**, *232*, 607–614.

71. Creanza, N.; Kolodny, O.; Feldman, M.W. Cultural evolutionary theory: How culture evolves and why it matters. *Proc. Natl. Acad. Sci. USA* **2017**, *114*, 7782–7789. [CrossRef]

72. Ferraro, P.J.; Hanauer, M.M. Protecting ecosystems and alleviating poverty with parks and reserves: 'win-win' or tradeoffs? *Environ. Resour. Econ.* **2011**, *48*, 269–286. [CrossRef]

73. Murillo-Sandoval, P.J.; Hilker, T.; Krawchuk, M.A.; Van Den Hoek, J. Detecting and attributing drivers of forest disturbance in the Colombian Andes using landsat time-series. *Forests* **2018**, *9*, 269. [CrossRef]

 *sustainability*

Article

Geomorphological and Geochronological Analysis Applied to the Quaternary Landscape Evolution of the Yeltes River (Salamanca, Spain)

Iván Martín-Martín *, Pablo-Gabriel Silva , Antonio Martínez-Graña and Javier Elez

Department of Geology, Faculty of Sciences, University of Salamanca, Plaza de la Merced s/n., 37008 Salamanca, Spain; pgsilva@usal.es (P.-G.S.); amgranna@usal.es (A.M.-G.); j.elez@usal.es (J.E.)
* Correspondence: ivan_martin96@usal.es

Received: 1 September 2020; Accepted: 22 September 2020; Published: 23 September 2020

Abstract: This paper aims to study the Quaternary geomorphological evolution of the Yeltes river-valley (Duero Basin, Central Spain) primarily based on the study of the Late Neogene piedmont dissected by the river and its Quaternary terrace sequence, since fluvial terraces are excellent archives to study the landscape and climate evolution during this period. Detailed geomorphological mapping implemented in GIS-based digital elevation models was used to the further applications of existing fluvial chronofunctions (relative terrace height-age transfer functions) to establish a numerical geochronology to the sequence of fluvial terraces in the zone. The obtained theoretical ages points to an onset of fluvial incision in the zone after 2.0–2.5 Myr ago, with the dissection of the *"Raña surface"* (a Gelasian alluvial piedmont widely developed in Central Spain). The obtained terrace ages coincide, in most cases, with warm isotopic stages (MIS) or mainly with the transit of cold to warm MIS. Additionally, this study suggests that the full connectivity of the Yeltes drainage (Ciudad Rodrigo Basin) with the Atlantic drainage was not completely effective until MIS 9 (c. 0.29 Myr). The new reported data allows for the exploration of the timing and processes involved in the capture of inland sedimentary basins (Ciudad Rodrigo, Duero basins) by the Atlantic drainage during the early Quaternary.

Keywords: fluvial terraces; chronology; quaternary landscape; Yeltes river; Duero basin (Spain)

1. Introduction

During the last million years, the Cenozoic basins in the center of the Iberian Peninsula opened to the Atlantic, produced the so-called Atlantic Capture [1]. This process represented the present drainage network development and this region's landscape modeling, through the dissection of the Plio-Quarternary alluvial piedmont known as the "Raña surface" in central Spain [2–7]. Fluvial dissection into the Raña surface allowed us to analyze the development of river valleys and their correlative terrace sequences in central Spain, as well as their relationships with the Quaternary climate changes.

The river terraces are excellent archives on the relationships of landscape evolution and climate during the Quaternary [8–10]. They record low-frequency climatic changes of astronomical origin, but also climatic fluctuations correlated with marine isotopic stratigraphy [11]. Valley aggradation occur during glacial periods (cold periods) and dissection during the end of interglacial periods (warm periods) [12]. However, river-response to climate change is complex, in terms of aggradation and incision periods and in the morpho-climatic characteristics of each area, which require specific studies [13]. Consequently, this work deals with the geomorphological analysis and evolution of the

Yeltes river-valley (Duero Basin, Central Spain; Figure 1) pondering all the geomorphological processes that shape the relief.

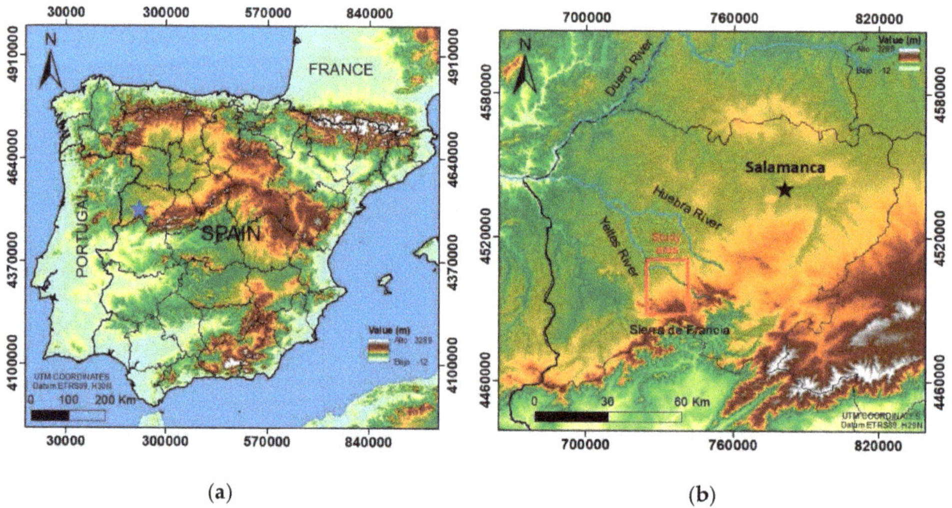

Figure 1. Location of the study area in the (**a**) Iberian Peninsula and the (**b**) Salamanca province.

This work is focused on a detailed geomorphological mapping of the Yeltes river basin, and more specifically of the sequence of existing fluvial terraces. These will be used as the main elements to set the temporal sequence of sedimentation-erosion processes behind valley development, but also to infer the timing of the Atlantic capture of the drainage in this southern zone of the Duero basin. The theoretical timing of the geomorphological evolution will be explored by the application of fluvial chronofunctions developed for the Atlantic river basins of the Iberian Peninsula, which calculate theoretical terrace ages from the relative height of fluvial terraces. Once established, the theoretical geochronology of valley development, the geomorphological, tectonics, and climatic factors operating in the area during the Quaternary will be explored, making it possible to advance a more accurate interpretation of the evolution of physical environment and the landscape of the Yeltes river valley.

2. Geology and Geography of the Study Area

The Yeltes River area is located in the central-western part of the Iberian Peninsula in the province of Salamanca (Castilla y León, Spain; Figure 1), occupying an area of 387 km². The southern part has reliefs belonging to the Sierra de Francia, a range-hill of the Spanish Central System, with a W-E orientation, and elevations around 1700 m. To the north of these landforms, there is a large plain that constitutes the "Raña" piedmont, where the altitude decreases from 1000 m in the range front to 753 m, which is the lowest elevation of the study area. The average altitude of the piedmont is about ± 800–900 m. All the drainage network of the area belongs to the Duero drainage-basin, which eventually flows into the Atlantic Ocean (Figure 1). The Yeltes river is the main drainage system in the area and the main objective of study for this work. The Yeltes originates in the Quilamas mountain range and flows into the Huebra river to then flow into the Duero river at the end of the "Arribes Canyon." When crossing the study area, the Yeltes river carves an important escarpment on its right bank and presents a well-developed terrace system on its left bank, which is an essential part of the analysis done in this work.

The study area is within the Paleozoic Iberian Massif of Central Spain and more specifically of the Central-Iberian Zone [14]. This is an ancient geological zone occupying the western part of the Iberian Peninsula and constitutes a large outcrop of the old European Variscan Chain. This zone is featured by the notable extension of Ordovician and Precambrian metamorphic outcrops (Schists, Shales, and Quartzites) as well as subsidiary late Palaeozoic granitic intrusions [15,16] (Figure 2). The oldest rock successions belong to the Precambrian being part of Neoproterozoic "Complejo Esquisto-Grauváquico" (Schist-Greywacke Complex) and Ordovician materials are constituted by the Armorican Quartzites, both typical of the European Variscan units [14]. These metamorphic units are verticalized and strongly folded featuring old structural units of the Tamames-Ahigal Syncline and the Sierra de Francia-Torralba Syncline [15,16]. Quarzite crests are the main landforms within the Sierra de Francia Range featuring the main intermontane water-divides (Figure 2). These folded units also display important fault systems, the dominant ones being those oriented at N40–55E y N10–20E with reverse and sinistral kinematics, otherwise typical of the Spanish Iberian Massif [17]. To the north, all these metamorphic and subsidiary granitic rocks are buried by Paleogene, Neogene, and Quaternary detrital deposits linked to the sedimentary infilling of the Ciudad Rodrigo Basin (Figure 2). This basin is a sub-basin of the largest Duero Basin, one of the more important Cenozoic basins within the Iberian Peninsula (Figure 1).

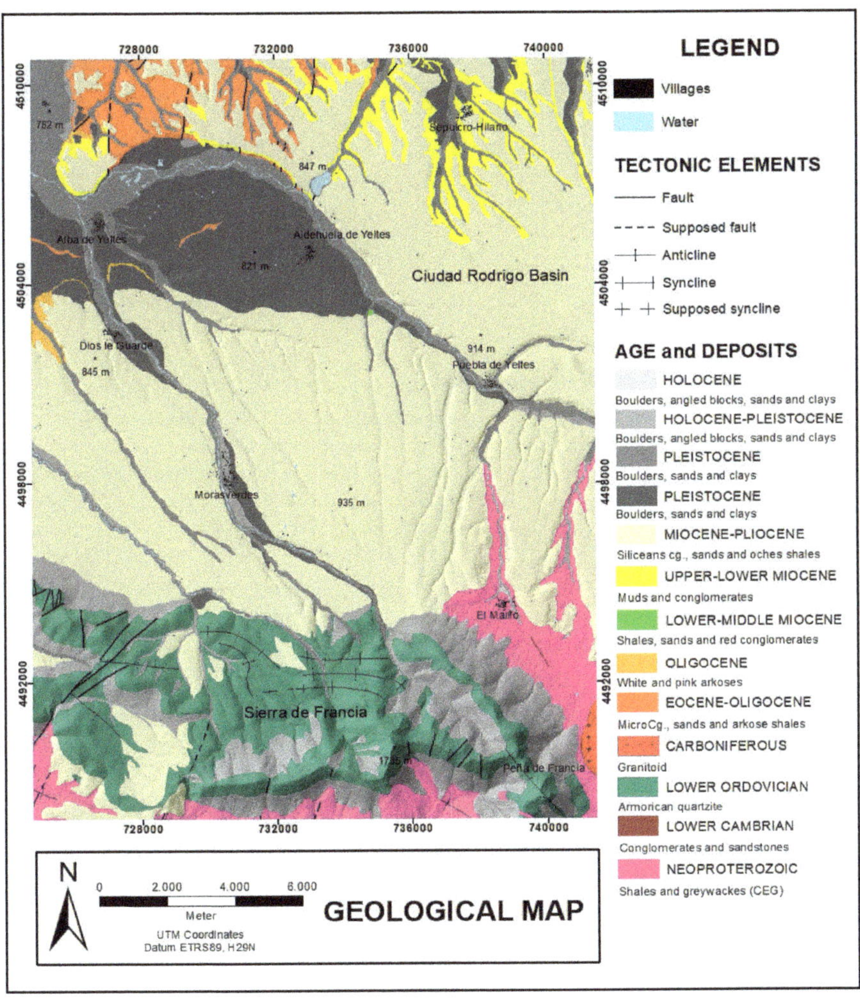

Figure 2. Geological map of the study area.

Ciudad Rodrigo Basin formation occurred during the Alpine orogeny and their sedimentary infilling records different phases linked to different tectonic pulses [18,19]. The first sedimentary phase occurred during the Paleogene (Eocene-Oligocene) and was featured by coarse-grained fluvial systems generating the so-called "arkosic filling" [18]. The second phase was dominated by the deposit of large alluvial fans fed by the relief of the Sierra de Francia (Figure 2), constituting the so-called post-arkosic Neogene sedimentary filling of the basin [18]. Faulting occurred during the last stages of the Variscan orogeny and had a great impact on the later Alpine evolution of the zone. These old fault systems conditioned the structure and geometry of basins as well as ranges throughout the entire Paleozoic Massif and were reactivated as normal or reverse faults from the Paleogene [19,20]. Therefore, the Ciudad Rodrigo Basin has a tectonic origin and has an asymmetrical structure with a greater sediment thickness at their southern zone [19,21].

3. Materials and Methods

A multidisciplinary methodology has been used, based mainly on field research and mapping, the compilation of existing geological information, remote sensing, geomatics, and statistics. We use field research and Geographic Information System (GIS)-based geomorphological mapping to perform a detailed geomorphological analysis (Section 3.1) and develop a theoretical geochronological approach (Section 3.2) to the age of the different fluvial terraces within the Yeltes river-valley. For this last issue, we used the existing fluvial chronofunctions proposed for the Atlantic river basins of the Iberian Peninsula [22], the models of fluvial evolution of the zone [23,24], as well as more recent papers on fluvial valley evolution for the Duero and Tagus river basins in Portugal [1] and Spain [25,26] containing modern geochronological data (Optically Stimulated Luminescence -OSL and Cosmogenic data) for the corresponding terrace systems.

3.1. Geomorphological Mapping

Geomorphological analysis was performed considering all the existing cartographic information for the studied area. Moreover, complementary photointerpretation and a fieldwork checking were accomplished. The geological cartographic information from the Spanish Geological Survey 1:50.000 maps [15–17,19–21] covering the studied area were digitized to obtain a geological map (Figure 2) as the previous background for the geomorphological analysis. All the cartographic information was homogenized, scaled, and implemented in a Geographic Information System (GIS) using the software ArcGIS v10.5. The processed digital information was: (1) topographic maps; (2) orthophotos of the Program PNOA 2017; (3) air-photos of the 1956–57 American flight; (4) 1977–78 Interministerial orthophotos; and (5) the geological cartography of the Spanish Geological Survey. In addition, a Digital Terrain Model (DTM) with 5 m/pixel resolution was generated to obtain the hillshade and slope DTMs. For the detailed geomorphological analysis, a higher resolution DTM (1 m/pixel) was generated from a LIDAR (Light Detection and Ranging or Laser Imaging Detection and Ranging) point cloud in LAS format.

The image generated from the 1 m/pixel DTM–LIDAR in hillshade model provided valuable geo-spatial information to produce a detailed photointerpretation capable to differentiate geomorphological units and individual landforms. Both were classified according to their morphogenetic origin following previous classifications for the zone [27] based on modern issues on geospatial digital geomorphological mapping [28,29]. An initial legend was prepared based on previous morphogenetic classifications for the zone [23] following classical guidelines differentiating landforms according to their endogenic (i.e., structural landforms) or exogenic origin (i.e., landforms generated by surface processes) [28]. A description of each landform starts from the morphogenetic classification and is summarized in Table 1. This description is based on the characteristics of its morphology (shape), genetics (process), composition, structure (background geology), and relative chronology. The exogenic landforms are mainly depositional elements whereas endogenic landforms are mainly erosional features or old elements inherited from the tectonic evolution of the area. The geomorphological classification

(Table 1) and mapping (Figure 3) follow the overall guidelines proposed for digital geomorphological mapping [29], previously applied to different sectors of the Duero basin [23,24,27,30] or to other zones all over the globe [31]. Once the overall geomorphology of the zone and individual landforms were checked and validated on the field, photointerpretation was revised, and the final geomorphological map was produced by the use of various manual and digital cartographic techniques [28]. The obtained digital map (Figure 3) constitutes a key-tool to accomplish the geomorphological analysis of the zone and reconstruct the temporal process-sequence behind the landscape evolution of the Yeltes river-valley.

Table 1. Summary of the morphogenetic system (M.S.), domains, and landforms of the Yeltes river-basin.

M. S.	Domain	Landforms	Description	Age
M. S. Structural	Structural Landforms	Fault	Fractures on materials that condition the structural reliefs and drainage orientations	Carboniferous (Variscan orogeny)
		Suspect fault	Probable fractures from photointerpretation not verified in the field	Carboniferous (Variscan orogeny)
		Crest	Narrow ridges (hogback-type) mainly developed on resistant quartzites	Paleogene-Neogene (Alpine orogeny)
		Summits	Water-divides within the mountain area, usually coincides with quartzite crests	Paleogene-Neogene (Alpine orogeny)
M. S. Fluvial	Fluvial-Alluvial Landforms	Drainage network	Present set of river and stream channels allowing water erosion and sediment transport	Holocene
		Flood plain	Alluvial surface adjacent to a river subject to periodic/episodic flooding	Holocene
		Terrace scarp	Topographic steps (slope-ruptures) between the different terrace levels	Holocene-Pleistocene
		Fluvial Terrace	Ancient floodplains hang up several meters above the present river thalweg, resulting from consecutive cycles of river downcutting and sedimentation	Holocene-Pleistocene
		Fluvial escarpments	Slope ruptures generated by the erosive action of the rivers	Holocene-Pleistocene
		Alluvial fan	Fan-shaped alluvial sedimentary bodies generated at the toe of large escarpments which extends radially downslope	Holocene-Pleistocene
		Erosional landforms	Areas excavated and shaped by river action in Neogene sediments	Holocene-Pleistocene
M. S. Lacustrine	Endorheic Landforms	Small lakes, ponds	Endorheic lakes of hydro-eolian origin favored by differential erosion along structural elements	Pleistocene
M.S. Gravitational	Slope Landforms	Talus cones	Debris accumulations related to gully erosion and escarpments retreat	Holocene
		Hillslopes	Steep slopes in a mountainous area, where various gravitational processes occur	Holocene-Upper Pleistocene
		Terrace wash-slopes	Accumulations developed with the erosion of the terrace escarpments	Holocene-Pleistocene
M. S. Periglacial	Periglacial Landforms	Talus-scree (Canchales)	Talus-slope accumulations of angular clasts in elevated zones at the foot of large quartzite crests and rock walls	Holocene-Upper Pleistocene
M.S. Polygenic	Polygenic surfaces and Landforms	Glacis	Glacis-type deposits coming from the reworking of raña deposits composed of relatively angular clasts. It could be considered as the first terrace	Lower Pleistocene
		Raña surface (Piedmont)	Extensive alluvial sedimentary platforms and deposits unconformably placed on top of the Neogene sedimentary filling of the basin	Pliocene-Pleistocene
		Pediment	Exhumed relic rocky surface free of the older overlaying weathering profile (removed by erosion)	Pliocene
M. S. Anthropogenic	Anthropogenic Elements	Roads	Main road network	Modern
		Quarry	Aggregate extraction quarries (Gravelly)	Modern
		Villages	Localities (sites)	Historic

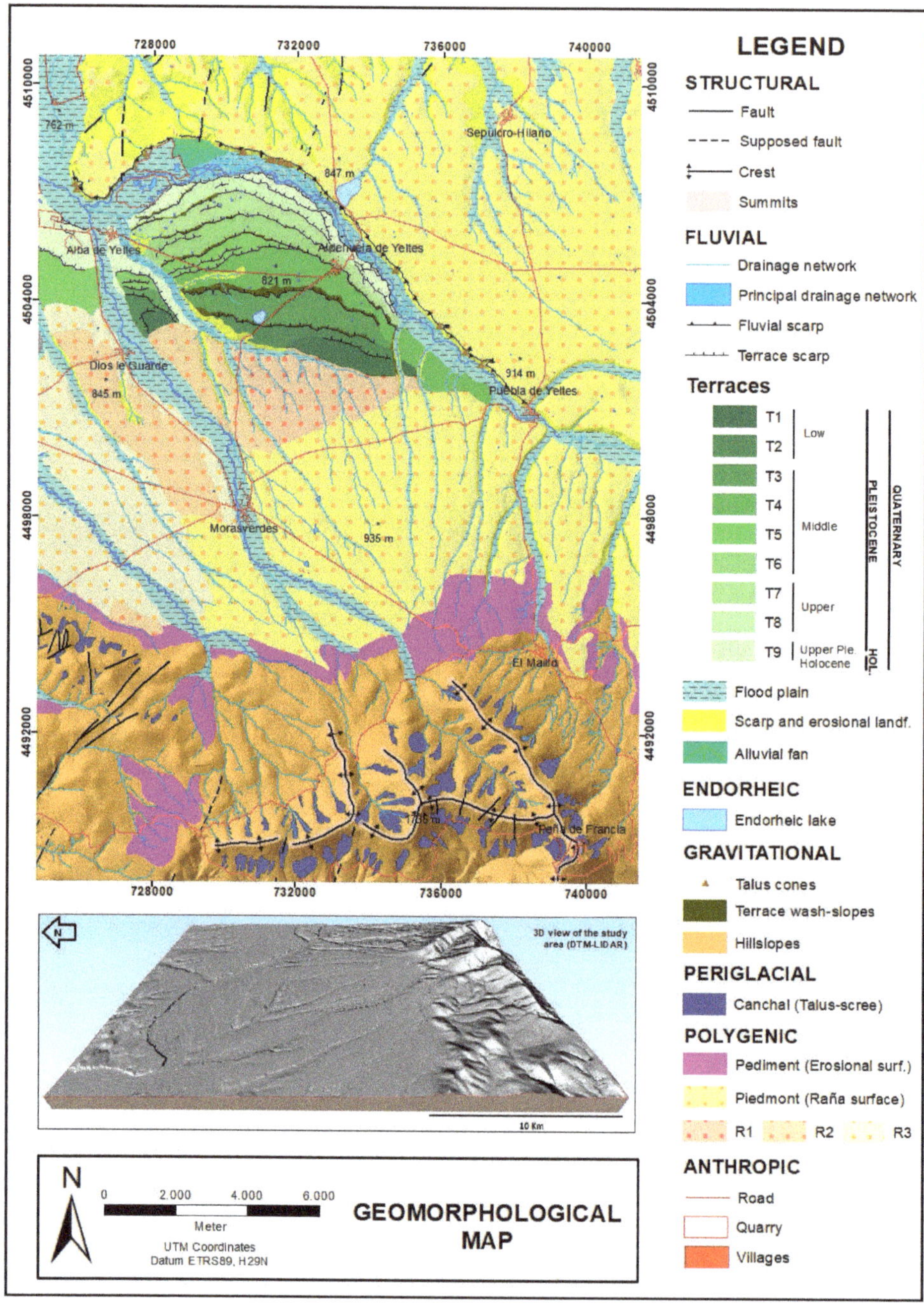

Figure 3. Geomorphological map of the study area.

3.2. Geochronological Analysis

In the geochronology section, theoretical ages were calculated and discussed for the fluvial terraces of the river Yeltes by the applications of the height-age transfer chronofunctions developed for the Atlantic river basins of Central Spain [22]. The used chronofunctions are based on the existing published dataset of numerical dates, coming from different dating methods:(^{14}Carbon -^{14}C-, Thermoluminescence -TL-, OSL, Thorium/Uranio -Th/U-, Electron Spin Resonance -ESR-, AminoAcids Racemization -AAR-, and Paleomagnetic determinations), for the Duero and Tagus basin till the year 2017 [22]. These equations describe the river incision and terrace development during the Quaternary from the large battery of numerical ages existing for these two Atlantic river-basins. This method offers a statistical approach to calculate the theoretical ages of the terraces in the Atlantic side of Iberia, suitable to be applied to the Yeltes river-basin, located nearly in the limit of the Duero and Tagus basins. The mathematical approach is based on multiple power and polynomial correlations between the relative height of the terraces with respect to the river thalwegs and the numerical ages obtained by different numerical dating methods (Table 2) [22]. These numerical models relate terrace relative height to age, transforming relative heights into numerical ages with correlation coefficients (R^2) above 0.92 (Table 2). The application of the four chronofunctions listed in Table 2 indicate that the two second order polynomial chronofunctions [a, d] are the best-fit for the study area [23]. These chronofunctions will provide us with the general numerical age (y) of the different terrace levels as a function of their relative height above the thalweg (x). As we will see, the obtained terrace ages fit quite well with recent geochronological studies in the Duero [25] and Tagus basins [26]. These chronofuntions are of use and applicable to the Atlantic river basins of the Iberian Peninsula, which underwent a similar geological history and sea-level changes; for other areas (i.e., Northern Europe, Mediterranean, etc.) similar equations have to be developed from local data [22]. For a deep discussion on the development and application of the listed fluvial chronofunctions, consult the work of Silva et al. [22].

Table 2. Best-fitted fluvial chronofunctions for Central Spain applied to the study area [22]. The parameter [x] represents the relative height of the terrace with respect to the present river thalweg, and the parameter [y] is the resulting theoretical numerical age for the terrace deposit. R^2 represents the correlation coefficient for each equation.

Chronofunction name	Equation (R^2)
2nd Order polynomial function (Duero Basin) [a]	$y = 0.098x^2 + 8.057x - 16.38$ (R^2 0.96)
Power Function for Arlanzón valley (Duero Basin) [b]	$y = 2.942x^{1.37}$ (R^2 0.92)
3rd Order polynomial function (Duero + Tagus Basin) [c]	$y = -0.001x^3 + 0.199x^2 + 4.38x - 21.16$ (R^2 0.98)
2nd Order polynomial function (Duero + Tagus Basin) [d]	$y = 0.085x^2 + 8.05x - 38.59$ (R^2 0.96)

4. Results and Discussion

The geomorphology section (Section 4.1) shows the resulting digital map, the classification of all the landforms and deposits in both morphogenetic systems, and domains summarized in Table 1. The distinguished terrace sequence is explained and discussed in Section 4.2 in relation to large polygenetic landforms (raña surfaces) developed before the fluvial dissection of the zone. The geochronology section (Section 4.3) is focused on the theoretical determination of the ages for the fluvial terraces of the river Yeltes by the application of the height-age transfer equations listed in Table 2. Finally, Section 4.4 presents the geomorphological analysis of all the information focused on the geomorphological evolution of the mapped area and its relationships with other nearby river basins around the Spanish Central System.

4.1. Geomorphology

The Yeltes River is located on the border between the northern plateau (Meseta norte) and the beginning of the mountainous relief of the Central System, linking the relief of the Sierra de Francia range with the Paleogene and Neogene filling of the Ciudad Rodrigo Basin. This results in a transition from a mountainous relief in the south to an almost flat area in the north, passing through an interesting piedmont made up of Neogene alluvial fan deposits. The uplift of the Central System and the subsequent incision of the drainage triggered the fluvial dissection of the piedmont, generating an important set of landforms and deposits of mainly fluvial origin (Figure 3 and Table 2) [23,25].

The main geomorphological and structural elements within the southern mountain ranges are the faults, crests, and summits. These last elements usually constitute the water-divides within the range and usually follow the outcrops of the resistant Armorican Quarzite. The orientation of valleys, drainage, and summits is strongly conditioned by the folded structure of the quartzites and by the fracture systems originated during the Variscan. In other words, the subjacent Variscan structure strongly controls the present landscape within the mountain range area.

Within this southern mountain area are also typical geomorphological elements generated by periglacial processes during the Late Pleistocene. Periglacial landforms are restricted to several colluvial deposits of the talus-scree type, called "canchal" in Spanish. These occur in the most elevated areas, linked to the steep slopes flanking the crests of Armorican Quartzite within the Sierra de Francia and Sierra de Valdefuentes ranges (Figure 3). As in other zones of the Spanish Central System, cryoclastic activity and gelifraction were important processes during the Last Glacial Cycle, generating important talus-scree formations [30].

The polygenic landforms correspond to those generated by the participation of different erosive and/or depositional morphogenetic processes (e.g., fluvial-aeolian; fluvial-alluvial, etc.) or those for which the main morphogenetic process is doubtful. Within this category of landforms, erosive pediments, sedimentary piedmonts, and complex large escarpments occur in the Yeltes River (Table 1). The pediment is located at the foot of the southern ranges constituting a narrow fringe around the reliefs. It constitutes a relic rocky surface that has been exhumed and is free of sediments. In the study area, it corresponds to a pre-paleogene etchplain-type surface constituting the ancient substrate of old and thick weathering profiles [19]. These weathered materials were removed by differential erosion during the process of denudation and exhumation of the zone before the deposit of the extensive piedmont of the "Raña surface" [4,5].

The "Raña" is the main polygenic landform featuring the piedmont areas around Central Spain and have been subject of numerous studies [2–6]. These are extensive sedimentary gravelly formations placed in unconformity on top of the old Neogene alluvial sediments within the Ciudad Rodrigo Basin [19]. The "rañas" constitute broad alluvial piedmonts composed of quartzite gravels pasted by a reddish clayey matrix coming from the torrential erosion of the weathered metamorphic materials (quartzites, schists, and shales) within the southern mountain areas. These formations do not display evident stratigraphic structures (mainly massive) but support well developed fersialithic red-soils testifying old near-tropical climates dominating Central Spain during the end of the Neogene [2–4]. All around the Spanish Central System, the raña formations mark the transit between the old Neogene endorheic conditions of the Duero and Tagus basins and their Quaternary fluvial dissection, that is, the onset of the present fluvial network [4]. They have been traditionally considered to have a generic Plio-Quaternary age before the incision of the present the fluvial valleys [2–4]. In the mapped area, the "raña" formation is found north and south of the Yeltes river-valley [23], which axially dissect this ancient piedmont (Figure 3). On the northern bank of the valley, the Yeltes river carved a prominent scarp on the raña deposits about 12 km length and 60–50 m high (Figure 3), being one of the more prominent polygenic landforms of the zone.

A last polygenic element in the area is the set of gentle slope accumulation glacis zones connecting the raña surface with the terrace system of the Yeltes river (Figure 3). These sedimentary features are constituted by sub-angular quartzite clasts embedded in a reddish clayey matrix, clearly derived from the erosion the raña deposits with no clear sedimentary structures [23]. The detailed mapping performed in this work allowed us to recognize that the surface of this accumulation glacis is gently stepped, similarly to the raña surfaces in Tagus basin [2]. Consequently, three stepped surfaces have been newly differentiated (R1, R2, and R3; Figure 3) connecting the Plio-Pleistocene piedmont (Raña surface) with the Quaternary terrace system. The "raña surface" is at +134–138 m above the river thalweg, and the stepped surfaces at +118–122 m (R1), +100–105 m (R2), and +89–92 m (R3) (Figure 4).

These deposits would be considered as the product of the early dissection episodes of the raña surface, with a similar meaning that the "rañizo terraces" dissecting the raña formations in the Tagus basin [22], that is, fanhead trench terraces indicating the early stage of dissection of the raña surface soon before the "Atlantic Capture" of the zone.

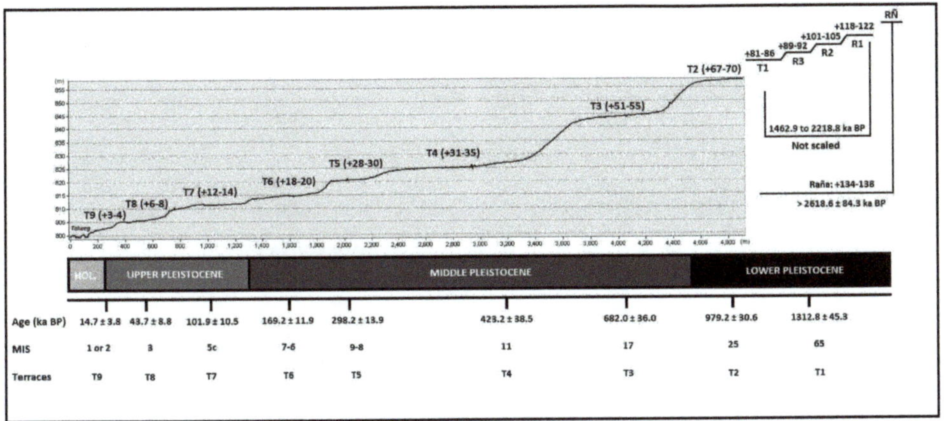

Age (ka BP)	14.7 ± 3.8	43.7 ± 8.8	101.9 ± 10.5	169.2 ± 11.9	298.2 ± 13.9	423.2 ± 38.5	682.0 ± 36.0	979.2 ± 30.6	1312.8 ± 45.3
MIS	1 or 2	3	5c	7-6	9-8	11	17	25	65
Terraces	T9	T8	T7	T6	T5	T4	T3	T2	T1

Figure 4. Topographic Profile of Yeltes river terraces between T2 and T9. T1 and stepped piedmont surfaces (R1 to R3) are indicated out of the profile. The horizontal time-bar is not scaled and theoretical terrace ages (kyr = 1000 years) and corresponding warm isotopic stages (MIS) are shown (See explanation in Section 4.3).

Fluvial landforms and deposits are the main geomorphological elements in the zone, mainly featured by the terrace system of the Yeltes river, constituted by nine staircased levels developed between +82–86 m (T1) and +3–4 m (T9) above the present river thalweg (Figure 3). As aforementioned, this terrace sequence is inset on the stepped piedmont of the raña surface and rañizo terraces between +134 and +89 m (Figure 4). The terrace levels T1 to T5 are terraces with a thickness of 2–3 m, whilst the younger terraces (T6–T9) are cut and fill terraces but with no relevant thickness (<5 m). Terrace deposits are specially developed in the southern slope of the valley and are constituted by sub-rounded quartzite and quartz clasts with a sandy matrix (Figure 5). This features the Yeltes valley as an asymmetric river valley with large terrace surfaces on the left bank, while the present flood plain and river-bed are located in the northern bank at the toe of the previously mentioned large escarpment carved in the raña deposits. Erosion is still an active process operating in this large escarpment, leading the development of gullies, slope deposits (colluvial slopes and small talus cones), and alluvial fans along their toe (Figure 3). The main alluvial fan occurs in the NW corner of the mapped zone (1 km long and 1.5 km wide), but is subject to anthropic modification by mining operations for the extraction of aggregates [23]. The largest floodplain in the Yeltes river develops a clear braided system with multiple channels (abandoned and active) and many sand and gravel bars in between. This floodplain can reach near 1.5 km wide in the NW corner of the mapped area (Sancti-Spiritus plain). The main tributaries of the Yeltes river, like the Morasverdes and Tenebron rivers, also display well-developed braided systems (Figure 3). These braided systems present a torrential activity, the flash floods within the floodplain being a common hazardous process [32]. The last important flooding occurred just this same year on 31st May 2020, affecting to the locality of Martín de Yeltes downstream the mapped area. These main rivers and most of the secondary drainage network seems to be adapted to NW-SE and N-S fault systems. These fractures clearly affect Neogene and Paleogene materials, but no fault offset has been observed in the more modern raña and fluvial deposits.

Figure 5. Stratigraphic profiles of the Yeltes river terraces. (**A**) Interbedded silt-sandy and gravel levels (T3). (**B**) Detail showing the poor classification and sub-angular nature of quartzite clasts.

The endorheic lacustrine elements are represented by two small lakes (ponds) on the T3 terrace surfaces near the village of Aldehuela de Yeltes. The Largest pond occurs just above the escarpment of the Yeltes river on the raña flat surface (El Cristo Lake). The development of small lakes or ponds in the raña surface is a common geomorphological element in the zone. These ponds have an eolian origin, since they are located in small depressions originated by the wind action in zones subject to strong deflation during the Late Pleistocene. Therefore, these endorheic elements can be labelled as hydro-eolian landforms [23,27] originated under arid to semi-arid climate with dominant NE-SW wind directions and favored by structural elements (faults, fractures) with similar orientations [33]. These climatic conditions were characteristic during the end of the last glacial period in the whole northern piedmont of the Spanish Central System, where dune fields and deflation actions occurred in the absence of water (retained in the glacier systems of the ranges) [34].

The gravitational morphogenetic system is mainly featured by talus slope deposits located in the southern mountain area of the mapped zone, previously featured as periglacial talus scree. Within the basin, the small talus cones developed along the toe of the mentioned Yeltes escarpment are also characteristic gravitational landforms (Figure 3). In this escarpment, minor earth-slides and creeping, out of the resolution of the map scale, also occur, evidencing that gravitational processes are still active along this landform, controlling its evolution. Within the terrace system, small wash-slope debris-slope deposits develop at the toe of the terrace scarps, especially in the older terraces up to +35 m (>T4). They are also included in this morphogenetic system (Figure 3).

The anthropogenic elements are summarized in Table 2. It is only necessary to highlight the morphological modifications of the ground surface generated by the intense quarrying of gravels and sand (aggregates) in the floodplain and lower terrace levels (Figure 3).

4.2. Fluvial Terrace Sequence of the Yeltes River

Fluvial terraces represent the ancient floodplains of a river-valley presently hanged several meters above the river thalweg because of pulses of river incision and sedimentation during the Quaternary [35]. Climatic sea-level changes, tectonics, or uplift cause relative drops of the river base-level, allowing the downcutting of river channels in their active floodplains, leaving these abandoned and hanging above the new incised river bed, constituting successive terrace levels [36]. The alternation of aggradation and incision processes occurred during warm and cold stages along the Quaternary produced terrace sequences in fluvial valleys, providing a record of river flow changes, sediment supply fluctuations, and base-level variations over the period of valley development [35]. A recent analysis in Central Spain indicates that terrace sedimentation mainly occurs during the transit of cold to warm Oxygen Isotopic Stages (OIS) [37]. Floodplains stabilized during warm OIS, where soil development starts and form the end of warm OIS (odd-numbered OISs) and the complete subsequent cold OIS (even-numbered OISs) river downcutting dominates in response to glacier sea-level drops [37,38]. Consequently, the set of terraces in a fluvial valley represent a chronological sequence of incision, stabilization, soil formation, and sedimentation within the river basin, allowing the study the geomorphological and climatic evolution of the area [38,39].

Within the mapped area, the terrace sequence of the Yeltes river displays a geometry of asymmetrical fan-splay incised in the raña surface giving place to a kind of asymmetrical valley [39]. The entire terrace system only develops on the southern bank of the river, creating a very large flat but gently stepped area of about 75 km^2 (Figure 3). In this sector, the terraces appear as cartographically narrow and elongated bands from WNW (T1) to N-S (T9) orientation, displaying a gentle slope towards NNW [27]. This particular "fan-splay" geometry of the terrace system indicates a progressive displacement towards the NNE of the river channel during the Quaternary downcutting of the valley. This fact, together with the N-S asymmetrical filling of the basin [39], strongly suggests continuous uplift of the Spanish Central System and correlative northwards tilting of the Plio-Pleistocene piedmont (raña surface) during the fluvial dissection of the zone.

Usually, it is difficult to find the scarps among the different terraces, since many times the topographic steps have a small height (1–2 m) and are normally very degraded. In many times, terrace scarps are nearly buried by the occurrence of wash-slope and debris-slope accumulations adjacent to the scarps. In basis to the texture and grain size of detritic materials incorporated in the fluvial terraces, it is possible to assess that the Yeltes river has sustained a braided fluvial system from its initial stages of development. The stratigraphy of the upper terraces displays numerous interbedded levels of floodplain facies (sands and silts) and gravel bars (quartzite clasts) (Figure 5A). In other outcrops, these braided gravels were originated by a large amount of matrix-less quartzite clasts, within river channels with insufficient capacity of transport, as presently occurs in the present braidplain of the river. This is one of the main reasons for the poor classification and sub-rounded nature of the gravels (Figure 5B) of almost the whole terrace sequence.

As aforementioned, the terrace system of the Yeltes river presents nine terrace levels (T1–T9). These have been discriminated by means of GIS and photointerpretation using the 1 m/pixel DTM created for this study [23]. Terrace 1 (T1) is the oldest fluvial level at +82–86 m above the river thalweg and T9 is the most modern one at +3–4 m, just above the present floodplain (braidplain) of the river, which is the younger fluvial landform of the mapped area (Figure 3). As already indicated, the glacis-deposits morphologically connecting the raña surface with the terrace sequence was identified as a unique polygenic unit in previous studies [23]. Detailed mapping allowed to differentiate a stepped piedmont with three distinct surfaces (R1, R2, and R3) inset within the raña surface. These pre-fluvial surfaces developed between +89 to +122 m inset in the raña surface which stands at a mean relative altitude of +134–138 m above the present thalweg of the Yeltes river. These relative elevation values fit well with the relative elevation of the highest fluvial terrace of the Duero river (+144 m), indicating the Atlantic Capture of the Duero Basin [22,25]. However, more recent studies place the final development of the raña until the Gelasian Period previous to the Olduvai normal paleomagnetic subchron prior to 2.0–1.9 Myr [22]. From this period on started the true fluvial dissection (i.e., the Atlantic Capture) of the Duero and Tagus basins [22]. More recent studies based on OSL and Cosmogenic dating profiles (e.g., ^{10}Be–^{26}Al) suggest that the endhoreic-exorheic transition in the Tagus and Duero basins occurred between 2.42 and 2.36 Myr [25,26].

The relative heights of the different terrace levels and raña surfaces were measured in a perpendicular NW-SE section from the locality of El Maillo to near the village of Alba de Yeltes (Figure 3). Despite T9 (+3–4 m) being elevated above the present floodplain, it is presently subject to active flooding during episodic strong storms in the zone [32]. The partial terrace profile (T2–T9) obtained by means GIS tools on the 1 m/pixel MDT is displayed in Figure 4.

4.3. Geochonology

This section is focused of the geochronological analysis of the terrace sequence based on the chronofunctions proposed for Central Spain [22] listed in Table 2. As mentioned in the methodological Section 3.2, the proposed equations are based on multiple correlations between the relative altimetry of fluvial terraces in the Tagus and Duero basins and the numerical ages obtained by different dating methods. Consequently, this supposes a statistical approach to assess the theoretical numerical ages of the terrace sequence for the Yeltes river-valley. Additionally, the same chronofunctions have been applied to the early incision surfaces (R1 to R3) inset in the raña and to the raña surface itself (Table 3). The obtained values for this last alluvial piedmont surface indicate the age in between its eventual deposition and its early dissection. As indicated for the Tagus basin, the early dissection of the raña surface are not properly linked to the Atlantic Capture of the zone, but to intrinsic controls common to the late evolution of alluvial fan systems (i.e., proximal to distal trenching) [22]. The Atlantic Capture occurred after an important degree of intrabasinal dissection of the old Neogene basins of Central Spain, which favored intensive soil development (raña red soils), as well as their eventual endorheic-exoreic transition [2]. Modern cosmogenic and relative dating of raña and rañizo surfaces [1,25,26] indicate that this transition was not a synchronous process in Central Spain as traditionally considered [4].

Table 3. Age-values for fluvial terraces and stepped piedmont (raña and rañizo surfaces) obtained from different chronofunctions [a, b, c, d] for fluvial terraces in Central Spain [22] listed in Table 2.

Terraces	Relative Height (m)	[a] 2nd Order Polynomial Duero (ky)	[b] Power Function Arlanzón Valley (ky)	[c] 3rd Order Polynomial Duero + Tagus (ky)	[d] 2nd Order Polynomial Duero + Tagus (ky)
T9	+3–4	13.04	16.45	−3.39	−9.35
T8	+6–8	44.92	42.53	19.09	22.01
T7	+12–14	105.02	98.95	67.37	80.51
T6	+18–20	172.18	166.28	127.18	145.13
T5	+28–30	299.79	296.66	248.94	266.43
T4	+31–35	424.18	422.24	370.74	383.35
T3	+51–55	686.32	677.73	627.17	627.17
T2	+67–70	995.59	962.92	891.19	911.87
T1	+82–86	1352.29	1273.41	1157.99	1237.71
R3	+89–92	1515.64	1410.20	1263.71	1386.30
R2	+100–105	1839.69	1672.63	1440.97	1680.10
R1	+118–122	2362.05	2075.64	1641.40	2151.75
Raña	+134–138	2822.95	2414.22	2822.95	2822.95

The analysis of the values obtained by means of the four chronofunctions proposed for Central Spain [22] offers very similar values to each other (Table 3). However, the data resulting from the 3rd Order [c] and 2nd Order polynomial [d] functions for the complete data set (Duera + Tagus basins) offer slightly lower values for the whole data set and even negative values for the lowest terrace (T9). This is due to the inclusion of palaeomagnetic data in these functions, which introduce great uncertainties in the computed values [22]. On the contrary, the values coming for 2nd Order polynomial function for the Duero basin [a] and the power function for the Arlanzón valley [b] (Table 3) offer more congruent and similar values and have been selected for this study. Their results are best suited to the geographical area of the Duero Basin [22]. To homogenize the theoretical chronologies of the terraces, we present the average ages resulting from the age values obtained by the two equations [a, b]. Table 4 displays the obtained average ages (kyr) but also the age uncertainties (± kyr) resulting to calculate the age of the maximum and minimum relative height for each terrace level; this is the difference between maximum and minimum calculated age "per case." The same methodology was applied to calculate average ages and errors for the stepped piedmont surfaces of the Raña and the Raña surface itself (Table 4). In this way, it is important to note the mean age obtained for the Raña surface (2.62 ± 0.84 Myr) will correspond to the youngest possible age for their deposition, since it is a pre-incision surface not truly related to the theoretical fluvial downcutting described by the equations.

Table 4. Average ages and errors (± ky) of the fluvial terraces and raña surfaces (stepped piedmont) for the Yeltes river valley. Numerical ages and errors obtained was obtained from mean values of the best-fit polynomial [a] and power [b] equations listed in Table 3.

Terraces	Relative Height (m)	Numerical Age (ky)	Error (± ky)	Epoch/Subepoch	MIS
T9	+3–4	14.75	3.8	Holocene-Upper Pleist.	2–1
T8	+6–8	43.72	8.8	Upper Pleistocene	3
T7	+12–14	101.98	10.5	Upper Pleistocene	5
T6	+18–20	169.23	11.9	Chibanian (Mid. Pleist)	7–6
T5	+28–30	298.22	13.9	Chibanian (Mid. Pleist)	9–8
T4	+31–35	423.21	38.5	Chibanian (Mid. Pleist)	11
T3	+51–55	682.02	36.0	Chibanian (Mid. Pleist)	17
T2	+67–70	979.25	30.6	Calabrian (Low. Pleist)	25
T1	+82–86	1312.85	45.3	Calabrian (Low. Pleist)	43
R3	+89–92	1462.92	35.4	Calabrian (Low. Pleist)	51
R2	+100–105	1756.16	63.1	Gelasian (Low. Pleist)	65
R1	+118–122	2218.84	55.3	Gelasian (Low. Pleist)	89
Raña	+134–138	2618.58	84.3	Late Pliocene	> 103

The obtained theoretical terrace ages are similar to those recently proposed for comparable tributary valleys of the Spanish Upper Duero Basin (i.e., Tormes valley) [24] and more recently for the terrace sequence of the Duero river between +117 m and +13 m [25]. However, the obtained values are slightly older than OSL-ESR ages obtained for terrace levels down to +30 m in the Lower Duero Basin near Barca d'Alva in Portugal [1]. As noted by these authors [25] the apparent difference of ages between the upper (Spain) and the lower (Portugal) sectors of the Duero basin can probably be linked to the important step of the longitudinal profile of the river introduced by "Los Arribes Canyon" (c. 200 m) between these two zones. All the tributary basins in Spain are located upstream of "Los Arribes Step," whilst the still scarce set of age determinations in Portugal are downstream it [1]. Similar age discrepancies occur with the Portuguese and Spanish sectors of the Tagus river basin, where a more robust set of numerical age determinations are available [22]. The obtained ages are listed in Table 4 and illustrated in the topographic section of Figure 4. As can be seen, all the fluvial sequence (stepped piedmont and terrace system) entirely develop during the Quaternary. The rañizo terraces (R1 to R3) developed during the Lower Pleistocene and the Lower-Middle Pleistocene transit occurs between the deposit of the T2 and T3 terraces (Table 4). The numerical approach suggests that an effective climatic sensitivity in terrace development occurs from the isotopic stage MIS 11 (T4) and specially from the MIS 9 (T5). In this sense, the obtained ages (Table 4) indicate in most of the cases terrace sedimentation mainly occurred during warm isotopic stages (odd MIS), and in many cases close or during the transit between cold to warm isotopic stages (i.e., glacier terminations). As indicated by other studies in the Iberian Peninsula [22,25,37], river downcutting occurs during cold stages (even MIS) in response to long-lasting sea-level drops, but sedimentation preferably took place during the transit between glacier-interglacial periods when the melting of the mountain glaciers promotes important sedimentation rates. The subsequent step progresses during the second half of warm stages (odd MIS), when terrace surfaces stabilized being dominant soil formation processes [37], before fluvial downcutting re-start again.

4.4. Discussion: Quaternary Landscape Evolution of Yeltes River

The onset of the present landscape evolution in the zone was promoted by the alpine tectonics, which disrupted the old peneplain imprinted on the Palaeozoic materials of the Iberian Massif. Alpine tectonics promoted the uplift of marginal reliefs around Spanish Central System (e.g., Sierra de Francia) and generated new sedimentary basins (e.g., Ciudad Rodrigo Basin) from the end of the Paleogene period. In many cases, basin formation was controlled by the reactivation of the old variscan faults cutting the Paleozoic massif. Cenozoic sedimentation was defined by the so-called arkosic (Paleogene) and post-arkosic (Neogene) cycles defined for this area, constituted by materials derived from the erosion of granitic materials of the southern reliefs [18]. These two cycles record a transit from fluvial (arkosic) to alluvial fan (post-arcosic) sedimentation indicating accelerated uplift of the Central System ranges and a replacement of source areas from granitic to metamorphic materials [18]. During the Neogene cycle, the superposition of large alluvial fans coming from the southern ranges generated a large piedmont system along the southern zone of the Ciudad Rodrigo Basin. Neogene sedimentation culminates with the deposit of the "raña" materials giving place to an extensive alluvial piedmont, which is a characteristic Plio-Quaternary landform around the Spanish Central System [2–4].

As traditionally considered, the deposit of the "raña" occurred as a time-transgressive event coupled to the endorheic-exoreic transition of the old Neogene basins in central Spain [4]. The Raña deposit constitutes the last sedimentary episode in endorheic conditions before the incision of the present fluvial system and the eventual "Atlantic Capture" of the Neogene basins [1,22]. This process occurred during and elapsed time of several thousands of years from the Late Pliocene to the Lower Pleistocene, but mainly during the recently recognized early stage of the Quaternary (i.e., Gelasian: c. 2.5–2.0 Myr ago), previously assigned to the Late Pliocene [22,40]. The "raña" deposit is formed by quartzite cobbles and pebbles embedded in a reddish clayey matrix (normally 2.5 YR in Munssel code [2]) coming from the reworking of Neogene weathering profiles in the Paleozoic metamorphic rocks within the ranges. They were deposited under a relatively humid climate, as testified by the red soils developed on top [1,2,22], but subject to torrential rainfall events favoring the mobilization of weathered materials downslope [3,27]. The increased intra-plate compression, the overall westward tilting of the Iberian Peninsula, and the important climate changes preceding the beginning of the Quaternary around the African-Eurasian plate boundary facilitated the so-called "Atlantic Capture" [1,6,7]. The high topography of the Iberian Massif peneplain in the western sector of the Iberian Peninsula allowed the isolation from the Atlantic base-level of the old sedimentary basin during the Neogene [41]. However, the climate deterioration (cooling) and sea-level fall occurred from the onset of the Quaternary produced a relevant base-level drop triggering the fluvial capture of the ancient sedimentary basins in central Spain [1,25,40]. Recent geomorphological approaches to this process in the Duero Basin points that the endorheic to exorheic transition in western Iberia started from about 3.7 Myr ago, testified by the development of a series of erosion surfaces inset in the Iberian massif peneplain [1]. However, fluvial dissection in the upper Tagus and Duero Late Neogene basins only occurs from c. 2.5–2.6 Myr [22,24,25].

The detailed geomorphological analysis carried out in this work allowed to characterize the raña surface in the studied area as a stepped piedmont, as already noticed in the southern slope of this Spanish Central System [2]. In this way, R1, R2, and R3 surfaces have been mapped inset in the raña piedmont in relation to its early incision stages (Figure 3). These stepped surfaces develop between +122 m and +89 m above the present river thalwegs and can be comparable to the upper rañizo terraces described in the Tagus basin [22,25,40]. In other words, large fanahead trench-like terraces linked to the intrinsic dissection of alluvial systems before the onset of proper fluvial downcutting by headward erosion linked to the "Atlantic Capture" [40].

The application of the fluvial chronofunctions (Table 4) to the stepped piedmont surfaces of the raña (+132 to +89 m) identified in this study indicates that the early dissection of the area occurred from c. 2.6 to 1.4 Myr (Figure 5). From this time starts the true fluvial incision and fluvial terrace development (T1, T2, T3) preserved upstream of Los Arribes Canyon (c. 1.3–0.68 Myr; Table 4). Climatic sensitivity to terrace development is clear from the isotopic stage MIS 11 (T4), from which fluvial sedimentation are clearly linked to the early phases of warm isotopic stages or to the transit of cold to warm stages (Table 4). This fact can be related to drainage connectivity problems between the lower and upper sectors of the Duero drainage basin due to the existence of important internal steps of the base-level. These internal steps would prevent the upstream propagation of headward erosion. In the study area, these steps are nicely represented by "Los Arribes Canyon" in the axial fluvial valley, but also by the series of gorges developed in all the tributaries in the eastern slope of the Canyon [36], as is the case of the Huebra river along which the Yeltes flows to the Duero (Figure 1). This hypothesis indicates that base-level induced headward erosion was limited in the Duero Basin before the development of the T4 (+31–35 m; Figure 5). Fluvial downcutting linked to the youngest terraces (<MIS 11) appears climatically controlled as internal thresholds theoretically disappeared. This hypothesis should be properly analyzed in further studies when the set of available numerical ages for the upper (Spain) and lower (Portugal) sectors of the Duero basin will be more robust than today.

Figure 6 illustrates terrace development of the Yeltes river-valley in relation to the 41 ka and 100 ka worlds defined by the Milankovitch orbital cycles and the so-called "Middle Pleistocene Transition" (MPT) [22,42]. The main curve describes terrace development for central Spain as described by the 3rd Order polynomial curve corresponding to the equation [c] listed in Table 2. In our case, is clear that the oldest rañizo surface (R1) and the raña surface itself show an important deviation from this general chronofuction curve (Figure 6). Since the calculation of these chronofunctions [22] only considered the rañizo terraces described in the Tagus basin, the observed deviation will indicate a younger evolution of the Duero basin in relation to the Tagus one, and therefore a more recent "Atlantic Capture" for the Duero as already suggested by previous authors [22,41]. On the other hand, it seems clear that the climatic control of fluvial terrace development occurs after the MPT, during the 100 ka world (Figure 6). T3 roughly coincides with the end of the MPT and the subsequent development of T4. T5 to T9 is clearly linked to climatic pulses (Figure 6), indicating that from T3 (c. 0.68 Myr) the Atlantic base-level changes could freely propagate upstream Los Arribes and the Huebra canyons towards the studied area. The extreme glaciations and base-level drops (c. ±120 m) depicting the 100 ka world [42] could largely help to eliminate probable internal thresholds working in the past.

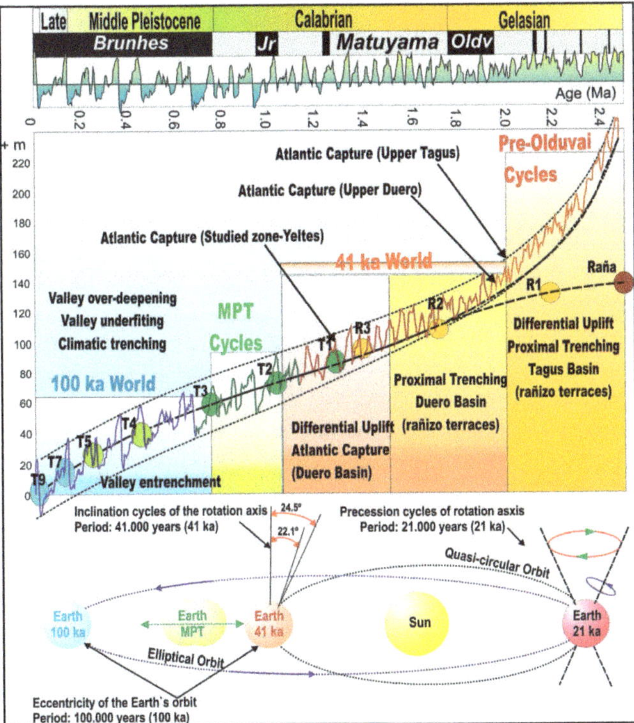

Figure 6. Theoretical behavior of the river under conditions of dynamic equilibrium in the readjustment of the classical MIS curve to the 3rd order polynomial function describing terrace development in Central Spain [c] in relation to the Milankovicht cycles. The circles mark the terrace age-data obtained for the Yeltes River. Modified and adapted from [22].

After terrace development, the subsequent significant impact on the landscape of the zone occurred during the last glacial cycle. The development of the ephemeral lakes of el Cristo and La Cervera on the highest terrace levels and the widespread development of "canchales" (talus-scree) in the southern mountain zone (Sierra de Francia) are related to periglacial conditions from the end of the Pleistocene to the present. These formations of matrix-less angular rock debris generate an important water reservoir within the mountain area, forming a quaternary type aquifer with numerous springs [25].

5. Conclusions

The geomorphological evolution of the studied zone began with the Alpine orogeny that triggered the uplift of the Spanish Central System and the generation of piedmont basins. One of these basins is the Ciudad Rodrigo basin filled by Paleogene and Neogene detritic materials of fluvial and alluvial origin. The final filling of the basin is featured by the alluvial deposit of the raña as occurs in most of the old Neogene basins in central Spain sourced by quartzite-schist metamorphic reliefs [2,3]. The performed geomorphological analysis has been largely supported and improved by the implementation of LIDAR data in high-resolution digital elevation models (1 m/pixel DTMs) allowing a detailed photointerpretation and facilitating fieldwork. The elaboration of a detailed geomorphological map (Figure 3) allowed the establishment of the spatial and temporal distribution of the different geomorphological units and systems operating in the area. Seven morphogenetic systems have been identified (Table 2): Structural (tectonic structure), Fluvial, Alluvial, Gravitational (slope processes), Periglacial, Polygenetic, and Anthropogenic (human activity). All these systems have been classified in the map legend differentiating among erosive and depositional landforms (Figure 3). The detailed geomorphological analysis developed in this work allowed to differentiate nine terrace Levels (T1 and T9), between +82–86 m and +3–4 m above the present river thalweg, but also three higher stepped piedmont surfaces (R1, R2, R3) between +89 and +122 m inset in the piedmont of the raña surface (+134–138 m). For the first time a stepped raña piedmont in the northern slope of the Spanish Central System is described, which make possible for the comparison of these newly reported surfaces with the rañizo terraces described in the Tagus basin [2,22,26].

Chronological analysis has been implemented by means of the application of fluvial chronofunctions developed for central Spain (Table 1). The obtained theoretical ages indicate that the entire fluvial and alluvial terraces and surfaces inset in the raña piedmont entirely developed during the Quaternary (≤2.2 Myr). The raña deposit results older than c. 2.6 Myr with a Late Pliocene age and its early dissection (R1–R3) during the Gelasian (Table 4), being these ages comparable to those obtained by cosmogenic dating for similar surfaces in central Spain (2.42–2.36 Myr) [26]. The Atlantic capture of the Ciudad Rodrigo (T1) basin occurs about 1.4–1.3 Myr. during the end stages of the Calabrian, but complete drainage connectivity does not seem to happen until the development of terrace T4 during the isotopic stage MIS 11 (c. 0.42 M.a). It seems that base-level changes did not freely propagate upstream in the Duero basin for a period of about one million years, by the occurrence of internal thresholds (steps) in the river profile along the "Los Arribes Canyon" and the Huebra river gorges. These constitute transverse drainage systems [25,41] connecting the lower (Portugal) and upper (Spain) sectors of the Duero drainage basins, inducing a step of c. 200 m in the river profiles. Fluvial terrace development seems clearly climatically controlled from MIS 11 when terrace deposits mainly occur during the initial phases of warm isotopic stages or closely related to cold-warm stage transitions. The large sea-level drop linked to the extreme glaciations of the 100 ka world (Figure 6) seems to facilitate the removal of the probable internal thresholds existing within the Duero basin during the Middle and Lower Pleistocene. The obtained chronological scenario for the Yeltes river-valley provides an estimate for the timing of terrace formation due to river incision that seem to be in agreement with recent proposals based on cosmogenic dating for the Duero river upstream Los Arribes Canyon [25], especially for terraces down to +40–35 m. The chronology of the other landforms (Table 1) existing in the zone was assigned in function to their position in relation to the terrace chronology.

The landscape of the area links polygenic piedmont fossil landforms prior to 2.5 M.a., dissected by much younger fluvial landforms (terraces) of Quaternary age. These last dominate the geomorphological evolution of the area during the last 1.4 M.a., but especially since the isotopic stage MIS 11 (C. 420 ka).

Author Contributions: Conceptualization, I.M.-M., P.-G.S., and A.M.-G.; methodology, I.M.-M., P.-G.S., and A.M.-G.; software, I.M.-M., A.M.-G., and J.E.; validation, I.M.-M., P.-G.S., and A.M.-G.; formal analysis, I.M.-M., P.-G.S., and A.M.-G.; investigation, I.M.-M., P.-G.S., and A.M.-G.; resources, I.M.-M., A.M.-G., and J.E.; data curation, I.M.-M., A.M.-G., and J.E.; writing—original draft preparation, I.M.-M., P.-G.S., and A.M.-G.; writing—review and editing, I.M.-M., P.-G.S., A.M.-G., and J.E.; visualization, I.M.-M., A.M.-G., and J.E.; supervision, P.-G.S. and A.M.-G.; project administration, P.-G.S. and A.M.-G.; funding acquisition, A.M.-G. All authors have read and agreed to the published version of the manuscript.

Funding: This research received no external funding.

Acknowledgments: This research was methodologically helped by Junta Castilla y León Project SA044G18 and supported by the MINECO-FEDER Project CGL2015–67169–P (USAL).

Conflicts of Interest: The authors declare no conflict of interest.

References

1. Cunha, P.P.; Martins, A.A.; Gomes, A.; Stokes, M.; Cabral, J.; Lopes, F.C.; Pereira, D.; De Vicente, G.; Buylaert, J.P.; Murray, A.S.; et al. Mechanisms and age estimates of continental-scale endorheic to exorheic drainage transition: Douro River, Western Iberia. *Glob. Planet. Chang.* **2019**, *181*, 102985. [CrossRef]

2. Pérez-González, A.; Gallardo, J. La Raña al sur de la Somosierra y Sierra de Ayllón; un piedemonte escalonado del Villafranquiente medio. *Geogaceta* **1987**, *2*, 29–32.

3. Martín Serrano, A. Sobre la positión de la raña en el contexto morfodinámico de la Meseta. Planteamientos Antiguos y Tendencias Actuales. *Bol. Geol. Min.* **1988**, *99*, 855–870.

4. Martín Serrano, A. La definición y el encajamiento de la red fluvial actual sobre el Macizo Hespérico en el marco de su geodinámica alpina. *Rev. Soc. Geol. Esp.* **1991**, *4*, 337–351.

5. Martín Serrano, A. El relieve del Macizo Hespérico; genesis y cronologia de los principales elementos morfologicos. *Cuader. Lab. Xeol. Laxe* **1994**, *19*, 37–55.

6. Gutierrez Elorza, M.; Garcia Ruiz, J.M.; Goy, J.L.; Gracia Prieto, F.J.; Gutierrez Santolalla, F.; Marti, C.; Martin Serrano, A.; Pérez-González, A.; Zazo, C.; Aguirre, E. Quaternary. In *The Geology of Spain*; Gibbons, W., Moreno, T., Eds.; Geological Society: Bath, UK, 2002; pp. 335–366.

7. Pereira, D.I.; Pereira, P.; Brilha, J.; Cunha, P.P. The Iberian Massif Landscape and Fluvial Network in Portugal: A geoheritage inventory based on the scientific value. *Proc. Geol. Assoc.* **2015**, *126*, 252–265. [CrossRef]

8. Bridgland, D.; Westaway, R. Climatically controlled river terrace staircases: A worldwide Quaternary phenomenon. *Geomorphology* **2008**, *98*, 285–315. [CrossRef]

9. Westaway, R.; Bridgland, D.R.; Sinha, R.; Demir, T. Fluvial sequences as evidence for landscape and climatic evolution in the Late Cenozoic: A synthesis of data from IGCP 518. *Glob. Planet. Chang.* **2009**, *68*, 237–253. [CrossRef]

10. Macklin, M.G.; Lewin, J.; Woodward, J.C. The fluvial record of climate change. *Philos. Trans. R. Soc.* **2012**, *A370*, 2143–2172. [CrossRef]

11. Bridgland, D.R. River terrace systems in north-west Europe: An archive of environmental change, uplift and early human occupation. *Quat. Sci. Rev.* **2000**, *19*, 1293–1303. [CrossRef]

12. Chorley, R.; Schumm, S.A.; Sudgen, D.E. *Geomorphology. Earth Sciences and the Past*; Methuen: London, UK, 1984; p. 605.

13. Stokes, M.; Cunha, P.P.; Martins, A.A. Techniques for analysing Late Cenozoic river terrace sequences. *Geomorphology* **2012**, *165–166*, 1–6. [CrossRef]

14. Julivert, M.; Fonboté, J.M.; Ribeiro, A.; Conde, L. *Mapa Tectónico de la Península Ibérica y Baleares. Escala 1:1.000.000*; Instituto Geológico y Minero de España (IGME): Madrid, Spain, 1974; p. 113.

15. Ugidos, J.M.; Rodríguez Alonso, M.D.; Albert Colomert, V.; Martín Herrero, D. *Cartografía y Memoria de la Hoja n° 552 (Miranda del Castañar). Escala 1:50.000*; Plan MAGNA; ITGE: Madrid, Spain, 1990; p. 81.

16. Ugidos, J.M.; Rodríguez Alonso, M.D.; Martín Herrero, D.; Bascones Alvira, L. *Cartografía y Memoria de la Hoja n° 575 (Hervás). Escala 1:50.000*; Plan MAGNA; ITGE: Madrid, Spain, 1988; p. 86.

17. Bascones Alvira, L.; Rodriguez Alonso, M.D. *Cartografía y memoria de la Hoja n° 526 (Serradilla del Arroyo), Escala 1:50.000*; Plan MAGNA; ITGE: Madrid, Spain, 1990; p. 72.

18. Cantano, M. Evolución Morfodinámica del Sector Suroccidental de la Cuenca de Ciudad Rodrigo, Salamanca. Ph.D. Thesis, Universidad de Huelva, Huelva, Spain, 1997; p. 277.

19. Martín-Serrano García, A.; Monteserín López, V.; Mediavilla López, R.; Rubio Pascual, F.; Santiesteban Navarro, J.I. *Cartografía y Memoria de la Hoja n° 501 (La Fuente de San Esteban). Escala 1:50.000*; Plan MAGNA; ITGE: Madrid, Spain, 1991; p. 76.

20. Martín Herrero, D.; Ugidos Meana, J.M.; Nozal Martín, F.; Pardo Alonso, M.V. *Cartografía y Memoria de la Hoja n° 527 (Tamames). Escala 1:50.000*; Plan MAGNA; ITGE: Madrid, Spain, 1990; p. 98.

21. Martín-Serrano García, A.; Monteserín López, V.; Mediavilla López, R.; Nozal Martín, F.; Diez Balda, M.A.; Pardo Alonso, M.V.; Carral González, M.P.; Rubio Pascual, F. *Cartografía y Memoria de la Hoja n° 502 (Matilla de los Caños del Río). Escala 1:50.000*; Plan MAGNA; ITGE: Madrid, Spain, 1993; p. 76.

22. Silva, P.G.; Roquero, E.; López-Recio, M.; Huerta, P.; Martínez-Graña, A.M. Chronology of fluvial terrace sequences for large Atlantic Rivers in the Iberian Peninsula (Upper Tagus and Duero drainage basins, Central Spain). *Quat. Sci. Rev.* **2017**, *166*, 188–203. [CrossRef]

23. Martín-Martín, I. Evolución Geomorfológica del Río Yeltes (Salamanca) Durante el Cuaternario. Master's Thesis, Universidad de Salamanca, Salamanca, Spain, 2019; p. 45.

24. Goy, J.L.; Rodriguez-López, G.; Martínez-Graña, A.M.; Cruz, R.; Valdés, V. Geomorphological analysis applied to the evolution of the quaternary landscape of the Tormes River (Salamanca, Spain). *Sustainability* **2019**, *11*, 7255. [CrossRef]

25. Rodríguez-Rodríguez, L.; Antón, L.; Rodés, A.; Pallàs, R.; García-Castellanos, D.; Jiménez-Munt, I.; Struth, L.; Leannig, L.; ASTER Team. Dates and rates of endo-exorheic drainage development: Insights from fluvial terraces (Duero River, Iberian Peninsula). *Glob. Planet. Chang.* **2020**, *193*, 103271. [CrossRef]

26. Karampaglidis, T.; Benito-Calvo, A.; Rodés, A.; Braucher, R.; Pérez-González, A.; Pares, J.; Stuart, F.; Di Nicola, L.; Bourles, D. Pliocene endoreheic-exhoreic drainage transition of the Cenozoic Madrid Basin (Central Spain). *Glob. Planet. Chang.* **2020**, *194*, 103295. [CrossRef]

27. Martínez Graña, A.M. Estudio Geológico Ambiental Para la Ordenación de los Espacios Naturales de "Las Batuecas-Sierra de Francia" y "Quilamas". Aplicaciones Geomorfológicas al Paisaje, Riesgos e Impactos, Análisis Cartográfico Mediante SIG. Ph.D. Thesis, Universidad de Salamanca, Salamanca, Spain, 2010; p. 806.

28. Bishop, M.P.; James, L.A.; Shroder, J.F.; Walsh, S.J. Geospatial technologies and digital geomorphological mapping: Concepts, issues and research. *Geomorphology* **2012**, *137*, 5–26. [CrossRef]

29. Gustavsson, M.; Kolstrup, E.; Seijmonsbergen, A.C. A new symbol-and-GIS based detailed geomorphological mapping system: Renewal of a scientific discipline for understanding landscape development. *Geomorphology* **2006**, *77*, 90–111. [CrossRef]

30. Martínez-Graña, A.M.; Goy, J.L.; Zazo, C.; Silva, P.G.; Santos-Francés, F. Configuration and evolution of the landscape from the geomorphological map in the Natural Parks Batuecas-Quilamas (Central System, SW Salamanca, Spain). *Sustainability* **2017**, *9*, 1458.

31. Quesada-Román, A.; Zamorano-Orozco, J.J. Geomorphology of the Upper General River Basin, Costa Rica. *J. Maps* **2019**, *15*, 95–101. [CrossRef]

32. Veleda, S.; Martínez-Graña, A.; Santos-Francés, F.; Sánchez-San Roman, J.; Criado, M. Analysis of the Hazard, Vulnerability and Exposure to the Risk of Flooding (Alba de Yeltes, Salamanca, Spain). *Appl. Sci.* **2017**, *7*, 157. [CrossRef]

33. Jambrina, M.; Corrochano, A.; Armenteros, I. Sedimentología e hidrogeoquímica de las lagunas de "El Cristo" y de "La Cervera" (Salamanca, España). *Stud. Geol. Salmanticiensia* **2010**, *46*, 25–45.

34. Bernat Rebollal, M.; Pérez-González, A. Campos de dunas y mantos eólicos de Tierra de Pinares (sureste de la cuenca del Duero, España). *Bol. Geol. Min.* **2005**, *116*, 23–38.

35. Leopold, L.B.; Wolman, M.G.; Miller, J.P. *Fluvial Processes in Geomorphology*; Freeman: San Francisco, CA, USA, 1964; p. 522.

36. Harden, C. Terrace, River. In *Encyclopedia of Geomorphology*; Goudie, A.S., Ed.; Routledge: London, UK, 2004; pp. 1039–1043.

37. Roquero, E.; Silva, P.G.; Goy, J.L.; Zazo, C.; Massana, J. Soil evolution indices in fluvial terrace chronosequences from central Spain (Tagus and Duero fluvial basins). *Quat. Int.* **2015**, *376*, 101–113. [CrossRef]

38. Gutiérrez-Elorza, M. *Geomorfología*, 1st ed.; Pearson-Prentice Hall: Madrid, Spain, 2008; p. 920.

39. Martín-Serrano, A.; Cantano, M.; Carral, P.; Rubio, F.; Mediavilla, R. La degradación cuaternaria del piedemonte del río Yeltes (Salamanca). *Cuat. Geomorfol.* **1998**, *12*, 5–17.

40. Silva, P.G.; Bardají, T.; Roquero, E.; Baena-Presley, J.; Cearreta, A.; Rodríguez-Pascua, M.A.; Rosas, A.; Zazo, C.; Goy, J.L. El Periodo Cuaternario: La Historia Geológica de la Prehistoria. *Cuater. Geomorfol.* **2017**, *31*, 7–24. [CrossRef]

41. Antón, L.; De Vicente, G.; Muñoz-Martín, A.; Stokes, M. Using river long profiles and geomorphic indices to evaluate the geomorphological signature of continental scale drainage capture, Duero basin (NW Iberia). *Geomorphology* **2014**, *206*, 250–261. [CrossRef]

42. Clark, P.U.; Archen, D.; Pollard, D.; Blum, J.D.; Rial, J.A.; Brovkin, V.; Mix, A.C.; Pisisas, N.G.; Roy, M. The mid-Pleistocene transition: Characteristic mechanisms and implications for long-term changes in atmospheric pCO2. *Quat. Sci. Rev.* **2006**, *25*, 3150–3184. [CrossRef]

 *sustainability*

Article

Mycological Indicators in Evaluating Conservation Status: The Case of *Quercus* spp. Dehesas in the Middle-West of the Iberian Peninsula (Spain)

Prudencio García Jiménez [1], Abel Fernández Ruiz [1], José Sánchez Sánchez [1,2] and David Rodríguez de la Cruz [1,2,*] 

1   Spanish-Portuguese Agricultural Research Centre (CIALE), Universidad de Salamanca, Río Duero 12, 37185 Villamayor, Spain; prudenciogarjim@gmail.com (P.G.J.); abel@usal.es (A.F.R.); jss@usal.es (J.S.S.)
2   Department of Botany and Plant Physiology, Universidad de Salamanca, Licenciado Méndez Nieto s/n, 37007 Salamanca, Spain
*   Correspondence: droc@usal.es; Tel.: 0034-677-584-172

Received: 11 November 2020; Accepted: 11 December 2020; Published: 14 December 2020

Abstract: The use of bioindicators to assess the conservation status of various ecosystems is becoming increasingly common, although fungi have not been widely used for this purpose. The aim was to use the analysis of the macromycetes fruiting bodies in the area of a natural reserve and the degree of preservation of its different zones combined with the use of geographical information systems (GIS). For this purpose, quantitative and qualitative fungal samples were carried out in plots of the middle-west of the Iberian Peninsula previously delimited and characterised thanks to GIS during the springs and autumns of the 2009–2012 period. In addition, the lifestyles of the fungal species were analysed as well as the influence of the main meteorological parameters on fungal fruiting. A total of 10,125 fruiting bodies belonging to 148 species were counted on 20 plots with four vegetation units (holm oak dehesas, mixed holm oaks and Pyrenean oak dehesas with different abundance and grasslands). The distribution of the different species, their lifestyles and the number of fruiting bodies in the different plots of the reserve indicated that the eastern part was best conserved, showing that the combination of fungal diversity studies and the use of GIS could be useful in the management of areas with environmental relevance.

Keywords: fungal indicators; GIS; conservation; dehesas; MW Spain

1. Introduction

The sustainable management of forests and resources that can be obtained through them, such as wood, fruit, or even ecosystem services (CO_2 capture), is becoming of considerable interest in recent decades [1,2]. One of the most important aspects that must be taken into account before dealing with the management of the forests and their resources is assessing their state of conservation. In this way, it will be possible to know the initial stage before starting any kind of action. To this end, throughout this century various methods have been proposed to evaluate the degree of preservation in various woodlands [3,4], in which distinct factors are evaluated through different indices. However, these methods did not, or only very partially, consider the contribution of one of the groups of organisms with a fundamental role in forest dynamics, fungi [5]. Traditionally, three types of life forms are considered in the Fungi kingdom, symbiotic with various plant species, saprophytes decomposing material of various types and parasites on other species in various kingdoms [6]. The relevance of these organisms and their lifestyles in the dynamics of different plant formations, especially forests, is well known, favouring the recycling of products resulting from the activity of the ecosystem itself [7,8], the growth of various tree and shrub species [9], many of which are fundamental in the physiognomic and ecological composition of the forest itself [10], and even with a potential role in the bioremediation of various pollutants [11]. It should not be overlooked that a good number of the fruiting bodies of these organisms also constitute a natural resource [12], nor the fact that 74% of threatened fungal species are found in woody formations compared to 9% present in different types of pasture [13], also highlighting the environmental importance of this type of vegetation. Most of the studies carried out on fungal fruiting bodies as indicators of the conservation status of different ecosystems were based on the occurrence of fruiting bodies of different species [14], or assessing the connectivity between tree formations and myceliums of saprophytic species in wood remains [15]. No assessments were developed through these studies, therefore, that could be transferred to technical teams managing forest areas as a further factor to be taken into account when estimating the preservation of different ecosystems.

The aim of this paper was the use of fungal sporocarps produced in a forest ecosystem as a relevant factor to assess the conservation status of these plant formations, and in the context of a characteristic habitat type of the central-western Iberian Peninsula, the dehesas (in Spanish). These ecosystems, also known as montado (in Portuguese), are the most widespread complex agrosilvopastoral systems in Europe and are dominated by several species of the genus *Quercus* L. [16]. It was also intended to examine the variation in the production of macroscopic fruiting bodies belonging to Ascomycota and Basidiomycota, both quantitatively (in a number of sporocarps) and qualitatively (species diversity) and the possible influence of meteorological parameters, extending the preliminary results shown in previous work within the study area [17].

2. Materials and Methods

2.1. Study Area

This work was carried out in the "Campanarios de Azaba" Natural Reserve located in the Middle-West of the Iberian Peninsula (N 40° 29.769 W 6° 47.551). This area has an area of 522 hectares located about 800 m.a.s.l. with a Mediterranean oceanic pluviseasonal bioclimate (Rivas-Martínez et al., 2001). The dominant ecosystem in the area was an open agrosilvopastoral ecosystem present in the Iberian Peninsula called "dehesa" (in Spanish) or "montado" (in Portuguese), with a predominance of holm oak (*Quercus ilex* subsp. *ballota* (Desf.) Samp.) and Pyrenean oak (*Quercus pyrenaica* Willd.), coexisting with cork oak (*Quercus suber* L.) and gall oak (*Quercus faginea* Lam.) individuals. These formations covered approximately half of the area of the reserve, which has traditionally been used for livestock (cattle and pigs) and occasionally for agriculture.

2.2. Delimitation of Vegetation Units

A physiognomic delimitation on a 1:1.000 scale of the different plant formations present in the reserve was developed, obtaining a series of plots or tesserae, which were assigned to general vegetation units, through geographic information systems (GIS), with the ArcGIS® 10.1 software package. This theoretical and practical delimitation in different plots was conducted following physiognomic and ecological criteria by means of the main plant formations and making reference mainly to the tree types [18,19]. The predominance of holm oak or Pyrenean oak in the mixed dehesas was established on the basis of a greater abundance of one or the other species, which was defined in at least 60% of the trees present in each vegetation unit. The characterization of the territory from an ecological point of view was completed by assigning the different vegetation units to habitats defined by the European Union [19,20], for future management activities in the study area.

2.3. Mycological Surveys

The evaluation of macromycological diversity was carried out between autumn 2009 and spring 2012, through four visits per weather station during three autumns (years 2009, 2010 and 2011) and three springs (years 2010, 2011 and 2012) to collect the different fungal fruiting bodies. In total, they involved 24 samples (3 autumns + 3 springs × 4 visits per weather station) of 30 m × 30 m [21] conducted on each of the previously delimited plots. The Ascomycota and Basidiomycota sporocarps were subsequently identified in the laboratory by macroscopic and microscopic character analysis using a LEICA DMRD microscope attached to a LEICA DC100 video camera and the LEICA Qwin image software. Specific literature was used on taxonomic discrimination ([22–26], among others), following CABI Index Fungorum [27] for nomenclature. The influence of meteorological parameters on fructification, both qualitative (in species diversity) and quantitative (in a number of sporocarps) was performed by means of Spearman's non-parametric correlation statistic, taking into account that fructification does not follow a normal distribution throughout the years. The software applied was SPSS v.23. The meteorological parameters used were the averages of average, maximum, minimum and difference in temperature (in °C) and total rainfall for the previous seven days [28], giving a correlation coefficient with values between −1 and 1, representing a negative and positive correlation, respectively. The meteorological data of temperature and precipitation were provided by the Spanish State Agency of Meteorology (AEMET) via the weather stations located in Navasfrías (20 km from the study area) and Alberguería de Argañán (7 km away).

2.4. Assessment of the Conservation Status

The evaluation of the state of conservation of the different plots was performed according to three parameters (health status, species richness and production of fruit bodies), estimated for each plot, and then as a whole, in order to delimit the degree of preservation of each one of them and of the reserve as a whole. These parameters and the scores assigned to the different categories are shown in Table 1. Within this estimate, plots of land assigned to disturbed environments, whether agricultural crops, reforestation, buildings or roads, as well as temporary ponds, have been discarded because they do not constitute an adequate environment for the development of fruiting bodies. Firstly, the health status was evaluated, considering the way of life of each of the identified specimens, estimating in each case its proximity to the proportion reported by Moreno [29], as an indicator of a good state of health of each plot (51% saprophytes, 47% symbiotics and 2% parasites). The number of fungal species was also considered, as greater species richness leads to a better state of preservation of the various plots in a given area [30]. The number of specimens was taken into account with less weight than the two previous parameters since the number of fruiting bodies does not seem to be so important in assessing conservation [31]. The latter parameters were assessed by taking into account the respective maximum values reached in a plot, and scores were assigned on the basis of these values. The results that defined the scores in the different sections are indicated as follows. Health status: optimum (fungal symbiotic species >40%), Good (31–40% symbiotic species), Favorable (21–30% symbiotic species), Poor (<20%); Species richness: Very high (>60 species), High (51–60 species), Medium (21–50 species), Low (<20 species); Production fruit bodies: High (>800 carpophores), Medium (300–800 carpophores), Low (<300 carpophores).

Table 1. Criteria and scores taken into account to estimate the conservation status of the different plots present in the study area.

Health Status		Species Richness		Production Fruit Bodies		Preservation Status	
Value	Health	Value	N species	Values	Carpophores	Value	Estimation
3	Optimum	3	Very high	2	High	8–7	Optimum
2	Good	2	High	1	Medium	6–5	Good
1	Favourable	1	Medium	0	Low	4–3	Suitable
0	Poor	0	Low			2–1	Poor
						0	Not favourable

N species: number of species.

3. Results and Discussion

The GIS analysis of the study area and the field evaluation led to the identification of four main and vegetation units: holm oak dehesas (dominated by the holm oak), mixed dehesas of Pyrenean oak and holm oak (with a greater presence of Pyrenean oak accompanied by the holm oak), mixed dehesas of holm oak and Pyrenean oak (same as the previous case, but with a greater abundance of holm oaks), and grasslands of different types and composition. It should be noted that the different lakes of natural or artificial origin were not included in this generic synthesis of vegetation units, because no fruiting bodies were observed to develop in them, except in the case that they had a herbaceous component on their borders, and they were ascribed to grasslands (Table 2, Figure 1). In total, grouping all the above factors, 20 different plots in the study area were analysed. The delimitation in units revealed that almost a third (31%) of the Natural Reserve was made up of holm oak dehesas, followed by grasslands (21%), mixed dehesas of holm oak and Pyrenean oak (12%) and mixed dehesas of Pyrenean oak and holm oak (3%). The remaining plots, together with another third of the area analysed, were considered to be areas with a high degree of human intervention, such as crops and reforestation (15%), Mediterranean temporary ponds or natural eutrophic ponds (11%), and roads and built-up areas (7%). The application of GIS for the conservation of forest formations has proven to be an effective tool in the sustainable management of their resources [32], as well as for the elaboration of preservation plans associated with these formations and their associated habitats [33].

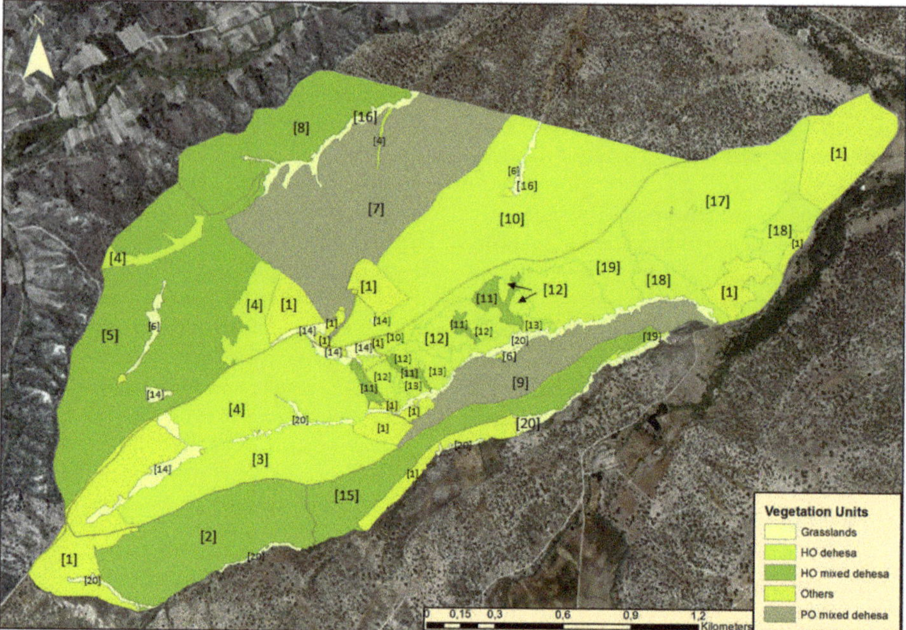

Figure 1. Vegetation units present in the "Campanarios de Azaba" Natural Reserve. HO: Holm oak; PO: Pyrenean oak. (The number of plots analysed in brackets).

Table 2. Vegetation units described in the Natural Reserve "Campanarios de Azaba" and habitats defined by the EU (Directive 92/43 CEE) that can be placed on them, with their code (EU_Code) and whether or not they have priority status (marked with an asterisk "*" and shaded in grey).

VEGETATION UNITS	EU HABITAT NAME	EU_CODE	PRIORITY
Holm oak dehesas	Pseudo-steppe with grasses and annuals of the *Thero-Brachypodietea*	6220	*
	Endemic oro-Mediterranean heaths with gorse	4090	Np
	Dehesas with evergreen *Quercus* spp.	6310	Np
Mixed dehesas of holm oak and Pyrenean oak	Pseudo-steppe with grasses and annuals of the *Thero-Brachypodietea*	6220	*
	Endemic oro-Mediterranean heaths with gorse	4090	Np
	Dehesas with evergreen *Quercus* spp.	6310	Np
	Galicio-Portuguese oak woods with *Quercus robur* and *Quercus pyrenaica*	9230	Np
Mixed dehesas of Pyrenean oak and holm oak	Pseudo-steppe with grasses and annuals of the *Thero-Brachypodietea*	6220	*
	Mediterranean tall humid herb grasslands of the *Molinio-Holoschoenion.*	6420	Np
	Endemic oro-Mediterranean heaths with gorse	4090	Np
	Dehesas with evergreen *Quercus* spp.	6310	Np
	Galicio-Portuguese oak woods with *Quercus robur* and *Quercus pyrenaica*	9230	Np
	Pseudo-steppe with grasses and annuals of the *Thero-Brachypodietea*	6220	*
	Mediterranean tall humid herb grasslands of the *Molinio-Holoschoenion.*	6420	Np

Np: No priority.

A total of 148 fungal species were identified in the reserve during the different visits in the period 2009–2012 (Table S1). Two species can be highlighted because they are included in the Red List of Endangered Mushrooms in Europe [34]: *Hericium erinaceus* (Bull.) Pers. in a mixed holm oak and Pyrenean oak dehesa, and *Torrendia pulchella* Bres. (=*Amanita torrendi* Justo) in a holm oak dehesa with sandy soil and some Pyrenean oaks [17]. The role of these red lists in conserving fungal species and the habitats in which they grow should cover more specific geographical or administrative areas, such as at country level [35]. In the Iberian Peninsula, there have been some proposals [36], but a red list has not yet been defined, and we think it would be a useful tool for the preservation of ecosystems [37]. Species richness is similar to that reported for other Mediterranean *Quercus* formations [38–40].

In the 2009–2010 period, almost half of the species of the whole catalogue were found, a proportion that increased to almost 70% by the end of spring 2011 (Figure 2). In autumn 2011 almost all the species were catalogued, with only three new species appearing in spring 2012. Seasonally, 77% of the species were collected during the autumn. This type of dynamic has already been identified in study plots with ecosystems similar to those in the area of analysis [41], suggesting that two years of study may be adequate to characterize fungal diversity through the analysis of fruiting bodies. However, the lack of sampling during the winter and summer could lead to a loss of information on climate change indicator species, since their sporocarps appear in adverse weather conditions [42]. The number of fruiting bodies that appeared in the set of plots analysed throughout the 24 weeks of study was 10,125, a number similar to that counted in other works also developed in Iberian dehesas [39]. The highest production of carpophores, as well as the fungal diversity, was obtained in the autumn weeks (71% of the total), particularly autumn of 2011 where almost half of the identified fruiting bodies were counted (Figure 2). Autumn has been reported as the season with the highest fungal and spore production diversity in other works carried out in Mediterranean coastal holm-oak forests [43]. The variability in sporocarp production from one year to another is very variable [44], as our study showed, since almost twice as many fruiting bodies were counted in 2011than in the other two years combined. This variability and inter-annual species diversity seem to be linked to differences in meteorological parameters [45], mainly temperature and precipitation prior to fruiting periods [46], especially in regions with a Mediterranean climate [47]. For this reason, the influence of these two meteorological parameters on the diversity of species and the number of sporocarps was evaluated using the Spearman correlation statistic (Table 3).

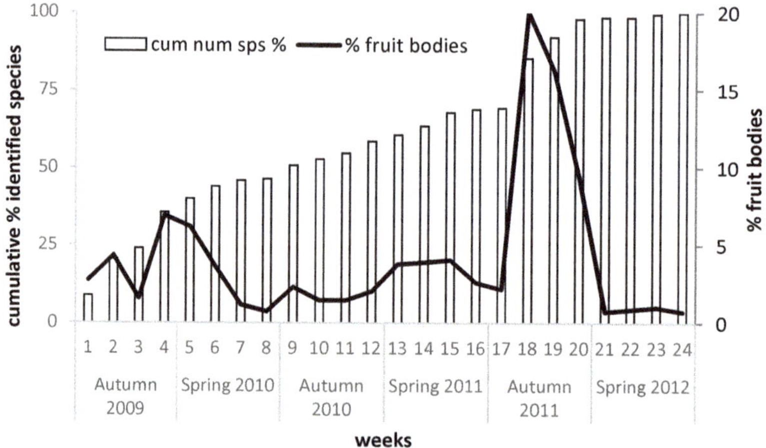

Figure 2. Number of species identified (cumulative %) and of fruiting bodies (in % of total) during 2009–2012.

Table 3. Influence of the average temperature and total rainfall of the previous 7 days on the diversity and production of carpophores recorded in the studied weeks.

	Number sps	Tmax	Tmin	Tmean	Dif. Temp	Rainfall
n carpophores	0.8437 **	−0.2133	0.1926	−0.1001	−0.3791	**0.4136 ***
Number sps		−0.4502 *	−0.0063	−0.3346	−0.5363 **	**0.5358 ****

Number sps: number of species; Tmax: maximum daily temperature; Tmin: minimum daily temperature; Tmean: daily mean temperature; Dif. Temp: daily temperature differences; temperature in °C. Rainfall (in mm). Significance levels: * 0.95%, ** 0.99%. In **black**, significant positive correlations. <u>Underlined</u>, significant negative correlations.

The results indicated a significant positive correlation between the number of species and sporocarps with rainfall, which was higher in the case of the number of species. For the number of species identified there was a negative correlation with the difference in daily temperature and the maximum daily temperature in the week prior to sampling. The positive effect of previous precipitation on fungal fruiting, both quantitative and qualitative, has already been reported in Mediterranean ecosystems [48]. In these environments, the difference in previous daily temperatures and the maximum temperature also had a negative effect on the diversity and abundance of sporocarps [49]. However, it would be advisable to increase the number of years analysed as well as to extend the frequency of study for a better evaluation of the influence of meteorological parameters on the diversity and production of fruiting bodies [41].

The species identified were mainly distributed in 69% of holm oak dehesas, followed by mixed holm oak and Pyrenean oak dehesas, Pyrenean oak and holm oak dehesas (15% and 12%, respectively), and grasslands (4%). One factor that could have influenced the lower location of specimens in the meadows, in addition to their smaller extension in the study area, was the previous agricultural use in some cases, which could even reduce the appearance of various fruiting bodies. In addition, the subsequent livestock load should be noted, as many of the carpophores were removed, damaged, or even ingested by the different animals for livestock use [50].

Analysis of the way of life in the area studied showed a notable predominance of saprophytic (72%) over symbiotic (27%) and, in a residual way, parasitic (1%) species. Other studies carried out in Mediterranean ecosystems dominated by various species of the genus *Quercus* [40,50] indicated lower percentages of saprophytic species, which could reflect the influence of livestock numbers on the presence of taxa with this lifestyle. This representation did not show great variations with respect to the different units of vegetation studied, with slight percentage variations in the holm oak dehesas and the mixed holm oak and Pyrenean oak in the presence of symbiotic species or otherwise in the oak grove, except in the grasslands, with a notable presence of saprophytic species (Figure 3). Centuries-old and extensive livestock use could have conditioned the large percentage of saprophytic taxa present in all the habitats considered, especially in the grasslands where livestock use was even greater, as indicated by the presence of various coprophilic species, such as the genera *Coprinus* Pers., *Coprinopsis* P. Karst., or *Panaeolus* (Fr.) Quél. [51].

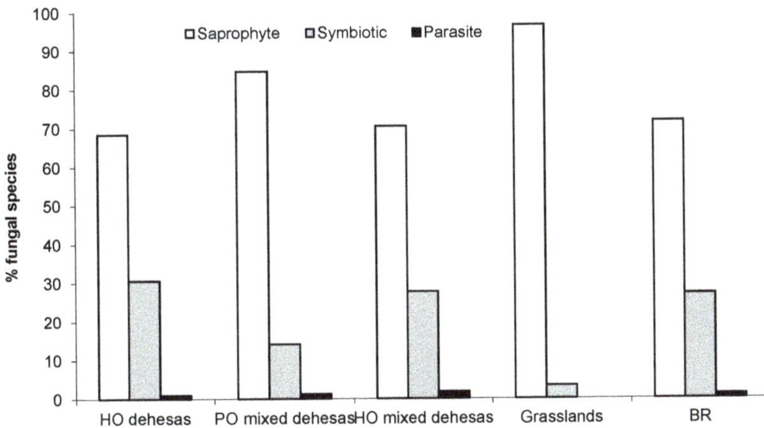

Figure 3. Average percentage representation of the different lifestyles considered in the vegetation units present within the territory under analysis (period 2009–2012). HO dehesas: holm oak dehesas. PO mixed dehesas: mixed Pyrenean oak and holm oak dehesas, dominated by the Pyrenean oak. **HO mixed dehesas**: mixed holm oak and Pyrenean oak dehesas, dominated by the holm oak. BR: the entire natural reserve "Campanarios de Azaba".

The evaluation of the conservation status within each of the physiognomically defined plots in the study area was based on three parameters. The results are summarized in Table 4. The first one, the health status was estimated taking into account the way of life of each of the species and the proximity to the indicators referred to by Moreno [29]. The evaluation of this parameter made it possible to estimate that more than half of the plots in the reserve presented a good or favourable health status (59.7% overall), although it should be noted that no plot was catalogued as "optimal" state (Figure 4). There could again be an influence of the livestock load in the years prior to the study on the increased presence of saprophytic taxa [52], and the low representativeness of good or optimal plots according to the lifestyle of the fungal species.

Table 4. Results of the parameters analysed to assess the state of conservation in the different plots.

Vegetation Unit	Plot	N sps	Fungal Lifestyle (%)			N Fruit Bodies
			Saprophte	Symbiotic	Parasite	
Culture	1	NE	NE	NE	NE	NE
Grasslands	6	16	59	41	0	240
	14	6	100	0	0	90
	16	4	100	0	0	60
	20	11	91	9	0	165
HO dehesas	2	13	62	38	0	195
	3	24	71	29	0	465
	4	33	76	24	0	555
	10	35	70	30	0	645
	12	21	54	46	0	360
	13	49	76	24	0	930
	17	82	68	31	1	1905
	18	56	51	46	3	1095
	19	36	69	29	2	675
HO mixed dehesas	5	39	89	9	2	645
	8	12	75	25	0	180
	11	23	78	22	0	345
	15	18	47	48	5	285
PO mixed dehesas	7	21	83	17	0	360
	9	51	85	13	2	930

N sps: number of species. Fungal lifestyle (%): percentage of each type of life in relation to the total. N fruit bodies: number of fruit bodies. HO dehesas: Holm oak dehesas. PO mixed dehesas: mixed Pyrenean oak and Holm oak dehesas, dominated by the Pyrenean oak. HO mixed dehesas: mixed holm oak and Pyrenean oak dehesas, dominated by the holm oak. NE: not evaluated.

Figure 4. Health status of the different plots in the "Campanarios de Azaba" Natural Reserve according to the lifestyle in macrofungal species.

Parasite species have been found mainly in Pyrenean oak specimens, which could be due to two main and even complementary reasons. The first of these refers to the limit of distribution of the Pyrenean oak in the area, as it usually appears on north-facing slopes and/or in places with greater soil humidity [53]. The appearance of a series of years in which a decrease in the total amount of precipitation or even in its seasonal distribution is noted, could have influenced a weakening of this tree species and the appearance of parasitic organisms [54]. This negative effect could also be due, to a greater or lesser extent, to the different uses derived from the debudding of the Pyrenean oak formations, and, in particular, the exercise of excessive and/or inadequate pruning [55]. This worse level of conservation shown by the mixed Pyrenean oak and holm oak dehesas was also reflected in a lower degree of symbiotic species (14%), which establish mycorrhizae mainly with this tree species, with respect to the holm oak dehesas (31%) and the mixed holm oak and Pyrenean oak dehesas (28%).

With regard to fungal diversity, based on the number of species, it was observed that only 7% of the plots had a very high or high number of species (Figure 5), with the percentage of plots with a medium amount (50.9%) being notably higher, and even for those with a low number of species (42.1%). The greatest diversity of taxa was found mainly in the eastern part of the reserve. The use of fungal diversity in plots of the same area to assess their state of conservation has been shown to be useful [56].

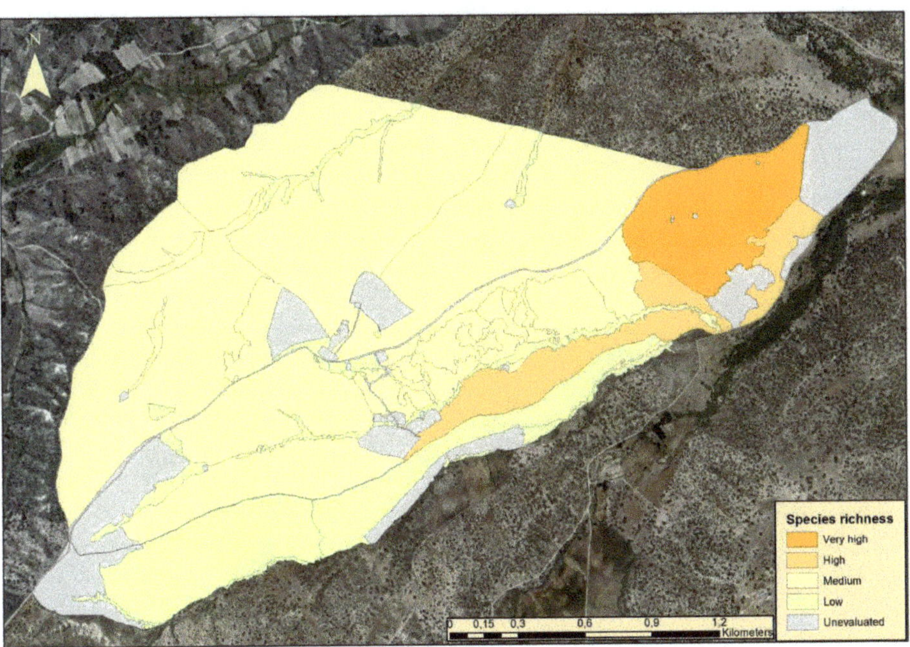

Figure 5. Number of fungal species identified in the period 2009–2012 within the different plots defined in the "Campanarios de Azaba" Natural Reserve.

The last parameter, the number of sporocarps counted, showed that approximately half of the plots analysed (49.1%) had a low number of specimens, only 10.5% with an abundance of fruiting bodies, leaving the rest with a medium number (Figure 6). These results generally indicated a greater number of fungal specimens in the holm oak dehesas, mainly in the eastern part of the area studied. This parameter was mainly used in ecological succession studies after fires [57], since in mature plant formations it is not usually considered a very relevant parameter [31].

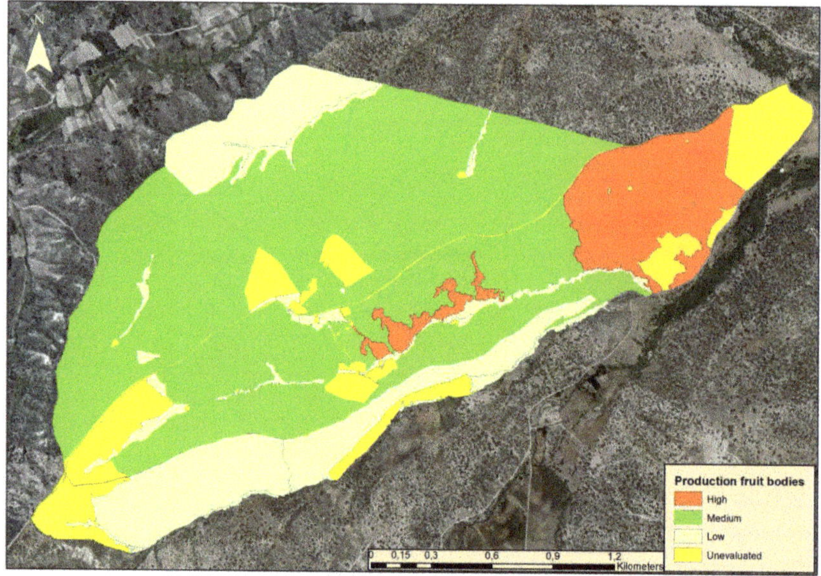

Figure 6. Number of fruit bodies found in the period 2009–2012 within the different plots defined in the "Campanarios de Azaba" Natural Reserve.

The combination of all these parameters was used to assess the state of conservation of the various plots in the Natural Reserve (Figure 7). Using the mycological indicators, it was considered that almost a third of the plots presented a suitable state of preservation, with only a small part (5.3%) having an optimum or good status, mainly the holm oak dehesas located at the East. Of the rest of the plots analysed, nearly half displayed poor status (8.8%) or not favourable (38.6%) status.

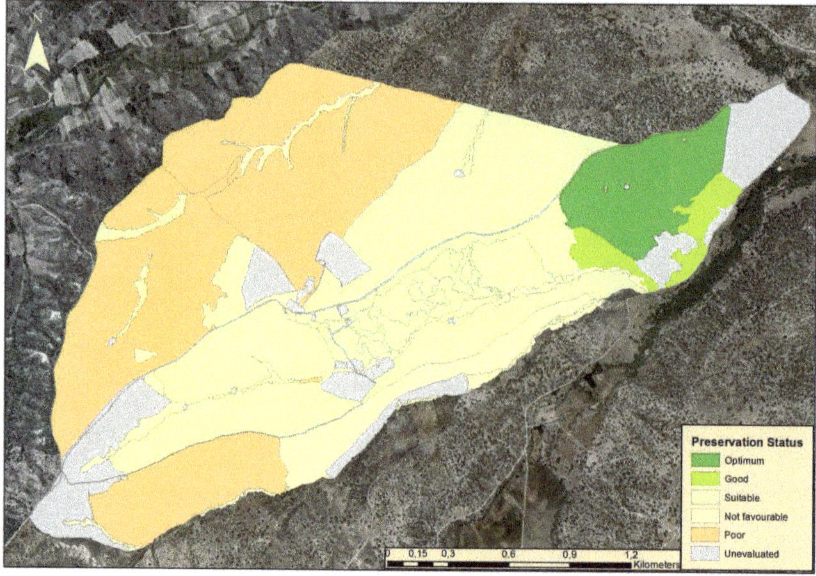

Figure 7. Preservation status of the different plots defined in the "Campanarios de Azaba" Natural Reserve according to fungal indicators (lifestyle, number of species and number of fruit bodies).

Macromycetes species have been used as indicators of the conservation status in various forest systems [58], focusing mainly on the dynamics of symbiotic species, mainly ectomycorrhizal [59] and sometimes with economic-gastronomic interest [60]. However, not enough is known about the fungal composition of forest formations to be able to use them widely as biological indicators of forest health [61]. In addition, some authors considered that joint studies with the dynamics and diversity of lichen-forming fungi [62] would be necessary for a more correct assessment of their degree of preservation, especially for forest habitats with older individuals [63]. In any case, the relevance of fungal species as a tool for assessing the health of habitats has been stressed [64], subject to further contributions that should be made in this field [65]. The use of GIS in the management of forest ecosystems has proven to be a very useful tool [66] and is now even employed in the planning of Mediterranean landscapes by linking economic and environmental perspectives [67]. However, studies combining fungal species as biological indicators and GIS were very scarce and related to the distribution of pathogenic fungal taxa [68]. This work aims to contribute to highlighting the relevance of macromycetes species as indicators of the degree of habitat conservation and the use of GIS for more efficient environmental management. In any case, a greater number of studies are needed in this area where possibly more years of study are contemplated in order to better evaluate mycological diversity and its dynamics, preferably evaluating possible changes in the distribution of species and their lifestyles.

Supplementary Materials: The following are available online at http://www.mdpi.com/2071-1050/12/24/10442/s1. Table S1. Mycological catalogue with the species identified in the period 2009–2012 (in alphabetical order).

Author Contributions: Conceptualization, P.G.J., D.R.d.l.C. and J.S.S.; methodology, P.G.J., D.R.d.l.C. and A.F.R.; software, D.R.d.l.C. and A.F.R.; validation, P.G.J., D.R.d.l.C., A.F.R. and J.S.S.; formal analysis, P.G.J., D.R.d.l.C. and A.F.R.; investigation, P.G.J., D.R.d.l.C., A.F.R. and J.S.S.; resources, P.G.J., D.R.d.l.C. and A.F.R.; data curation, P.G.J. and A.F.R.; writing—original draft preparation, P.G.J., D.R.d.l.C. and A.F.R.; writing—review and editing, P.G.J., D.R.d.l.C., A.F.R. and J.S.S.; visualization, P.G.J., D.R.d.l.C. and A.F.R.; supervision, P.G.J., D.R.d.l.C. and J.S.S.; project administration, D.R.d.l.C. and J.S.S.; funding acquisition, D.R.d.l.C. and J.S.S. All authors have read and agreed to the published version of the manuscript.

Funding: This research received no external funding.

Acknowledgments: The authors would like to thank José Ángel Sánchez Agudo for his technical support. This research was supported by LIFE 07/E/NAT/000762.

Conflicts of Interest: The authors declare no conflict of interest.

References

1. Gratton, L.; Hone, F. Les défis de la forêt privée: La conservation, l'utilisation durable de la forêt et l'écotourisme. *Téoros* **2006**, *25*, 30–35.
2. Goldmann, K.; Schöning, I.; Buscot, F.; Wubet, Y. Forest Management Type Influences Diversity and Community Composition of Soil Fungi across Temperate Forest Ecosystems. *Front. Microbiol.* **2015**, *6*, 1300. [CrossRef] [PubMed]
3. du Bus de Warnaffe, G.; Devillez, F. Quantifier la valeur écologique des milieux pour intégrer la conservation de la nature dans l'aménagement des forêts: Une démarche multicritères. *Ann. Sci.* **2002**, *59*, 369–387. [CrossRef]
4. Sacchis Lopes, M.; Kozhikkodan Veettil, B.; Saldanha, D.L. Assessment of Small-Scale Ecosystem Conservation in the Brazilian Atlantic Forest: A Study from Rio Canoas State Park, Southern Brazil. *Sustainability* **2019**, *11*, 2948. [CrossRef]
5. Baldrian, P. Forest microbiome: Diversity, complexity and dynamics. *FEMS Microbiol. Rev.* **2017**, *41*, 109–130. [CrossRef]
6. Perotto, S.; Angelini, P.; Bianciotto, V.; Bonfante, P.; Girlanda, M.; Kull, T. Interaction of fungi with other organisms. *Plant. Biosyst.* **2013**, *147*, 208–218. [CrossRef]
7. Lonsdale, D.; Pautasso, M.; Holdenrieder, O. Wood-decaying fungi in the forest: Conservation needs and management options. *Eur. J. For. Res.* **2008**, *127*, 1–22. [CrossRef]

8. Park, J.-H.; Pavlov, I.N.; Kim, M.-J.; Park, M.S.; Oh, S.-Y.; Park, K.H.; Fong, J.J.; Lim, Y.W. Investigating Wood Decaying Fungi Diversity in Central Siberia, Russia Using ITS Sequence Analysis and Interaction with Host Trees. *Sustainability* **2020**, *12*, 2535. [CrossRef]

9. Chávez, D.; Pereira, G.; Machuca, Á. Estimulación del crecimiento en plántulas de *Pinus radiata* utilizando hongos ectomicorrícicos y saprobios como biofertilizantes. *Bosque* **2014**, *35*, 57–63. [CrossRef]

10. Taylor, A.F.S.; Alexander, I. The ectomycorrhizal symbiosis: Life in the real world. *Mycologist* **2005**, *19*, 102–112. [CrossRef]

11. Kurniati, E.; Arfarita, N.; Imai, T.; Higuchi, T.; Kanno, A.; Yamamoto, K.; Sekine, M. Potential bioremediation of mercury-contaminated substrate using filamentous fungi isolated from forest soil. *J. Environ. Sci.* **2014**, *26*, 1223–1231. [CrossRef]

12. Martínez-Ibarra, E.; Gómez-Martín, M.B.; Armesto-López, X.A. Climatic and Socioeconomic Aspects of Mushrooms: The Case of Spain. *Sustainability* **2019**, *11*, 1030. [CrossRef]

13. Arnolds, E.; de Vries, B. Conservation of fungi in Europe. In *Fungi of Europe—Ivestigation Recording and Mapping*, 1st ed.; Pegler, D.N., Boddy, L., Ing, O., Kirk, P.M., Eds.; Royal Botanic Gardens: Kew, UK, 1993; pp. 231–238.

14. Richard, F.; Moreau, P.A.; Selosse, M.A.; Gardes, M. Diversity and fruiting patterns of ectomycorrhizal and saprobic fungi in an old-growth Mediterranean forest dominated by *Quercus ilex* L. *Can. J. Bot.* **2004**, *82*, 1711–1729. [CrossRef]

15. Abrego, N.; Bässler, C.; Christensen, M.; Heilmann-Clausen, J. Implications of reserve size and forest connectivity for the conservation of wood-inhabiting fungi in Europe. *Biol. Conserv.* **2015**, *191*, 469–477. [CrossRef]

16. Moreno, G.; Pulido, F.J. The Functioning, Management and Persistence of Dehesas. In *Agroforestry in Europe: Current Status and Future Prospects*, 1st ed.; Rigueiro-Rodríguez, A., McAdam, J., Mosquera-Losada, M.R., Eds.; Springer Science + Business Media, B.V.: Amsterdam, The Netherlands, 2009; pp. 127–160.

17. Sánchez-Martínez, C.; Benito-Peñil, D.; García de Enterría, S.; Barajas-Castro, I.; Martín-Herrero, N.; Pérez-Ruiz, C.; Sánchez-Sánchez, J.; Sánchez-Agudo, J.A.; Rodríguez-de la Cruz, D.; Galante-Patiño, E. *Manual de Gestión Sostenible de Bosques Abiertos Mediterráneos*; Castilla Tradicional: Valladolid, Spain, 2012; pp. 68–79.

18. Romero-Calcerrada, R.; Novillo, C.J.; Millington, J.D.A.; Gómez-Jiménez, I. GIS analysis of spatial patterns of human-caused wildfire ignition risk in the SW of Madrid (Central Spain). *Landsc. Ecol.* **2008**, *23*, 341–354. [CrossRef]

19. European Commission. Council Directive 92/43/EEC of 21 May 1992 on the conservation of natural habitats and of wild fauna and flora. *Off. J. Eur. Union* **1992**, *L 206*, 7–50.

20. VV.AA. *Bases Ecológicas Preliminares Para la Conservación de los Tipos de Hábitat de Interés Comunitario en España*; Ministerio de Medio Ambiente, y Medio Rural y Marino: Madrid, Spain, 2009.

21. Taylor, A.F.S. Fungal diversity in ectomycorrhizal communities: Sampling effort and species detection. *Plant. Soil.* **2002**, *244*, 19–28. [CrossRef]

22. Neville, P.; Poumarat, S. *Amaniteae. Fungi Europaei*; Candusso: Origgio, Italy, 2004.

23. García Jiménez, P.; Pérez Gorjón, S.; Sánchez Rodríguez, J.A.; Sánchez Sánchez, J.; Valle Gutiérrez, C.J. *Setas de Salamanca*; Diputación de Salamanca, Naturaleza y Medio Ambiente: Salamanca, Spain, 2005.

24. Muñoz, J. *Boletus s.l. Fungi Europaei*; Candusso: Origgio, Italy, 2005.

25. Parra, L.A. *Agaricus s.l. Fungi Europaei. Volume 1*; Candusso: Origgio, Italy, 2008.

26. Parra, L.A. *Agaricus s.l. Fungi Europaei. Volume 2*; Candusso: Origgio, Italy, 2012.

27. CABI Index Fungorum. Available online: http://www.indexfungorum.org/names/names.asp (accessed on 1 October 2020).

28. Baptista, P.; Martins, A.; Tavares, R.M.; Lino-Neto, T. Diversity and fruiting pattern of macrofungi associated with chestnut (*Castanea sativa*) in the Trás-os-Montes region (Northeast Portugal). *Fungal Ecol.* **2010**, *3*, 9–19. [CrossRef]

29. Moreno, G. *Setas Micorrrizógenas, Parásitas y Saprófitas; Una Forma de Valorar el Impacto Ambiental en Nuestros Bosques*; Congress Communication: Laredo, Spain, 1996.

30. Saitta, A.; Anslan, S.; Bahram, M.; Brocca, L.; Tedersoo, L. Tree species identity and diversity drive fungal richness and community composition along an elevational gradient in a Mediterranean ecosystem. *Mycorrhiza* **2018**, *28*, 39–47. [CrossRef]

31. Richard, F.; Millot, S.; Gardes, M.; Selosse, M.A. Diversity and specificity of ectomycorrhizal fungi retrieved from an old-growth Mediterranean forest dominated by *Quercus ilex*. *New Phytol.* **2005**, *166*, 1011–1023. [CrossRef]

32. Lehtomäki, J.; Tomppo, E.; Kuokkanen, P.; Hanski, I.; Moilanen, A. Applying spatial conservation prioritization software and high-resolution GIS data to a national-scale study in forest conservation. *Ecol. Manag.* **2009**, *258*, 2439–2449. [CrossRef]

33. Hernández-Lambraño, R.E.; Rodríguez de la Cruz, D.; Sánchez-Agudo, J.A. Spatial oak decline models to inform conservation planning in the Central-Western Iberian Peninsula. *Ecol. Manag* **2019**, *441*, 115–126. [CrossRef]

34. Dahlberg, A.; Croneborg, H. *33 Threatened Fungi in Europe*; Complementary and Revised Information on Candidates for Listing in Appendix I of the Bern Convention: Uppsala, Sweden, 2003.

35. Dahlberg, A.; Mueller, G.M. Applying IUCN red-listing criteria for assessing and reporting on the conservation status of fungal species. *Fungal Ecol.* **2011**, *4*, 147–162. [CrossRef]

36. Calonge, D. Apuntes para la futura Lista Roja de hongos españoles. *Bol. Soc. Micol. Madr.* **2004**, *28*, 391–395.

37. Heilmann-Clausen, J.; Barron, E.S.; Boddy, L.; Dahlberg, A.; Griffith, G.W.; Nordén, J.; Ovaskainen, O.; Perini, C.; Senn-Irlet, B.; Halme, P. A fungal perspective on conservation biology. *Conserv. Biol.* **2015**, *29*, 61–68. [CrossRef] [PubMed]

38. Ortega, A.; Lorite, J. Macrofungi diversity in cork-oak and holm-oak forests in Andalusia (southern Spain); an efficient parameter for establishing priorities for its evaluation and conservation. *Cent. Eur. J. Biol.* **2007**, *2*, 276–296. [CrossRef]

39. Azul, A.M.; Castro, P.; Sousa, J.P.; Freitas, H. Diversity and fruiting patterns of ectomycorrhizal and saprobic fungi as indicators of land-use severity in managed woodlands dominated by *Quercus suber*—A case study from southern Portugal. *Can. J. Res.* **2009**, *39*, 2404–2417. [CrossRef]

40. Fernández, A.; Sánchez, S.; García, P.; Sánchez, J. Macrofungal diversity in an isolated and fragmented Mediterranean Forest ecosystem. *Plant. Biosyst.* **2019**, *154*, 139–148. [CrossRef]

41. Fernández, A. Diversidad Macrofúngica del Monte de la Orbada (Salamanca, España), un Ecosistema Forestal Mediterráneo Aislado y Fragmentado. Ph.D. Thesis, Universidad de Salamanca, Salamanca, Spain, 2019.

42. Moreno, G.; Manjón, J.L.; Álvarez-Jiménez, J. Los hongos y el cambio climático. In *Los Bosques y la Biodiversidad frente al Cambio Climático: Impactos, Vulnerabilidad y Adaptación en España*; Herrero, A., Zavala, M.A., Eds.; Ministerio de Agricultura, Alimentación y Medio Ambiente: Madrid, Spain; pp. 129–135.

43. Perini, C.; Barluzzi, C.; De Dominicis, V. Seasonal fruit body production of macrofungi in Mediterranean vegetation. *Bocconea* **1996**, *5*, 359–373.

44. Boddy, L.; Büntgen, U.; Egli, S.; Gange, A.C.; Heegaard, E.; Kirk, P.M.; Mohammad, A.; Kauserud, H. Climate variation effects on fungal fruiting. *Fungal Ecol.* **2014**, *10*, 20–33. [CrossRef]

45. Andrew, C.; Heegaard, E.; Høiland, K.; Senn-Irlet, B.; Kuyper, T.W.; Krisai-Greilhuber, I.; Kirk, P.M.; Heilmann-Clausen, J.; Gange, A.C.; Egli, S.; et al. Explaining European fungal fruiting phenology with climate variability. *Ecology* **2018**, *99*, 1306–1315. [CrossRef] [PubMed]

46. Jarvis, S.G.; Holden, E.M.; Taylor, A.F.S. Rainfall and temperature effects on fruit body production by stipitate hydnoid fungi in Inverey Wood, Scotland. *Fungal Ecol.* **2017**, *29*, 137–140. [CrossRef]

47. Ágreda, T.; Águeda, B.; Olano, J.M.; Vicente-Serrano, S.M.; Fernández-Toirán, M. Increase evapotranspiration demand in a Mediterranean climate might cause a decline in fungal yields under global warming. *Glob. Chang. Biol.* **2015**, *21*, 3499–3510. [CrossRef] [PubMed]

48. Salerni, E.; Laganà, A.; Perini, C.; Loppi, S.; Dominicis, V.D. Effects of temperature and rainfall on fruiting of macrofungi in oak forests of the Mediterranean area. *Isr. J. Plant Sci.* **2002**, *50*, 189–198. [CrossRef]

49. Ágreda, T.; Águeda, B.; Fernández-Toirán, M.; Vicente-Serrano, S.M.; Òlano, J.M. Long-term monitoring reveals a highly structured interspecific variability in climatic control of sporocarp production. *Agric. For. Meteorol.* **2016**, *223*, 39–47. [CrossRef]

50. Azul, A.M.; Mendes, S.M.; Sousa, J.P.; Freitas, H. Fungal fruitbodies and soil macrofauna as indicators of land use practices on soil biodiversity in Montado. *Agroforest. Syst.* **2011**, *82*, 121–138. [CrossRef]

51. Doveri, F. Occurrence of coprophilous Agaricales in Italy, new records, and comparisons with their European and extraeuropean distribution. *Mycosphere* **2010**, *1*, 103–140.

52. Richardson, M.J. Diversity and occurrence of coprophilous fungi. *Mycol. Res.* **2001**, *105*, 387–402. [CrossRef]

53. Sánchez de Dios, R.; Benito-Garzón, M.; Sainz-Ollero, H. Present and future extension of the Iberian submediterranean territories as determined from the distribution of marcescent oaks. *Plant. Ecol.* **2009**, *204*, 189–205. [CrossRef]

54. Nieto-Quintano, P.; Caudullo, G.; de Rigo, D. *Quercus pyrenaica* in Europe: Distribution, habitat, usage and threats. In *European Atlas of Forest Tree Species*, 1st ed.; San-Miguel-Ayanz, J., de Rigo, D., Caudullo, G., Houston Durrant, T., Mauri, A., Eds.; Publ. Off. EU: Luxembourg, 2016; pp. 158–159.

55. Moreno, G.; López-Díaz, M.L. The Dehesa: The Most Extensive Agroforestry System in Europe. In *Agroforestry Systems as a Technique for Sustainable Land Management*, 1st ed.; Mosquera-Losada, M.R., Fernández-Lorenzo, J.L., Rigueiro-Rodríguez, A., Eds.; Unicopia: Lugo, Spain, 2009; pp. 171–183.

56. Zotti, M.; Pautasso, M. Macrofungi in Mediterranean *Quercus ilex* woodlands: Relations to vegetation structure, ecological gradients and higher-taxon approach. *Czech. Mycol.* **2013**, *65*, 193–218. [CrossRef]

57. Hernández-Rodríguez, M.; Oria-de-Rueda, J.A.; Martín-Pinto, P. Post-fire fungal succession in a Mediterranean ecosystem dominated by *Cistus ladanifer* L. *Ecol. Manag* **2013**, *289*, 48–57. [CrossRef]

58. Laganà, A.; Salerni, E.; Barluzzi, C.; Perini, C.; De Dominicis, V. Macrofungi as long-term indicators of forest health and management in central Italy. *Cryptogam. Mycol.* **2020**, *23*, 39–50.

59. Azul, A.M.; Nunes, J.; Ferreira, I.; Coelho, A.S.; Veríssimo, P.; Trovão, J.; Campos, A.; Castro, P.; Freitas, H. Valuing native ectomycorrhizal fungi as a Mediterranean forestry component for sustainable and innovative solutions. *Botany* **2014**, *92*, 161–171. [CrossRef]

60. Ágreda, T.; Cisneros, Ó.; Águeda, B.; Fernández-Toirán, L.M. Age class influence on the yield of edible fungi in a managed Mediterranean forest. *Mycorrhiza* **2014**, *24*, 143–152. [CrossRef] [PubMed]

61. Egli, S. Mycorrhizal mushroom diversity and productivity—An indicator of forest health? *Ann. For. Sci.* **2011**, *68*, 81–88. [CrossRef]

62. Jönsson, M.T.; Ruete, A.; Kellner, O.; Gunnarsson, U.; Snäll, T. Will forest conservation areas protect functionally important diversity of fungi and lichens over time? *Biodivers. Conserv.* **2017**, *26*, 2547–2567. [CrossRef]

63. Izzo, A.; Agbowo, J.; Bruns, T.D. Detection of plot-level changes in ectomycorrhizal communities across years in an old-growth mixed-conifer forest. *New Phytol.* **2005**, *166*, 619–630. [CrossRef]

64. Berglund, H.; Jonsson, B.G. Nested plant and fungal communities; the importance of area and habitat quality in maximizing species capture in boreal old-growth forests. *Biol. Conserv.* **2003**, *112*, 319–328. [CrossRef]

65. Dahlberg, A.; Genney, D.R.; Heilmann-Clausen, J. Developing a comprehensive strategy for fungal conservation in Europe: Current status and future needs. *Fungal Ecol.* **2010**, *3*, 50–64. [CrossRef]

66. Falcão, A.O.; Borges, J.G. Designing decision support tools for Mediterranean forest ecosystems management: A case study in Portugal. *Ann. Sci.* **2005**, *62*, 751–760. [CrossRef]

67. Molina, J.R.; Rodríguez y Silva, F.; Herrera, M.A. Integrating economic landscape valuation into Mediterranean territorial planning. *Environ. Sci. Policy* **2016**, *56*, 120–128. [CrossRef]

68. Lorestani, E.Z.; Kamkar, B.; Razaci, S.E.; Teixeira da Silva, J.A. Modeling and mapping diversity of pathogenic fungi of wheat fields using geographic information systems (GIS). *J. Crop Prot.* **2013**, *54*, 74–83. [CrossRef]

Publisher's Note: MDPI stays neutral with regard to jurisdictional claims in published maps and institutional affiliations.

 *sustainability*

Article

Puyango, Ecuador Petrified Forest, a Geological Heritage of the Cretaceous Albian-Middle, and Its Relevance for the Sustainable Development of Geotourism

Fernando Morante-Carballo [1,2,3,*], Geanella Herrera-Narváez [1,4,*], Nelson Jiménez-Orellana [4] and Paúl Carrión-Mero [1,4]

1   Centro de Investigación y Proyectos Aplicados a las Ciencias de la Tierra (CIPAT), ESPOL Polytechnic University, Campus Gustavo Galindo Km. 30.5 Vía Perimetral, 9015863 Guayaquil, Ecuador; pcarrion@espol.edu.ec

2   Facultad de Ciencias Naturales y Matemáticas (FCNM), ESPOL Polytechnic University, Campus Gustavo Galindo Km. 30.5 Vía Perimetral, 9015863 Guayaquil, Ecuador

3   Geo-recursos y Aplicaciones GIGA, ESPOL Polytechnic University, Campus Gustavo Galindo Km. 30.5 Vía Perimetral, 9015863 Guayaquil, Ecuador

4   Facultad de Ingeniería en Ciencias de la Tierra (FICT), ESPOL Polytechnic University, Campus Gustavo Galindo Km 30.5 Vía Perimetral, 9015863 Guayaquil, Ecuador; nmjimene@espol.edu.ec

*   Correspondence: fmorante@espol.edu.ec (F.M.-C.); geamiher@espol.edu.ec (G.H.-N); Tel.: +59-396-976-0276 (F.M.-C.)

Received: 7 July 2020; Accepted: 6 August 2020; Published: 14 August 2020

Abstract: Geodiversity treaties have multiplied and given rise to geological heritage as a singular value of protection and preservation for territories. The Puyango Petrified Forest (PPF) is a recognized Ecuadorian reserve, which was declared a National Heritage Treasure. It has an area of 2659 hectares, and it is located in the south of Ecuador, between the provinces of El Oro and Loja. The petrified trunks and trees were buried by volcanic lava, dating from the Cretaceous Period, 96 to 112 million years ago. Thus, silicification and carbonization, two important fossilization events, have produced hundreds of samples of paleontological wealth in Puyango. The objective of this work is to methodologically assess the geodiversity of a fraction of the PPF by registering its geological heritage and value for its preservation and sustainable development. The methodology is based on: (i) Analysis of information on the territory used for tourist visits, as a pilot study area. Presentation of paleontological components and their main sections to enhance their geotouristics value; (ii) Assessment of the geological heritage for its geotourism categorization with a recognized scientific methodology and one proposed by the co-authors propose; and (iii) Analysis of Strengths, Weaknesses, Opportunities, and Threats (SWOT) as a guide for protection and development strategies. Findings reveal the high geotourism potential for a Geopark Project in Puyango, since only 300 hectares are used for tourism and the remaining area is a virgin environment for research and improving knowledge of geodiversity and biodiversity.

Keywords: petrified forest; sustainable development; geodiversity; biodiversity; Puyango; geopark

---

1. Introduction

Geodiversity considers all the geological elements of the Earth's crust, from the landscape to its internal structure that constitutes the various materials such as rocks, minerals and fossils. It is an inanimate part of nature, but at the same time significant to sustaining biodiversity, since the soil and subsoil generated by a series of geological processes are what sustain it, and what together are part of the natural beauty of a site; however, geodiversity has been downplayed by historically giving greater prominence to biodiversity [1]. Thus, the term geodiversity refers to the quality, spectacularity, and beauty of a site of an abiotic nature, which is why it deserves conservation [2]. The protection and conservation of geodiversity is an issue that until recently was not understood. However, this failure is being overcome thanks to the close link that exists between biodiversity and geodiversity, which can be achieved through proper management to preserve biocenosis and the biotope in an integral way [3,4].

Currently, the geological heritage is promoted and protected adequately towards the sustainable development of geotourism; this term emphasizes a form of tourism to natural areas that focuses explicitly on geology and landscape, promoting geosite tourism and the conservation of geodiversity for a better understanding of the earth sciences through appreciation and learning [5]. In geological terms, a geosite is defined by its scientific value that demonstrates the importance of the geological heritage of a specific area, which must be relevant and of importance to science [6]. This is accomplished through independent visits to geological features, such as geo-trails and viewpoints, guided tours, geo-activities, and the sponsorship of geosite visitor centres [7,8]. The term Geoheritage considers particular elements of geodiversity (petrological, geomorphological, structural, mineralogical, paleontological, stratigraphic, hydrogeological, pedological, among others) with a high scientific value [9]. Paleontological heritage of a geosite is the study of its scientific value through fossils, which are testimonies of life in the past, and which also reflect certain events in the geological history of Earth. Around them and the deposits in which they were found, numerous scientific features converge that can be considered objectively, allowing their value to be established; such as fossil types, relative age, state of conservation, among others. Therefore, paleontological heritage is part of the geological and natural heritage [10].

Geodiversity is evident that the fossils and deposits have meaning and provide information on the history of Life and Earth. Morevoer, the scientific component is relevant to place fossils and sites on a theoretical scale of paleontological value [11]. Fossil forests or petrified forests register a remarkable geological-paleontological heritage characterized by the wood of trees that has been buried under sediments and preserved by the absence of oxygen [12]. This type of fossilization is known as permineralization and emphasizes the replacement of the body's molecules by minerals. In this case, petrified wood explains vegetal biodiversity of diverse historical times since it preserves its original structure to a microscopic level.

There are more than 20 specimens of petrified forests worldwide and only 3 belong to America—the Petrified Forest National Park in Arizona-United States, the Petrified Forest of Santa Cruz Natural Monument in Patagonia, Argentina, and the Puyango Petrified Forest in Ecuador. All these forests are remarkable for their age, surface and paleontological wealth [13]. In Ecuador, forests cover 42% of the country's total area; half of the area is used for production [14]. Dry forests are of particular importance, since they have less biodiversity than rainforests, but they are the habitat of more than 130 species of birds. These forests are located in two different areas: (a) on the central Pacific coast, which corresponds to the provinces of Esmeraldas, Manabí, Santa Elena and Guayas; and (b) in the southern coast and western foothills of the Andes in El Oro and Loja that comprise the equatorial dry forest with a unique ecosystem in the world [15]. In the past, 35% of western Ecuador was covered by dry forest. However, 75% of the area has disappeared due to deforestation pressures and growth of the agricultural and livestock frontier.

Furthermore, PPF is recognized for its heritage as a true paleontological jewel that contains a large number of petrified trees of approximately 100 million years old, where the largest collection of petrified wood in the world is located. One of its largest specimens is the Petrino with dimensions of 2 m in diameter and 15 m in length. Its magnificence is given by a large number of petrified trunks that open the door to family, educational, scientific, geological-paleontological and naturist tourism. PPF is a unique beauty in the region with remnants of trees such as trunks and petrified leaves of Mesozoic flora, and fossils of invertebrates such as bivalves, ammonites, echinoderms, among others. Furthermore, its biodiversity coexists protected by the great slopes and breaks of the area [16,17].

In this regard, this study aims to respond the question: Could we define PPF as a geological heritage and a driving force for geotourism for its spectacularity, good exposure, paleontology, biostratigraphy, and stratigraphic content? Hence, the objective is to methodologically assess the geodiversity of a fraction of PPF as a pilot project, register its outstanding components (geological heritage) and value it for the preservation and sustainable development of the entire forest. For these purposes, an assessment is carried out using the scientific method of the Spanish Inventory of Places of Geological Interest (IELIG, *acronym in Spanish*) [18]. Moreover, some research-level experiences are recorded to assess sites of geological interest that could be considered as geosites [19,20].

2. Overview of the Study Area

Puyango forest is in the southern region of Ecuador, located in the canton Las Lajas, province of El Oro and Puyango-Alamor, canton-parish of the province of Loja. The cantons are divided by the Puyango river, located 7 km from the border with Peru. The forest occupies an extension of 2658 hectares (Figure 1), and in 1971, it was discovered by the academic staff of the Huaquillas night school that named it "Petrified Forest of Puyango" [20]. In 1973, PPF was declared Cultural Heritage On 9th January 1987 through Ministerial Agreement No. 22, it was declared a protected forest and vegetation due to the efforts of El Oro Cultural Development Center and the Central Bank of Ecuador. Later, in March 1988, it was declared part of the Natural Heritage of Ecuador [21,22].

This region is mountainous, with heights between 360 and 500 m.a.s.l. The current flora and fauna represent a group of transitional forms between the Pacific lowlands and the Andean elevations [23]. The biodiversity of the area corresponds to a tropical dry forest, which is one of the most threatened ecosystems in the world, and its species belong to the Tumbesino Center of endemism owned by Ecuador and Peru [8]. Due to the geomorphological conditions of the area, caused by tectonic events and modeled by the erosion of rivers, PPF presents a great potential for geotourism appeal in this area for the exposure of petrified logs, a product of the fossilization process [24,25].

Puyango Petrified Forest (PPF) is considered one of the few remnants of tropical dry forest in the southwest of the country, where steep slopes and streams such as El Guineo, Las Concreciones, El Chirimoyo, El Limón, Sábalos and Cochurco have preserved the endemic vegetation of its ecosystem and other areas of secondary forest in recovery. In fact, stratigraphic units reveal a fossil richness of invertebrates of the phylum Mollusca and microfossils of foraminifera and calcareous nanofossils [26]. Puyango is an open book of geological succession and paleontological information of high scientific value for society, since it is one of the most representative and relevant forests in South America. PPF has been compared to the Petrified Forest National Park in Arizona, United States—the largest in the world with more than 20,000 hectares. It has petrified tree trunks belonging to the Araucarioxylon arizonicum that are preserved from a conifer corresponding to the Late Triassic Period, already extinct in our times. Moreover, it has fern plant fossils, animals such as the Chinle frogs of the Chinle Geological Formation, and it is the habitat of a great variety of mammals, fauna, birds, reptiles, and amphibians in a desert environment [27].

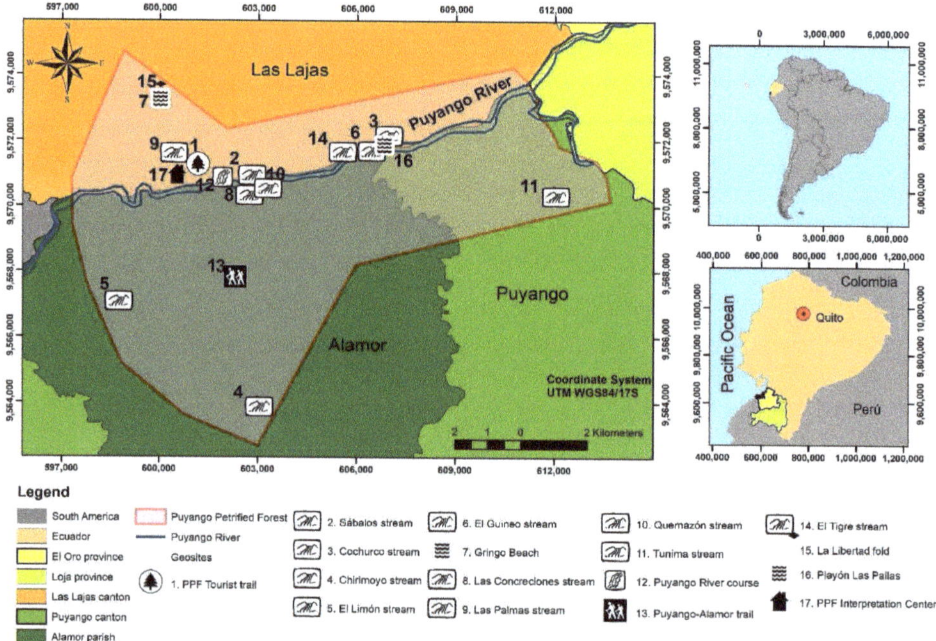

Figure 1. Study area of the Puyango Petrified Forest and its geosites in the provinces of El Oro and Loja. Modified from [28].

Before becoming a tropical dry forest, Puyango was a sea, which dried up and transformed into large hectares of forests and animals. Due to natural cataclysms, geological movements and time, the organisms buried underground arose on the surface forming a trace of the planet's remote past and a transcendent number of ancient and representative fossils that correspond to marine organisms that currently oscillate between 60 and 120 million years [26]. In this regard, Puyango emerged under coastal and terrestrial marine conditions in a relatively narrow basin caused by the continuous erosive tectonism, deformation, displacement, and deposits of pyroclastic materials during the Cretaceous. Before the Andes rose to the end of the Cretaceous 65 million years ago, gymnosperm forests originated on a relatively flat area like the lithified sediments that currently lean in different directions due to folding and tectonism. During this period, the area had a warm temperature. However, with the floods and volcanic activity east of the Andes, the forests were destroyed and deposits of buried trunk layers in alternating sequences of siltstones, sandstones, graywackes and conglomerates were reestablished by erosion [29]. The Andes mountain range slowly submerged the forest, placing it south of its current position in a subtropical climate along with calcium carbonate deposits and the presence of marine fossil invertebrates belonging to shallow water. The geological formation of the site is defined as a sequence of sedimentary rocks of the Late Cretaceous period formed by thin layers of calcareous black shales, crystalline, and massive black limestones, volcanic agglomerates, gray-green siliceous shales, brown shales, and volcano-sedimentary shales (Figure 2). In this type of Cretaceous (Albian-Middle) age materials, the petrified wood is found and in the upper part of the Coquina volcanic material, brown calcareous clays with mollusks and ammonites. In addition, a decreasing sedimentary volcanic grain sequence, clayey sandstones, tobaceous silt with petrified wood and reddish clay can be observed [30]. The trunks were dragged in an aqueous medium of remains of pyroclastic materials and deposited together with the sediments in a marine environment where they were covered by sediments as a result of the erosion of the Andes. Sediments are divided into four geological formations such as the Zapotillo, Cazaderos, Ciano, Ambín and Progreso. Marine invertebrate fossils such as gastropods, ostracods, and bivalves, which are associated with microfossils of the order of the foraminifera and calcareous nanophosiles of the Cretaceous (Middle Albian) age, are found in the layers of sedimentary materials [31]. The Puyango river is the limit between the provinces of El Oro and Loja and represents a great geological fault in the East–West direction. It has caused the folding of rocks at the gorges of El Chirimoyo, Cochurco and El Limón streams [32,33].

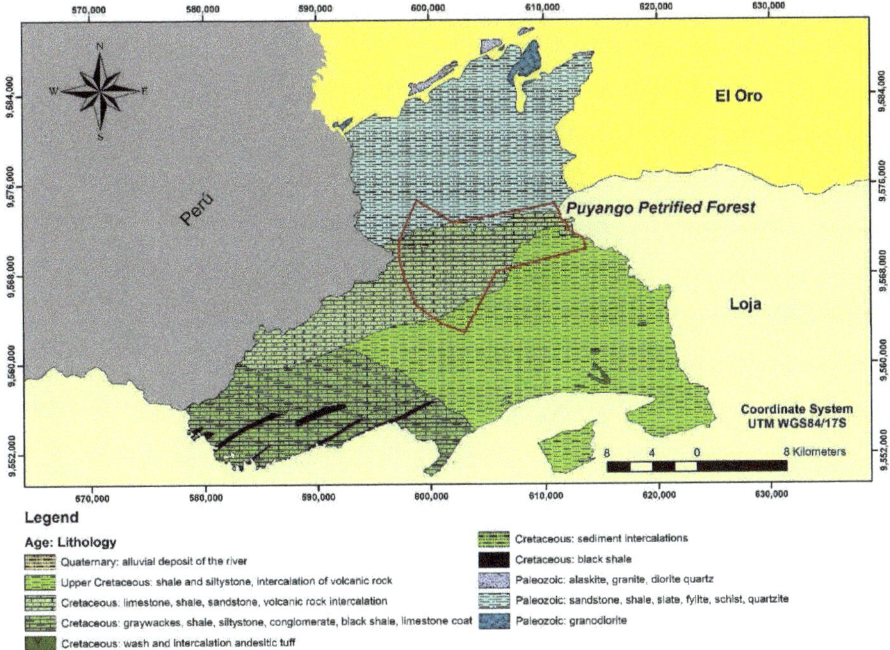

Figure 2. Geological map of the Las Lajas and Puyango cantons of the provinces of El Oro and Loja respectively. Modified from [28].

3. Materials and Methods

The proposed methodology for this study comprised three phases (Figure 3): (i) analysis of information from the pilot study area, the case of the Puyango Petrified Forest, scientific information regarding its geological interest, the presentation the paleontological components and their main geotouristic sections; (ii) assessment of the heritage site, using the methodology of the Spanish Inventory of Places of Geological Interest (IELIG, *acronym in Spanish*) and another proposal by authors to assess its geotouristics potential; and (iii) the results were confirmed with a Strengths, Weaknesses, Opportunities, and Threats (SWOT) analysis, to assess the heritage site status, and propose sustainable development strategies in the SWOT matrix [34].

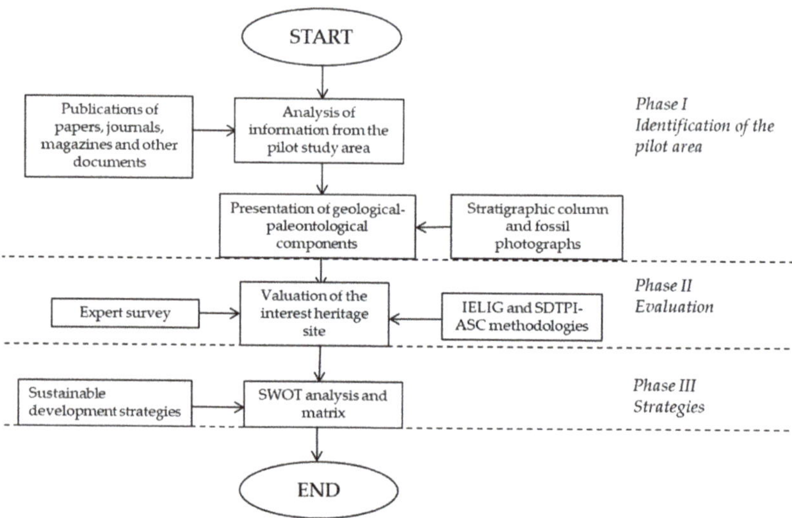

Figure 3. Flowchart of the methodology used for research.

3.1. First Phase: Identification of the Pilot Area

In the first phase, the researchers realized a technical visit about the outstanding characteristics of the studied forest, which was carried out with a tourist guide and some national tourists. The route of the place consists of six stations of great geological and paleontological interest, in addition to the collection of historical and scientific information carried out in the sector [26,29,30] through the configuration of a database in scientific publications, works outreach, project reporting and data collection through expert interviews as the basis for further evaluation. This route included the use of database in Geographic Information Systems (GIS) to obtain a map of the tourist route recorded in the Puyango forest.

Finally, this phase included the description geological, paleontological aspects and relevant observations of the geotouristics potential of seventeen possible geosites. We also presented its paleontological components that give the forest an important scientific interest for geologists and paleontologists. Much of its evidence was recorded in fossils such as petrified wood samples and remains of mollusc shell moulds that were found in some streams belonging to the provinces of El Oro and Loja, giving great geotouristics interest to the forest. Besides that, a stratigraphic column belonging to the Las Concreciones stream was described, and finally, it showed pictures of marine fossil invertebrates of the classes of pelecypod and cephalopod molluscs.

3.2. Second Phase: Quantitative and Qualitative Evaluation

Moreover, the applied methodology comprises two factors: (1) intrinsic value of the Site of Geological Interest (LIG, *acronym in Spanish*), highlighting the geological aspects, and (2) use-value of the LIG. Therefore, with the collaboration of paleontologist Nelson Jiménez and archaeologist Jorge Marcos, experts who evaluate the scientific, didactic and touristic premises value, surveys are collected. The experts assign weights to each premise of the methodology developed in [18] (Table 1), where the degrees of Scientific interest (Si), Didactic interest (Di) and Touristic interest (Ti) are obtained. The scores of each parameter vary from 0, 1, 2 and 4, where 0 is the lowest score and 4 the highest, scores that are multiplied by the weight of the individual interests given in the methodology (Table 1).

Table 1. Indicators and weights used for the quantitative assessment of Sites of Geological Interest (LIGs). Modified from [18].

IELIG Methodology							
Scientific Weight (Sw), Didactic Weight (Dw), Tourist Weight (Tv), Fragility Weight (Fw) and Vulnerability due to Anthropogenic Threats Weight (Vw)							
Indicators/Parameters	**Punctuation**	**Weight**					
		Sw	**Dw**	**Tw**	**Fw**	**Vw**	
Representativeness		30	5				
Prototype location character		10	5				
Degree of scientific knowledge of the place		15					
State of conservation		10	5				
Observation conditions		10	5	5			
Rarity		15	5				
Geological diversity		10	10				
Didactic content			20				
Logistics infrastructure			15	5			
Accessibility			15	10			
Association with other elements of the natural and/or cultural heritage			5	5			
Magnificence or beauty			5	20			
Population density (potential aggression)	0–4		5	5		5	
Proximity to recreational areas (immediate potential demand)				5		5	
Informative content/Informative use detected				15			
Potential to carry out touristic and recreational activities				5			
Socioeconomic environment				10			
LIG size				15	40		
Vulnerability to plunder					30		
Natural threats					30		
Proximity to anthropic activities (infrastructure)						20	
Interest for mining						15	
Site protection regime						15	
Physical or indirect protection						15	
Accessibility (potential assault)						15	
Place ownership regime						10	
Total		100	100	100	100	100	

Then, the total provides the value of Si, Di and Ti. If the LIG exceeds 266 points, it is considered a place of "Very high" interest. Hence, the scores between 134 and 266 will be of "High" interest, and those lower than 134 points will be considered to be of "Medium" interest. The following equations are defined for the value of each interest:

$$Si = \sum_{i=parameter}^{n\ parameters} Puntuation \times Scientific\ weight \tag{1}$$

$$Di = \sum_{i=parameter}^{n\ parameters} Puntuation \times Didactic\ weight \tag{2}$$

$$Ti = \sum_{i=parameter}^{n\ parameters} Puntuation \times Tourist\ weight \tag{3}$$

Furthermore, a qualitative evaluation about conservation of the site is achieved, where the Susceptibility of Degradation (SD) of the site is evaluated based on the Fragility (F) and the vulnerability due to anthropogenic Threats (T). With the SD, researchers can obtain the Protection Priority (PP) for Si, Di, Ti of the LIG and rank the interests according to the PP value in its different vertices: scientific (SPP), didactic (DPP), touristic-recreation (TPP), and global (PP). Taking the threshold of the pilot project in the Iberian Cordillera [18] for a given value of the SD as reference, if the value is higher than 26, the PP degree "High" of the LIG. If the SD is equal to 26, the PP degree is "Medium". The "Medium-high" and "Medium-low" PP will be around 8. To obtain the values of fragility (F) and threat (T) that allow prioritizing and monitoring of the conservation status, where it is susceptible to degradation, and to quantify the priorities SPP, DPP, TPP, and PP, the researchers used the following equations:

$$F = \sum_{i=parameter}^{n\ parameters} Puntuation \times Fragility\ weight \tag{4}$$

$$T = \sum_{i=parameter}^{n\ parameters} Puntuation \times Threat\ weight \tag{5}$$

$$SD = ((F \times T) \times 1/400) \tag{6}$$

$$SPP = \left(\left(Si^2 \times SD\right) \times 1/400^2\right) \tag{7}$$

$$DPP = \left(\left(Di^2 \times SD\right) \times 1/400^2\right) \tag{8}$$

$$TPP = \left(\left(Ti^2 \times SD\right) \times 1/400^2\right) \tag{9}$$

$$PP = [(Si + Di + Ti)/3]^2 \times SD \times 1/400^2 \tag{10}$$

Moreover, the authors include a completed form of the proposed methodology that includes the Scientific, Didactic, Tourist and Popular Interest factors with Accessibility, Sensitivity and Conservation status (SDTPI-ASC), considering the 17 LIGs of relevant geological-paleontological importance. This approach strengthens the evaluated geological heritage. For this purpose, the interest categories are divided into Scientific interest (Si), Didactic interest (Di) and Touristic interest (Ti) and Popularization interest (Pi), with its sections on accessibility to the place, sensitivity to plunder and current conservation status, as a basis for future studies.

3.3. Third Phase: Strategies

In this phase, based on expert judgment, the SWOT matrix configuration was done, with experiences from technical visits and studies in the Puyango Petrified Forest (BPP). This matrix allows establishing opportunities for geotourism development, identifying strengths, weaknesses and threats. The particular singularity of the territory in its geological and paleontological heritage has been highlighted, with a high interest for national and international geologists interested in investigating the geological record of the place, the scientific, educational and recreational use to strengthen the local economy.

There is a high fossil content, pleated structures, sedimentary, volcanic, metamorphic rocks with a natural, integral and biodiverse perspective, with samples of petrified trunks, pelecypods, macrofossils and ammonites. In the analysis of experts, the limitations or problems of the geosites are also considered, to turn them into new strategies for local development.

A SWOT analysis was carried out to assess the conservation status of the heritage site and to propose sustainable development strategies in a SWOT matrix [34], where the Puyango geopark project has great potential to promote the development of geotourism in the zone.

4. Results

4.1. Identification of the Pilot Area

The identification of the pilot area includes the visit of the tourist trail carried out in the Puyango Petrified Forest (PPF), in the province of El Oro. This guided route lasts approximately 45 min and observation stops are made at the stations: (1) Lava flows, (2) Path of the Araucarias, that is a genus of evergreen coniferous trees in the family Araucariaceae, (3) Deposit of petrified logs, (4) Carboniferous zones, (5) Path of the Giant, and (6) Giant Petrine (Figure 4).

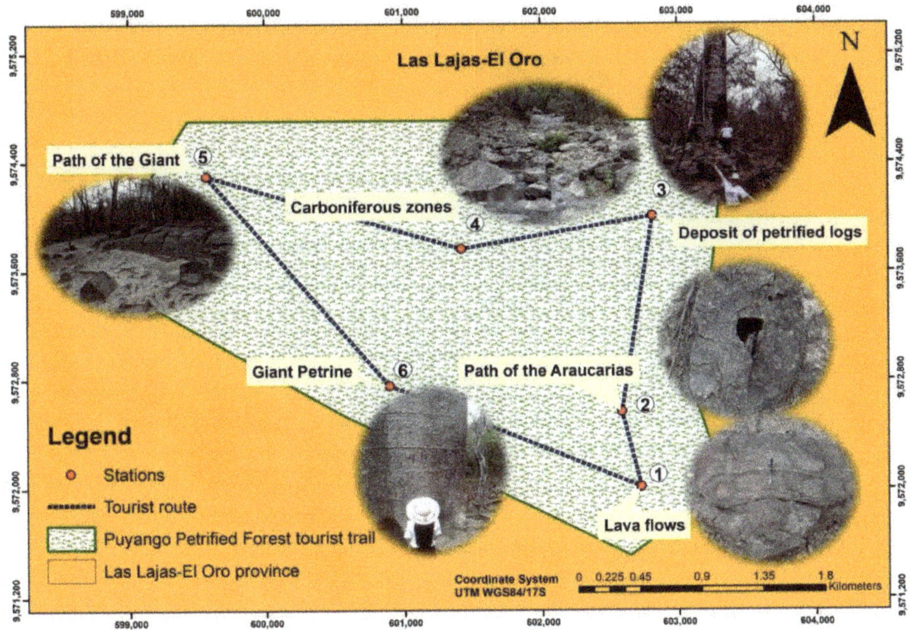

Figure 4. Map of the Puyango Petrified Forest tourist trail, El Oro Province [28].

For a better scientific reference about LIGs, Table 2 details the important geological aspects such as geological structures and type of rocks found in the geosite, as well as paleontological aspects that stand out for their fossil content, the type of fossilization and the number of petrified trunks found in each geosite which makes it an excellent geotouristic remnant.

Some of the LIGs mentioned in Table 2 belong to the province of El Oro, such as the streams Sábalos, Las Palmas, El Tigre, El Guineo, El Gringo, Quemazón and in the province of Loja are the El Chirimoyo, Cochurco, El Limón, Las Concreciones and Tunima (Figure 5). In the Cochurco, El Chirimoyo and Las Concreciones streams, remains of mollusk shell molds such as pelecypods and ammonites have been found in the limestones [26].

Table 2. Geological and paleontological aspects of the LIGs.

N°	LIGs	Geological Aspects	Paleontological Aspects	Observations
1	PPF Tourist trail	Volcano-sedimentary rocks.	It contains abundant remains of well-preserved petrified trunks.	It is a guided tourist route in the tropical dry forest (fauna and flora).
2	Sábalos stream	Geological domain: Alamor, Lancones Basin. At the base, some ortho-quartzites and conglomerates rest discordantly on the Amotape, Tahuin massif. Black marl, massive layers of limestone and banded black shales, clays, siltstones and sandstones with lava and sedimentary volcano rocks.	It contains abundant remains of petrified trunks corresponding to *Araucariaceas*.	The trunks are well preserved, practically intact of large dimensions, the only fossil specimens resulting from the fossilization process of petrification, product of the replacement of organic matter by silica.
3	Cochurco stream	Volcanic materials and sedimentary limestones and shales.	Remains of fossil roots (charred), remains of petrified trunks. Fossil invertebrates, present as internal and external moulds.	A petrified trunk of *Araucariaceas* well preserved, in a significant vertical position.
4	Chirimoyo stream	Geological deformations (folds) in sedimentary rocks represented by clay.	Remains of fossil invertebrates: Pelecypods, Inoceramus and Ammonites; and microfossils: planktonic and benthic foraminifera, calcareous nanofossils and Palinomorphs.	Invertebrate marine fauna is very frequent.
5	El Limón stream	Geological deformations, fold (anticline) in volcano-sedimentary rocks. Calcareous rocks (Coquina type).	Petrified trunks well preserved. Invertebrate fossils: foraminifera and calcareous nanofossils.	Petrified (carbonized) trunks, in which the cellulose of the trees was transformed into anthracite due to the loss of methane, water and carbon dioxide. This geosite has a potential area to recreation (crystal clear lagoon).
6	El Guineo stream	It shows stratifications by the alternative deposit of sediments of different composition.	Fossil invertebrates: a *phylum* of molluscs, bivalves and ammonites and petrified trunks.	Pelecypod prints.
7	Gringo Beach	Volcano-sedimentary rocks, limestones, shales and clays.	Fossil invertebrates: a phylum of molluscs, bivalves and ammonites.	Vegetable fossil remains, from fossilization processes (carbonization).
8	Las Concreciones stream	Limestone outcrop associated with a core of Ammonites. Shales, marls and thin layers of clays. The limestones inside have pyrite.	Ammonite concretions. Molluscs, foraminifera and calcareous nanofossils and ostracods.	A unique geosite: can be shown calcareous concretions.
9	Las Palmas stream	Volcano-sedimentary rocks.	Remains of allochthonous petrified trunks.	Little diversity of petrified trunks.
10	Quemazón stream	Geological deformations, folding and thrust in limestone rocks.	Appears a few remains of petrified (charred) trunks.	Fossilization processes (carbonization).
11	Tunima stream	Volcano-clastic rocks, with levels containing petrified trunks.	Petrified trunks well preserved.	Petrified trunk in volcano-clastic rocks with a diameter smaller than found in Sábalos and Chocurco streams.
12	Puyango River course	Geographical reference whose cause has an altitude of 200 m.a.s.l, is the water axis of importance for the ENE-WSW direction according to the structural limits of the sector.	Allochthonous petrified trunks.	Appears remains of petrified trunks and fossil invertebrates.

Table 2. *Cont.*

N°	LIGs	Geological Aspects	Paleontological Aspects	Observations
13	Puyango, Alamor trail	None.	Allochthonous petrified trunks.	None.
14	El Tigre stream	Volcano-sedimentary sediments reddish color due to the presence of iron oxides.	Vegetable fossil remains (fossilized leaf) by fossilization processes (carbonization).	The only geosite with a petrified leaf specimen in the fossilization process (carbonization).
15	La Libertad fold	Metamorphic deformations. Anticline fold.	Geosite without paleontological remains.	Geological deformations due to compressive stresses.
16	Playón Las Pailas	Appears deposits of visible material of sedimentary rocks along the river bed in the dry season.	Sedimentary rocks carbonated, with remains of invertebrates.	Recreational tourist place for organized activities.
17	PPF Interpretation left	Exists geological information (samples of rocks and calcareous concretions).	Petrified trunks, fossils. Recreation area, location maps of the main gorges, paleontological information.	Disclosure area.

Figure 5. Streams of the province of El Oro (**a**) Sábalos, (**b**) Cochurco, (**c**) Las Palmas and (**d**) Las Concreciones.

Moreover, a stratigraphic column of the Las Concreciones stream is shown (Figure 6) as a result of the scientific research carried out in. The stream adopted that name because at the beginning of the stream, precisely upstream, there are concretions of limestone rolled from the upper part, whose outcrop is made of sedimentary material. On the shales, limestones, marls and thin layers of clay are observed. Limestone concretions of 0.80 m in diameter are observed in the rock with a light gray hue. The limestones are crystalline with pyrite as a mineral. In addition, the concretion of two Ammonite molds lie, and a large limestone outcrop approximately 7 m thick stands out at the end of the station [27].

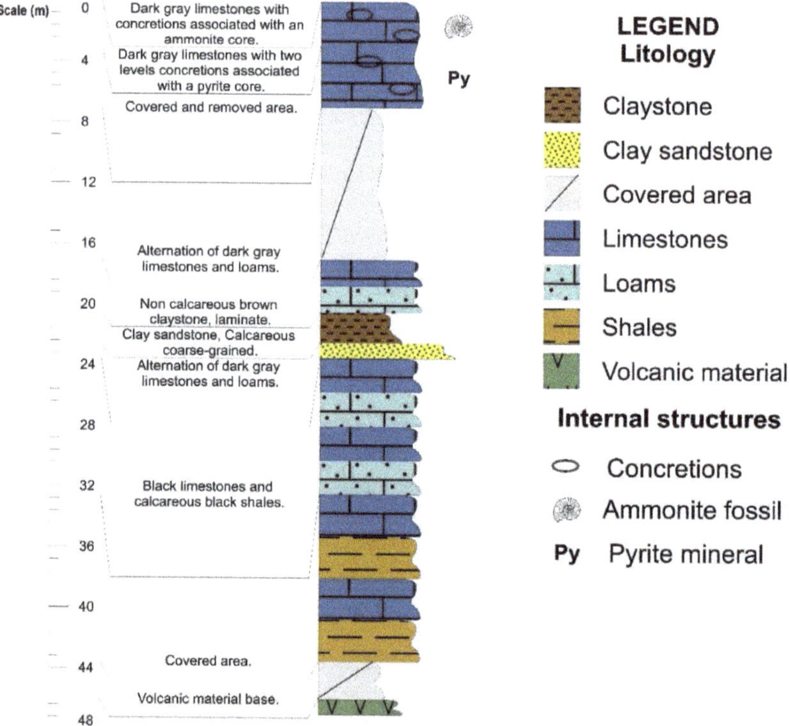

Figure 6. Stratigraphic Column of the Las Concreciones stream. Modified from [29].

The fossil invertebrates are marine and belong to the pelecypods and cephalopod classes of mollusks (Figure 7). The cephalopod class is represented by ammonoids represented by the internal and external molds as well as petrified shells in the Cochurco, El Chirimoyo, El Limón, Las Concreciones, and El Guineo streams. The internal molds of mollusks with *Nucula spp*, *Inoceramus concentricus*, *Astarte spp* and *Heterodontido* (See Supplementary Figures S1–S4) [26].

Figure 7. Macro-fossils in the province of Loja: (**a**) Pelecipodo *heterodontido*, (**b**) *Ostrea sp.*, (**c**) *Astarte sp.*, (**d**) *Peltoceras sp.*, (**e**) Fragment of return of *Ammonitido* and (**f**) *Schoenbachia sp.*

4.2. Evaluation

The results of phase ii with the scientific, didactic and touristic interests are presented according to the IELIG methodology with its justification in Table 3.

To obtain the value of each item of interest, Equations (1)–(3) were used. Results reveal that the Puyango Petrified Forest presents a "Very high" and "High" global degree of interest in the Scientific (Si), Didactic (Di) and Touristic (Ti) aspects. Although some of them have a "Middle" public interest (Figure 8), the reason is that the lack of strategies that promote the geodiversity of the area through geotourism.

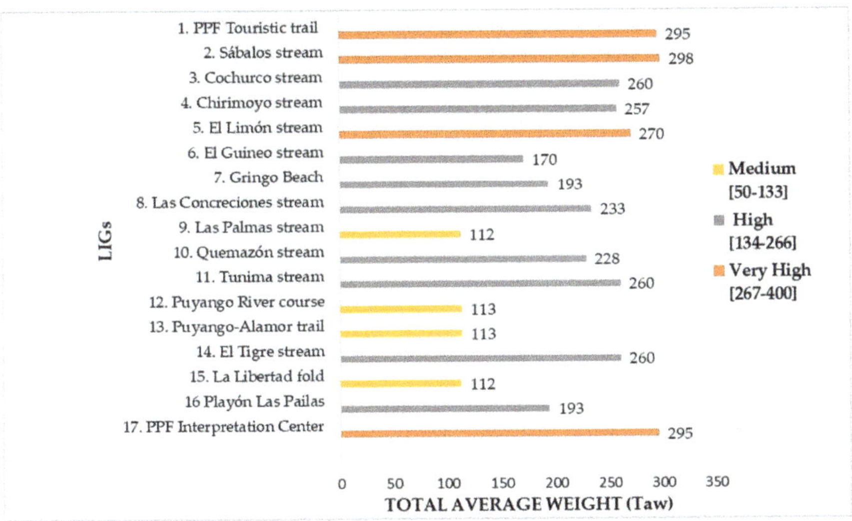

Figure 8. Assessment of the degree of geological interest of the LIGs in the Puyango Petrified Forest according to IELIG.

Table 3. Description of the main indicators evaluated by the IELIG methodology.

Value	Indicators/Parameters	Description
Scientific value	Representativeness	The Puyango Petrified Forest (PPF) is a source of knowledge and well-being and it is the unique place with the best paleontological exposure in Ecuador. The area shows a sequence of volcano-sediment from the Cretaceous age (Albian-Middle). A collection of petrified trunks, fossilized leaves of common gymnosperms and remains of mollusk shell molds of the Southern Cretaceous have been found in PPF deposits.
	Prototype location character	PPF is a site of geological interest and a good example of a paleontological deposit, where specimens of petrified trunks and other allochthonous from Aptian-Middle and Jurassic-Cretaceous are found in situ.
	Degree of scientific knowledge of the place	There are several works published in national and international journals performed by geoscientists and theses carried out by students and professionals on geological, paleontological, archaeological, botanical, cultural, economic and tourism topics.
	State of conservation	PPF presents a regular state of conservation. Some deteriorated areas prevent the observation of geological characteristics (landslides). Some information panels are partially deteriorated.
	Observation conditions	The observation of several silicified trunks of *Araucariaceas* and *Metapodocarpoxylon* specimens, marine invertebrate fossils of the phylum Mollusca of the pelecypods class and cephalopod *(ammonoids)* possible. The fossilized flora corresponds to the *subphylum* of the mid-Mesozoic gymnosperms of the *Zamites, Dioonites, Nilssonia, Otozamites, Podozamites, Carpites,* and other genera. It is also a dry-tropical forest ecosystem that preserves species (fauna and flora) existing today.
	Rarity	The petrified trunks of Puyango represent one of the largest collections of petrified wood in Ecuador and probably in the world. The site has unique characteristics from the Cretaceous Period.
	Geological diversity	The paleontological aspect stands out as the main geological interest, followed by the stratigraphic, sedimentological, structural and geological history. Secondary features are the historical/archaeological, biodiversity, cultural and landscape.

Table 3. *Cont.*

Value	Indicators/Parameters	Description
Educational potential and touristic use	Didactic content	Educational visits and excursions from schools, colleges, and universities take place considering the protection of the geological-paleontological heritage of the place.
	Logistics infrastructure	It lacks nearby accommodations and restaurants for tourist groups, but it has an interpretation and information left for tourists.
	Accessibility	Using a national highway there is the access Guayaquil-Machala-Arenillas-Puyango with a 276 km route. From Huaquillas (border with Peru), there are 62 km. Another access road goes from Loja-Veracruz- Catacocha's city through a state highgway and then El Empalme-Celica-Alamor-Puyango with a distance of 213 km [26].
	Association with other elements of the natural and/or cultural heritage	Numerous archaeological remains (petroglyphs) found in different parts of the Puyango canton.
	Magnificence or beauty	Landscape, river course, remains of plant fossils and invertebrates as evidence of ancient times.
	Informative content/Informative use detected	Limited and without a tourist information department.
	Potential to carry out touristic and recreational activities	It has touristic trails, fossil deposits and streams.
	Proximity to recreational areas (immediate potential demand)	There is a camping area less than 500 m from the forest.
	Socioeconomic environment	The most common economic activities are the short cycle crops sowing and coffee, cattle and pig raising.
Fragility	LIG size	Area of 2.6 ha. with 17 LIGs.
	Vulnerability to plunder	A paleontological site of great value, with numerous specimens and easy plunder.
	Natural threats	Possible landslides, flooding of rivers, weakening of the soil and climatic variations due to severe droughts.

Table 3. *Cont.*

Value	Indicators/Parameters	Description
Vulnerability due to anthropogenic threats	Proximity to anthropic activities (infrastructure)	Place not threatened.
	Interest for mining	No mining interest in the area.
	Site protection regime	Cultural Heritage, Ordinance for the Declaration of the Bi-provincial Protected Area, in the Category of Ecological Conservation Area.
	Physical or indirect protection	Protected area with access to tourists.
	Accessibility (potential assault)	It is directly accessible through unpaved and passable track for tourism.
	Place ownership regime	Location in restricted access areas declared as natural heritage.
	Population density (potential aggression)	The town is located at the entrance to the forest tourist complex that belongs to the province of El Oro, Las Lajas canton. It has an economically active population of approximately 200 people, mostly farmers [26].

The results of the applied evaluation of the IELIG methodology are represented in Table 4, noting that there are four LIGs with Geological Interest (IG) "Very High", nine of interest "High" and four "Medium", reflecting the great relevance of the Puyango Petrified Forest. The Global Protection Priority (PPG) values are also shown, finding most of the LIGs with the rating of "Medium-low".

The fossilized trunks resulted from the fossilization process, especially petrification by replacing organic matter such as cellulose and limenine with silica [33,35]. The paleontological importance with the highest concentration of logs has been found in the old Puyango-Alamor highway and El Chirimoyo, El Limón, and Cochurco streams, where specimens of logs of up to 26 m long by 2.2 m in diameter have been located (Figure 9).

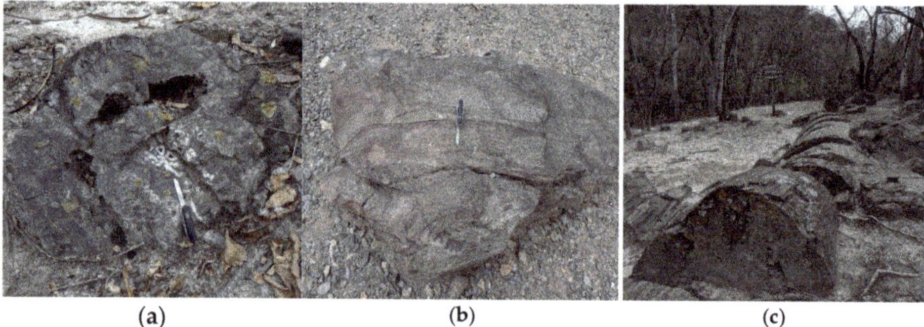

<table>
<tr><td align="center">(a)</td><td align="center">(b)</td><td align="center">(c)</td></tr>
</table>

Figure 9. (a,b) Logs petrified with silica. (c) Giant petrified trunk.

Table 4. Quantitative assessment of parameters Scientific interest (Si), Didactic interest (Di), Touristic interest (Ti), Total Average Weight (Taw), Degree of geological interest (GI), Susceptibility of Degradation (SD), Scientific Protection Priority (SPP), Didactic Protection Priority (DPP), Touristic Protection Priority (TPP), Protection Priority (PP) and Global Protection Priority (PPG), according to the IELIG methodology.

N°	LIGs	Si	Di	Ti	Taw	GI	SD	SPP	DPP	TPP	PP	PPG
1	PPF Touristic trail	380	275	230	295	Very High	50	45.13	23.63	16.53	27.20	High
2	Sábalos stream	360	250	285	298	Very High	4.50	3.65	1.76	2.28	2.50	Medium-low
3	Cochurco stream	320	225	235	260	High	8.25	5.28	2.61	2.85	3.49	Medium-low
4	Chirimoyo stream	330	215	225	257	High	18	12.25	5.20	5.69	7.41	Medium-low
5	El Limón stream	335	215	260	270	High	0	0	0	0	0	Medium-low
6	El Guineo stream	185	150	175	170	High	0	0	0	0	0	Medium-low
7	Gringo Beach	240	165	175	193	High	9.75	3.15	1.66	1.87	2.27	Medium-low
8	Las Concreciones stream	270	175	255	233	High	5.63	2.56	1.08	2.29	1.91	Medium-low
9	Las Palmas stream	110	80	145	112	Medium	9	0.68	0.36	1.18	0.70	Medium-low
10	Quemazón stream	305	170	210	228	High	0	0	0	0	0	Medium-low
11	Tunima stream	320	225	235	260	High	8.25	5.28	2.61	2.85	3.49	Medium-low
12	Puyango River course	110	90	140	113	Medium	82.50	6.24	4.18	10.11	6.62	Medium-low
13	Puyango-Alamor trail	110	90	140	113	Medium	82.50	6.24	4.18	10.11	6.62	Medium-low
14	El Tigre stream	320	225	235	260	High	8.25	5.28	2.61	2.85	3.49	Medium-low
15	La Libertad fold	110	80	145	112	Medium	9	0.68	0.36	1.18	0.70	Medium-low
16	Playón Las Pailas	240	165	175	193	High	9.75	3.15	1.66	1.87	2.27	Medium-low
17	PPF Interpretation Center	380	275	230	295	Very High	50	45.13	23.63	16.53	27.20	High

From the point of view of the conservation of LIG, it is necessary to determine the susceptibility of degradation. This analysis is performed through the fragility and natural threats components. Hence, mitigation actions can be taken to reduce as far as possible the vulnerability of the geosite and the damage caused by an event and anthropic threats identified in the sector. The data obtained as a result of the implementation of Equations (6)–(10) were presented in Table 4. With this values, Protection priorities "Medium-low" were identified for each of the protection priority parameters SPP, DPP, TPP and PP, which depend on the value calculated in the LIG Degradation Susceptibility (SD), this parameter is calculated independently of the others. The Figure 10 shows the sections of SPP, DPP and TPP with a global Protection Priority (PP) "Medium-low", except for two LIGs with a PP "High".

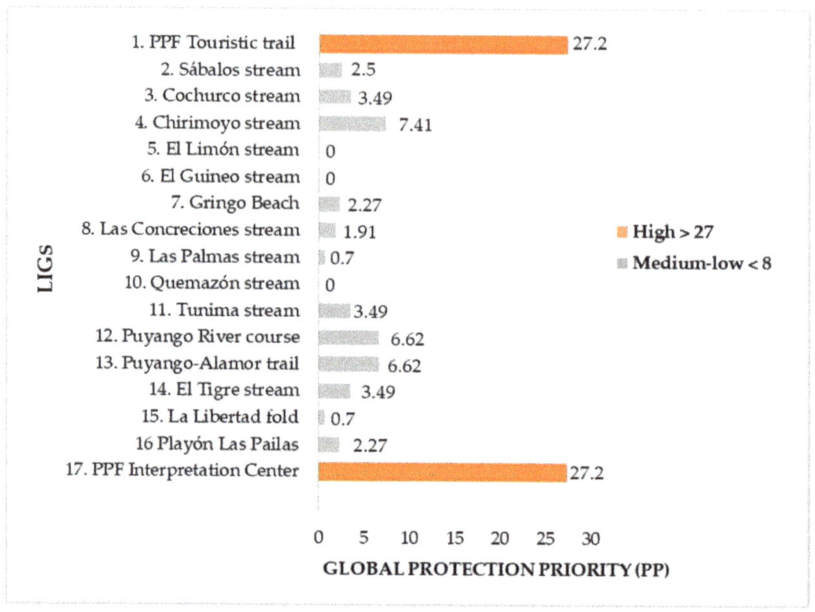

Figure 10. Assessment of the Protection Priority of the LIGs.

Additionally, all 17 LIGs which strengthen the entire system of geological-paleontological heritage in this sector are marked in the PPF. With the application of the Scientific, Didactic, Tourist and Popular Interest with Accessibility, Sensitivity and Conservation Status methodology (SDTPI-ASC), the authors identified the different types of interest: Scientific (Si), Didactic (Di), Tourist (Ti) and Popularization (Pi). The SDTPI-ASC was supported by the criteria of experts Nelson Jiménez, Jorge Marcos and Irina Xomchuk, based on [26]; Table 5 illustrates the SDTPI-ASC methodology.

Table 5. Referential assessment of the LIGs by the SDTPI-ASC methodology evaluated by experts in the Puyango Petrified Forest. It has a type of interest (✓).

N°	LIGs	UTM Coordinates	Type of Interest				Accessibility	Sensitivity	Conservation Status
			Si	Di	Ti	Pi			
1	PPF Tourist trail	601735E, 9560324S	✓	✓	✓	✓	Good	Average	Good
2	Sábalos stream	602734E, 9570326S	✓	✓	✓	✓	Good	Average	Good
3	Cochurco stream	606966E, 9572143S	✓	✓	✓	✓	Good	Poor	Good
4	Chirimoyo stream	603047E, 9563869S	✓	✓	✓		Below average	Poor	Good
5	El Limón stream	606437E, 9571676S	✓	✓	✓	✓	Below average	Poor	Below average
6	El Guineo stream	598773E, 9567082S	✓		✓		Below average	Poor	Below average
7	Gringo Beach	600022E, 9573244S	✓		✓	✓	Below average	Poor	Good
8	Las Concreciones stream	602810E, 9570930S	✓	✓			Poor	Poor	Below average

Table 5. *Cont.*

N°	LIGs	UTM Coordinates	Type of Interest				Accessibility	Sensitivity	Conservation Status
			Si	Di	Ti	Pi			
9	Las Palmas stream	600442E, 9571602S	✓		✓	✓	Below average	Poor	Good
10	Quemazón stream	603298E, 9570534S			✓		Good	Poor	Below average
11	Tunima stream	612035E, 9570322S			✓		Below average	Poor	Good
12	Puyango River course	601926E, 9570859S		✓	✓		Regular	Poor	Below average
13	Puyango-Alamor trail	602305E, 9567823S		✓	✓		Good	Poor	Below average
14	El Tigre stream	605566E, 9571653S	✓	✓			Good	Below average	Average
15	La Libertad fold	600021E, 9573243S	✓	✓			Below average	Below average	Below average
16	Playón Las Pailas	606828E, 9571865S			✓	✓	Good	Below average	Average
17	PPF Interpretation Center	601354E 9568864S		✓	✓	✓	Good	Below average	Below average

4.3. Strategies

The results obtained through a SWOT analysis, which required extensive geological-paleontological fieldwork with experts and people from the community, are presented in Table 6. The authors took the Sustainable Development Goals (SDGs) into account to obtain strategies in terms of sustainable geotourism.

Table 6. Matrix of Strengths, Weaknesses, Opportunities, and Threats (SWOT) of the Puyango Petrified Forest.

		Strengths	Weaknesses
Internal environment		S_1. Puyango is the unique petrified forest of Cretaceous-Aptian age in the country and it has international relevance. S_2. It has an outstanding scientific interest due to the exceptional type of fossilization of the petrified trunks. S_3. It was declared the Ecuadorian a Cultural Heritage Treasure. S_4. The unique forest that combines petrified and natural trees. It is the habitat of 65 species of birds and other animals.	W_1. Lack of "social awareness" necessary in a geological-paleontological and cultural heritage. W_2. Lack of infrastructure and tourist care services (emergency ward, restaurants, and lodgings). W_3. Lack of a bilingual tourist information center that provides scientific and relevant information W_4. Lack of internet access and media communications.
External environment			
Opportunities		**Strategy:** **Strengths + Opportunities**	**Strategy:** **Weaknesses + Opportunities**

Table 6. *Cont.*

O₁. Geopark Project proposal under development by government entities and universities. O₂. Protection and restoration of LIGs through various strategies with national and international organizations. O₃. Geotourism development open to improvement. O₄. Declare PPF as a Biosphere Reserve (BR) in the dry forest of southern Ecuador. It is a strategic opportunity.	$S_1.O_1$. Complement studies to generate a database of scientific information on paleontological resources. $S_2.O_2$. Develop strategies for the sustainable management of cultural and natural resources: conservation, research, and development for the proposal of the Puyango Geopark project. $S_3.O_3$. Undertake conservation and cleaning work on fossil outcrops, protection of specimens in situ, an adaptation of trails and placement of information panels.	$W_1.O_1$. Participatory Geopark proposal that integrates geodiversity, biodiversity, territory and people. $W_2.O_2$. Promote Social Management programs (training and awareness) that involve rural families in the management of resources. $W_4.O_4$. Promote the management of information, production and popularization of knowledge and tourism promotion at national and international level, through ICTs.
Threats	**Strategy:** **Strength + Threats**	**Strategy:** **Weaknesses + Threats**
T_1. The burning of vegetation to obtain new land for planting. T_2. Deforestation and overgrazing that degrades the ecosystem. T_3. Collection of samples without approval from Authorities. T_4. Demand from the neighboring country Peru for the pollution of the Puyango River due to mining effects outside the PPF [36]. T_5. Possible landslides, river floods and climatic variations due to severe droughts.	$S_2.T_4$. Encourage the development of good bilateral relations with the neighboring country Peru. This strategy stabilizes the intra-border situation and encourages Peruvian tourism to the PPF area. $S_3.T_3.T_4$. Recovery of the paleontological deposit of fossil outcrops of trees and marine invertebrates from the PPF and nearby areas such as the Piedmont of the Tumbesina eco-region (Ecuador-Peru). $S_4.T_3$. Legal regulations with permanent control on all activities within the forest, such as trail management, conservation, staff performance, tour operators, transportation income, agricultural production, etc.	$W_1.T_1$. Promote sustainable tourism and mitigation of local anthropogenic threats. $W_2.T_2$. To plan and promote the reforestation in selected areas through a pilot program, all together with farm owners within the forest has protected area to establish a mining-environmental order. $W_3.T_3$. Find private investors for the construction of a tourist accommodation and recreation center. $W_4.T_4$. Build a permanent Information Center and administrative offices, according to the regulations to certify the area as an eco-touristic site.

5. Discussion

The results obtained based on the IELIG's methodology provided data for both a quantitative and qualitative analysis, where the experts in geological, paleontological and archaeological sciences highlighted the forest's value as a unique example of the Cretaceous Period at a national level, which requires the recovery and preservation of its importance of the PPF as an Ecuadorian unique example of the Cretaceous Period, which requires the recovery and conservation of its heritage for sustainable development [37]. Additionally, the analysis suggested that the geotourism alternative supports the "Puyango" Geopark project. Experts have considered a referential assessment from 17 LIGs taken in the forest which on average are described as "below average" accessibility, "poor" sensibility and a "below average" conservations state that supports the usage of strategies to enhance the development of this area destinated for the geotourism as a pedagogical tool; it also promotes the restitution and improvement of geological and paleontological informative panels. Thus, it is essential to incorporate a web page to strengthen and increase the influx of tourists in the area, where the site is scientific and cultural information is disseminated nationally and internationally and to keep a record of annual visits.

The methodological study allows considering the PPF as a Geological Interest Place (LIG, *acronym in Spanish*) for the geological, paleontological, historical and cultural environment that surrounds it. The results evidenced to "Very high" geological interest in the LIGs "PPF Touristic trail", "Sábalos stream", "El Limón stream" and "PPF Interpretation Center"; "High" geological interest in "Cochurco stream", "Chirimoyo stream", "El Guineo stream", "Gringo Beach", "Las Concreciones stream", "Quemazón stream", "Tunima stream", "El Tigre stream" and "Playón Las Pailas" highlighting the paleontology of the place. The LIG Protection Priority (PP) scores classified as "Medium-low" in most LIGs; this is because the Puyango Petrified Forest (PPF) is a protected natural reserve of public administration, which gives it the competence and authority to management and protection policies. The LIGs to "PPF Touristic trail" and "PPF Interpretation Center" have a "High" rating of Protection Priority since they have a high influx of national and international visitors [17]. In the assessment, vulnerability to plunder also responded to the anthropic threat caused directly by collectors as the paleontological appeal is an intrinsic characteristic of the geosite. Therefore, knowing its PP fosters the adoption of measures for the conservation of the LIG [38,39].

Despite having high values in the geological-paleontological interest, it is necessary to monitor PPF, implement the proposed strategies and analyze its tourist development since it has 13,000 visitors per year, 70% are domestic tourists and 30% foreign [40]. Furthermore, the present study identifies shortcomings in terms of the infrastructure, as the place demands services for tourists such as accommodation, restaurants, a health care center and internet access and media communication. Nevertheless, PPF, with just 2659 hectares, has great potential compared to other forests with a large concentration of petrified wood such as the Jaramillo Petrified Forest National Park in Santa Cruz-Argentina of the Middle-Upper Jurassic Period [41]. The latter has 15,000 hectares and an average of 4800 tourists a year. Another case is the Petrified Forest National Park in Arizona, United States, of the Triassic Period, that includes more than 20,000 ha and around 5800 visitors per year [42,43].

Therefore, the results obtained in this study were satisfactory and provided necessary information to evaluate possible threats that affect the forest and its heritage in a natural environment and to boost the execution and implementation of scientific and territorial ordering in projects according to the population's capacity and the reality of the site [44,45].

6. Conclusions

The assessment of the methodology of the Spanish Inventory of Places of Geological Interest (IELIG, acronym in Spanish) carried out on seventeen different stations in the Puyango Petrified Forest, proved that it is a Place of Geological Interest (LIG, *acronym in Spanish*). Thus, PPF is in the categories of Very high and High interest in the scientific, tourist and educational sectors, for its great potential and geological relevance to promote geotouristic development.

The geodiversity of the forest is the main strength of the area, since it is considered a geological-paleontological heritage site, for purposes of touristic interpretation. PPF is one of the few sites in the world where you can analyze the paleontological aspects of the prehistoric flora and relate it to the current plant landscape. These characteristics make PPF an icon of geotourism in the sector. Hence, the proposed methodology SDTPI-ASC assessed the 17 LIGs in a preliminary stage and gave PPF a great value as a place of geological-paleontological heritage. Moreover, this analysis is a significant contribution to academia and its application facilitates the recognition of LIGs. Thus, the geopark project generates an alternative for the scientific, cultural and economic development of the population through a territorial order that addresses the 2030 Agenda and contains 17 Sustainable Development Goals (SDGs).

The importance of anthropogenic threats (deforestation, overgrazing, sample collection, burning of vegetation for new land and planting) and natural threats (landslides, river floods, severe droughts) faced by the geosites urge the protection and conservation of the LIGs. Therefore, strategies will substantially improve environmental, geological and paleontological conservation, as well as community participation and dissemination in the short and long term.

Finally, the development of bilateral relations with the neighboring country Peru will stabilize the intra-border situation and encourage Peruvian tourism to the PPF area. In addition, it will promote relations with international organizations interested in supporting the proposal for the sustainable conservation and resource management of the Puyango geopark project.

Supplementary Materials: The following are available online at http://www.mdpi.com/2071-1050/12/16/6579/s1, in the book "PUYANGO: ENTRE PASADO Y PRESENTE" ("PUYANGO: BETWEEN PAST AND PRESENT") of the Valuation Study of the Petrified Forest of Puyango, Loja Province, Ecuador. First edition, Guayaquil-Ecuador, 2004, ISBN: 9978-310-07 [26], Figure S1: Macrofossils, Figure S2: Macrofossils, Figure S3: Microfossils, Figure S4: Microfossils.

Author Contributions: Conceptualization, P.C.-M., F.M.-C. and N.J.-O.; methodology, F.M.-C., G.H.-N., N.J.-O. and P.C.-M.; software, G.H.-N.; validation, P.C.-M., N.J.-O. and F.M.-C.; formal analysis, F.M.-C., G.H.-N., N.J.-O. and P.C.-M.; investigation, N.J.-O., F.M.-C. and G.H.-N.; resources, F.M.-C., G.H.-N., N.J.-O. and P.C.-M.; data curation, N.J.-O., G.H.-N. and F.M.-C.; writing—original draft preparation, N.J.-O. and G.H.-N.; writing—review and editing, P.C.-M., F.M.-C., N.J.-O. and G.H.-N.; visualization, F.M.-C., G.H.-N., N.J.-O. and P.C.-M.; supervision, F.M.-C. and N.J.-O.; project administration, N.J.-O and F.M.-C.; funding acquisition, N.J.-O. All authors have read and agreed to the published version of the manuscript.

Funding: This work was supported by ESPOL Research projects: "Estudio de Valoración y Diagnóstico de Paleontología, Botánica, Arqueología y Etnografía en Puyango, Celica y Paltas, Provincia de Loja" (Study of assessment and diagnosis of paleontology, botany, archaeology and ethnography in Puyango, Celica and Paltas, Loja province), "Propuesta de Geoparque Ruta del Oro y su incidencia en el desarrollo territorial" ("Ruta del Oro" Geopark proposal and its impact on territorial development) under grant nos CIPAT-02-2018 and "Registro del Patrimonio Geológico y Minero y su incidencia en la defensa y preservación de la geodiversidad en Ecuador" (Registry of Geological and Mining Heritage and its impact on the defense and preservation of geodiversity in Ecuador) under grant nos CIPAT-01-2018.

Acknowledgments: The authors wish to acknowledge Ivan Romero and Josue Briones for their support and recommendations in the preparation of this research.

Conflicts of Interest: The authors declare no conflict of interest.

References

1.	Štrba, L.; Kolackovská, J.; Kudelas, D.; Kršák, B.; Sidor, C. Geoheritage and geotourism contribution to tourism development in protected areas of Slovakia-theoretical considerations. *Sustainability* **2020**, *12*, 2979. [CrossRef]

2.	Williams, M.A.; McHenry, M.T.; Boothroyd, A. Geoconservation and Geotourism: Challenges and Unifying Themes. *Geoheritage* **2020**, *12*, 1–14. [CrossRef]

3.	Lopes, R.F.; Candeiro, C.R.A.; de Valais, S. Geoconservation of the paleontological heritage of the geosite of dinosaur footprints (sauropods) in the locality of São Domingos, municipality of Itaguatins, state of Tocantins, Brazil. *Environ. Earth Sci.* **2019**, *78*. [CrossRef]

4.	Shekhar, S.; Kumar, P.; Chauhan, G.; Thakkar, M.G. Conservation and Sustainable Development of Geoheritage, Geopark, and Geotourism: A Case Study of Cenozoic Successions of Western Kutch, India. *Geoheritage* **2019**, *11*, 1475–1488. [CrossRef]

5.	Ferreira, A.R.R.; Lobo, H.A.S.; de Jesus Perinotto, J.A. Inventory and Quantification of Geosites in the State Tourist Park of Alto Ribeira (PETAR, São Paulo State, Brazil). *Geoheritage* **2019**, *11*, 783–792. [CrossRef]

6.	Dowling, R.K. Geotourism's Global Growth. *Geoheritage* **2011**, *3*, 1–13. [CrossRef]

7.	Hurtado, H.; Dowling, R.; Sanders, D. An Exploratory Study to Develop a Geotourism Typology Model. *Int. J. Tour. Res.* **2006**, *8*. [CrossRef]

8.	Newsome, D.; Dowling, R. Geoheritage and geotourism. In Geoheritage: Assessment, Protection, and Management. *Elsevier. Inc.* **2018**, *17*, 305–321. [CrossRef]

9.	Dos Reis Polck, M.A.; de Medeiros, M.A.M.; deAraújo-Júnior, H.I. Geodiversity in Urban Cultural Spaces of Rio de Janeiro City: Revealing the Geoscientific Knowledge with Emphasis on the Fossil Content. *Geoheritage* **2020**, *12*. [CrossRef]

10.	Tavares, G.N.D.; Boggiani, P.C.; de Moraes Leme, J.; Trindade, R.I. The Inventory of the Geological and Paleontological Sites in the Area of the Aspirant Geopark Bodoquena-Pantanal in Brazil. *Geoheritage* **2020**, *12*. [CrossRef]

11. Morales, J.; Azanza, B.; Gómez, E. El Patrimonio Paleontológico Español. *Coloq. de Paleontol.* **1999**, *50*, 53–62. (In Spanish)

12. Gutiérrez, J. Determinación de la metodología límites aceptables de cambio como estrategia para el manejo del ecoturismo en el Bosque Petrificado Puyango. Third level. Bachelor's Thesis, Machala Technical University, Machala, Ecuador, October 2015. (In Spanish).

13. Meléndez, G. Definición y Valoración de la Geodiversidad, Análisis, Valoración y Protección Legal del Patrimonio Geológico y Paleontológico. Bachelor's Thesis, Zaragoza University, Bilbao, Spain, 13 May 2010.

14. Ordoñez, M.; Jiménez, N.; Suárez, J. *Micropaleontología Ecuatoriana*; Petroproducción, Guayaquil Geological Research Center, CIGG.: Guayaquil, Ecuador, 2006. (In Spanish)

15. Ministerio del Ambiente del Ecuador. 2007. Plan Estratégico del Sistema Nacional de Áreas Protegidas del Ecuador 2007–201 (SNAP-GEF). REGLA-ECOLEX. Quito. Available online: http://maetransparente. ambiente.gob.ec/documentacion/WebAPs/PLAN%20ESTRATEGICO%20DEL%20SNAP.pdf (accessed on 14 December 2019).

16. Yaguachi, B. Identificación y dinamización del Corredor de Endemismo Tumbesino, sector Bosque Petrificado de Puyango. Bachelor's Thesis, Loja National University, Loja, Ecuador, 2012. (In Spanish).

17. Instituto Espacial Ecuatoriano (IEE) y MAGAP (SINAGAP). *Generación de Geoinformación Para la Gestión del Territorio a Nivel Nacional*; Technical Memory (Geomorphology): Puyango Canton, Ecuador, 2013. (In Spanish)

18. García, A.; Carcavilla, L. *Documento Metodológico Para la Elaboración del Inventario Español de Lugares de Interés Geológico (IELIG)*; IGME Geological and Mining Heritage Research Area: Madrid, Spain, 2013. (In Spanish)

19. Aguirre, Z.; Lopez, G. Conservations status of the dry forests of the province of Loja, Ecuador. *Arnaldoa* **2017**, *24*, 207–228. [CrossRef]

20. Carrión, P.; Herrera, G.; Briones, J.; Caldevilla, P. Geotourism and Development Base on Geological and Mining Site Utilization, Zaruma-Portovelo, Ecuador. *Geosciences* **2018**, *6*, 205. [CrossRef]

21. Jaramillo, J.; García, T.; Bolaños, M. *Bosque Petrificado de Puyango y sus Alrededores: Inventario de Lugares de Interés Geológico*; GEO Latitud, Publisher: Quito, Ecuador, 2017; Volume 1. (In Spanish)

22. Gobierno Autónomo Descentralizado Municipal de Puyango. Plan de Desarrollo y Ordenamiento Territorial. 2014–2019. Available online: https://www.puyango.gov.ec. (accessed on 16 December 2019).

23. *Conservación de la Diversidad Biológica en Los Bosques Tropicales Bajo Régimen de Ordenación*; UICN: Gland, Suiza y Cambridge, Reino Unido, 1992–1995; Volume Xii, p. 272. ISBN 2-8317-0251-8.

24. Jiménez, H. Valoración de los servicios ecosistémicos de recreación y belleza escénica del Bosque Petrificado Puyango, 2015. Bachelor's Thesis, Loja Private Technical University, Loja, Ecuador, 2017. (In Spanish).

25. Puño, N. *Environmental Management of the Puyango-Tumbes River Basin in Ecuador and Perú. Securing Water and Wastewater Systems: Global Experiences*; Springer International Publishing: Cham, Switzerland, 2014; Volume 8, pp. 161–187. [CrossRef]

26. Xomchuck, I.; Jimenez, N.; Valverde, F. *PUYANGO: Entre pasado y presente. Estudio de Valoración del Bosque Petrificado de Puyango, Provincia de Loja, Ecuador*, 1st ed.; ESPOL: Guayaquil, Ecuador, 2004; pp. 9–60. (In Spanish)

27. Martz, J.W.; Parker, W.G. Revised lithostratigraphy of the Sonsela Member (Chinle Formation, Upper Triassic) in the southern part of Petrified Forest National Park, Arizona. *PLoS ONE* **2010**, *5*, e09329. [CrossRef] [PubMed]

28. National Secretary of Planning and Development and National Information System, Information for planning and territorial planning, National Geographic Atlas. 2013. Available online: http://sni.gob.ec/atlas-geografico-nacional-2013. (accessed on 19 January 2020).

29. Jiménez, N.; Ordóñez, M.; Suárez, J.; Tigreros, J. *Estudio de Valoración y Diagnóstico de Paleontología, Botánica, Arqueología y Etnografía en los Cantones de Puyango, Celica y Paltas de la Provincia de Loja, 2001–2002, Paleontología y Micropaleontología del Bosque Petrificado de Puyango Provincia de Loja*; Proyecto de Loja, Vol. 1; CEAA-ESPOL: Guayaquil, Ecuador, 2004. (In Spanish)

30. Xomchuk, I.; Ramos, G.; Avilés, G. *Estudio de Valoración y Diagnóstico de Paleontología, Botánica, Arqueología y Etnografía en los Cantones de Puyango, Celica y Paltas de la Provincia de Loja, 2001–2002, Valoración Socio-Etnográfica de los Cantones Puyango, Celica y Paltas*; Proyecto de Loja, Vol. 2; CEAA-ESPOL: Guayaquil, Ecuador, 2004. (In Spanish)

31. Zhunaula, V. Análisis de los Factores Potenciales que inciden en el hurto del patrimonio cultural en el Bosque Petrificado Puyango. Bachelor's Thesis, Machala Technical University, Machala, Ecuador, 2016. (In Spanish).

32. Cartuche, D.V.; Armijos, L.A.; Romero, C.S.; Ocampo, C.H. Evaluación del desarrollo turístico en el Bosque Petrificado de Puyango (BPP), sur de Ecuador. *Rev. Espac.* **2019**, *40*, 23.

33. Calva-Nagua, D.X. El desafío de las Fuentes arqueológicas para la educación ecuatoriana. *Maestros y Sociedad* **2018**, *15*, 393–408. (In Spanish)

34. Carrión-Mero, P.; Loor-Oporto, O.; Andrade-Ríos, H.; Herrera-Franco, G.; Morante-Carballo, F.; Jaya-Montalvo, M.; Aguilar-Aguilar, M.; Torres-Peña, K.; Berrezueta, E. Quantitative and Qualitative Assessment of the "El Sexmo" Tourist Gold Mine (Zaruma, Ecuador) as A Geosite and Mining Site. *Resources* **2020**, *9*, 28. [CrossRef]

35. García, J.; Falaschi, P.; Zamuner, A. Fungal-arthropod-plant interactions from the Jurassic petrified forest Monumento Natural Bosques Petrificados, Patagonia, Argentina. *Palaeogeogr. Palaeoclimatol. Palaeoecol.* **2012**, *329–330*, 37–46. [CrossRef]

36. Woodcock, D.; Meyer, H.; Dunbar, N.; Mclntosh, W. Geologic and taphonomic context of El Bosque Petrificado Piedra Chamana (Cajamarca, Peru). *Geol. Soc. Am. Bull.* **2009**, *121*, 1172. [CrossRef]

37. Tarras-Wahlberg, N.H.; Flachier, A.; Lane, S.N.; Sangfors, O. Environmental impacts and metal exposure of aquatic ecosystems in rivers contaminated by small scale gold mining: The Puyango River basin, souther Ecuador. *Sci. Total Environ.* **2001**, *278*, 239–261. [CrossRef]

38. Salazar, S. Propuesta de un Plan de comunicación para promocionar, localmente, al bosque petrificado Puyango, ubicado en las provincias ecuatorianas de El Oro y Loja, como una novedosa alternativa turística. Bachelor's Thesis, Las Americas University, Quito, Ecuador, 2010. (In Spanish).

39. Carcavilla. *Geodiversidad y Patrimonio Geológico*. *Instituto Geológico y Minero de España*; Edición Parques Nacionales: Madrid, Spanish, 2014; p. 21, NIPO: 474-11-012-3. (In Spanish)

40. Gómez, J.; Magnin, L. Cartography of Geomorphological Units of the Bosques Petrificados de Jaramillo National Park (Santa Cruz Argentina) for its Geo-Archaeological Implementation. *Inst. De Geogr. Unam* **2019**, *98*. [CrossRef]

41. Baranyi, V.; Reichgelt, T.; Olsen, P.E.; Parker, W.G.; Kurschner, W.M. Norian vegetation history and related environmental changes: New data from the Chinle Formation, Petrified Forest National Park (Arizona, SW USA). *Bolletin Geol. Soc. Am.* **2018**, *130*, 775–795. [CrossRef]

42. Núñez, P.; Vejsbjerg, L. El Turismo, entre la actividad económica y el derecho social en El Parque Nacional Nahuel Huapi, Argentina, 1934–1955. *Estud. Y Perspect. En. Tur.* **2010**, *19*, 930–945. (In Spanish)

43. Aragón, S.; Woodcock, D. Plant Community Structure and Conservation of a Northern Peru Sclerophullous Forest. *Biotropica* **2010**, *42*, 262. [CrossRef]

44. Herrera-Franco, G.; Carrión-Mero, P.; Alvarado, N.; Morante-Carballo, F.; Maldonado, A.; Caldevilla, P.; Briones-Bitar, J.; Berrezueta, E. Geosites and Georesources to Foster Geotourism in Communities: Case Study of the Santa Elena Peninsula Geopark Project in Ecuador. *Sustainability* **2020**, *12*, 4484. [CrossRef]

45. Štrba, Ľ.; Kršák, B.; Sidor, C. Some Comments to Geosite Assessment, Visitors, and Geotourism Sustainability. *Sustainability* **2018**, *10*, 2589. [CrossRef]

Article

Landscape Evaluation as a Complementary Tool in Environmental Assessment. Study Case in Urban Areas: Salamanca (Spain)

Marco Criado [1,*] , **Antonio Martínez-Graña** [2] , **Fernando Santos-Francés** [1] and **Leticia Merchán** [1]

1   Department of Soil Sciences, Faculty of Environmental Sciences, University of Salamanca, 37007 Salamanca, Spain; fsantos@usal.es (F.S.-F.); leticiamerchan@usal.es (L.M.)
2   Department of Geology, Faculty of Sciences, University of Salamanca, 37008 Salamanca, Spain; amgranna@usal.es
*   Correspondence: marcocn@usal.es; Tel.: +34-923-294-546

Received: 14 July 2020; Accepted: 6 August 2020; Published: 8 August 2020

Abstract: In recent years, the landscape has become another environmental resource, so it is important to incorporate it into planning actions. However, its broad sense of study has made it difficult to develop methodologies that precisely diagnose the state of the landscape and its management requirements, especially in dynamic spaces like urban areas. In order to develop a method capable of providing information that can be incorporated into environmental assessment and territorial planning tasks so that the needs of the landscape are taken into account in the decision-making stages, an objective methodology is presented based on the study of different parameters (biotic, abiotic and socioeconomic) analyzed in the field and subsequently geoprocessed through Geographic Information Systems according to their influence on the landscape. Through the proposed methodology it is possible to determine the quality, fragility and need of protection of the landscape, as well as to identify the diverse landscape units that form the landscape of a territory. Based on these results, a landscape diagnosis can be drawn up to quantify its overall and partial state, carry out monitoring analyses and make comparisons between different landscape units, so that management measures can be adopted according to the obtained scenarios.

Keywords: landscape; landscape quality; landscape fragility, need of protection; landscape diagnosis; GIS; environmental assessment

---

1. Introduction

The landscape is an area, as perceived by people, whose character is the result of the action and interaction of natural and/or human factors [1]. Thus, the landscape can be understood as an entity resulting from the interaction between ecology, vegetation, geology, geomorphology, hydrology, edaphology, climatology, fauna and anthropic activity, and its dynamism must also be taken into account due to the continuous action of natural agents and human activity, which is currently the most important in the alteration of landscapes [2–4]. However, the complexity of the relationships between the different components, the difficulty in studying them, the discrepancies in their interpretation and the analysis of the landscape exclusively through the visual aspect (VIA), without including other aspects (natural, social, cultural, etc.) made the analysis and inclusion of the landscape in the environmental assessment processes late in relation to the rest of the physical environment components [4]. Nowadays, the landscape is considered to be yet another resource of great importance within the set of environmental values demanded by society, since it is perceived as an important factor in the quality of life [5,6]. Furthermore, the search for highly valuable natural landscapes promotes tourism and is the economic driving force behind small rural municipalities [7,8]. All these aspects mean that the landscape is integrated into planning policies and actions as another element of the environment [9,10].

Today, anthropic activity is one of the main causes of landscape alteration, especially in urban environments [11], due to the agglomeration of population and services that rapidly and irreversibly consume the land [12,13]. The magnitude and speed of changes in land use [14] lead to the breakdown of the landscape, causing transformations in its initial value and identity [4]. Therefore, properly assessing the landscape in different development initiatives is essential to avoid the degradation of the most valuable landscapes. Planners and leaders need to have methodologies that allow for the preparation of detailed landscape inventories and the monitoring of changes in order to make appropriate decisions, a very important aspect in such dynamic, complex and multifunctional spaces as cities [15]. In addition, ensuring a diversity of well-preserved landscapes is necessary given the diversity of observers, with subjective assessments of them according to their tastes and perception [2]. This aspect becomes crucial in the urban environment due to the need to achieve landscape performance in line with sustainable development and in addition to performing economic and social functions, it also preserves ecological ones [16,17].

For landscape analysis, there are direct and indirect methodologies or combinations of both [18–20]. The first to appear were the direct methodologies (1970s), which are based on the subjective evaluation of landscape aesthetics based on sensory perceptions such as visual, sound or olfactory [21,22]. In that regard, studies were carried out that estimated the preferences of the population and groups of experts in relation to perception "in situ" or through photographs of the landscape [23–26], and these can currently be carried out using virtual reality and similar technologies [27,28]. However, some authors have begun to highlight the need for quantitative landscape measurements to define indicators for landscape assessment that would contribute to a conceptual framework for landscape planning [29–31]. In this way, indirect methods were developed, which analyze the distribution of the components of the landscape and their relationship with the different components of the environment (vegetation, orography, etc.) whose combination defines the elements of the landscape [32]. Indirect methods have increased since the advance of Geographical Information Systems (GIS) at the beginning of the 21st century [33–37], and they establish a series of study factors and evaluate their impact on the landscape [38].

The principles underlying the methodology proposed in this manuscript are in line with the European Landscape Convention: identification of existing landscapes, analysis of their characteristics and identification of transformation pressures [1]. Landscape units are spaces that share a unique and singular structural, functional or perceptually differentiated configuration, identified by their internal coherence and their differences from other units [2,39]. The quality of the landscape evaluates the different spaces according to their relevance, singularity and importance in the perception of the landscape [40,41]. The fragility of the landscape is related to its vulnerability to change through the action of an external impact that implies a deterioration of its values [42,43]. Based on these two concepts, the absorption capacity of a landscape can be estimated, which is the capacity of response that a landscape has in the face of an impact and, from this, the need for protection that the landscape requires in each location can be established [38,44]. Transformation pressures are often linked to socioeconomic dynamics and are often multiple and complex in urban and periurban environments [11,45], and alteration of the landscape can also affect the ecological structure [46–48].

There is therefore a clear need for adequate and up-to-date information on the state of the landscape. In view of this, the aim of the manuscript is to establish a methodology based on the study and processing of preestablished objective criteria using GIS techniques that will make it possible in a highly anthropized area (1) to characterize the diversity of existing landscapes, (2) to evaluate the quality and fragility of the landscape, (3) to determine the need for landscape protection, (4) to carry out a diagnosis of the landscape situation and (5) to create a methodology capable of developing appropriate information that can be incorporated into environmental assessment procedures.

2. Materials and Methods

2.1. Study Area

The study area, located in central-western Spain, includes the city of Salamanca and its periurban area of influence (Figure 1), where 200,000 inhabitants live [49]. Salamanca is the capital of the province to which it gives its name and is the largest urban agglomeration in its surroundings, eminently agricultural. One of the main attractions of Salamanca is the diversity and large number of elements of historical, artistic and cultural interest that it contains, which is why its old town is listed as a UNESCO World Heritage Site [50].

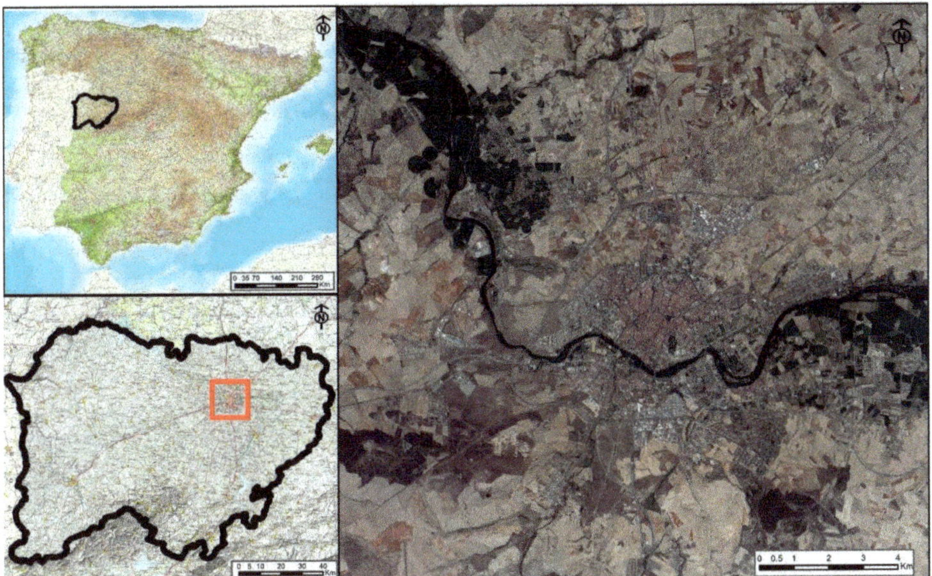

Figure 1. Study area location.

2.2. General Methodology

An indirect methodology has been followed for the study of the landscape, which allows an objective and quantitative assessment of the landscape based on the analysis of various components of the environment [38] (Figure 2). Firstly, the characteristic landscape units are defined. After this, the quality of the landscape is evaluated through the analysis of intrinsic and extrinsic quality, which consider different criteria for analysis, obtaining the quality mapping that determines the landscape value of each area. Next, the fragility of the landscape is studied in the face of possible impacts that would cause the alteration of its conditions through intrinsic and extrinsic fragility. Finally, the areas are classified according to the need for protection they present in each place and, superimposing this on the distribution of the landscape units, a landscape diagnosis can be made of each one of them.

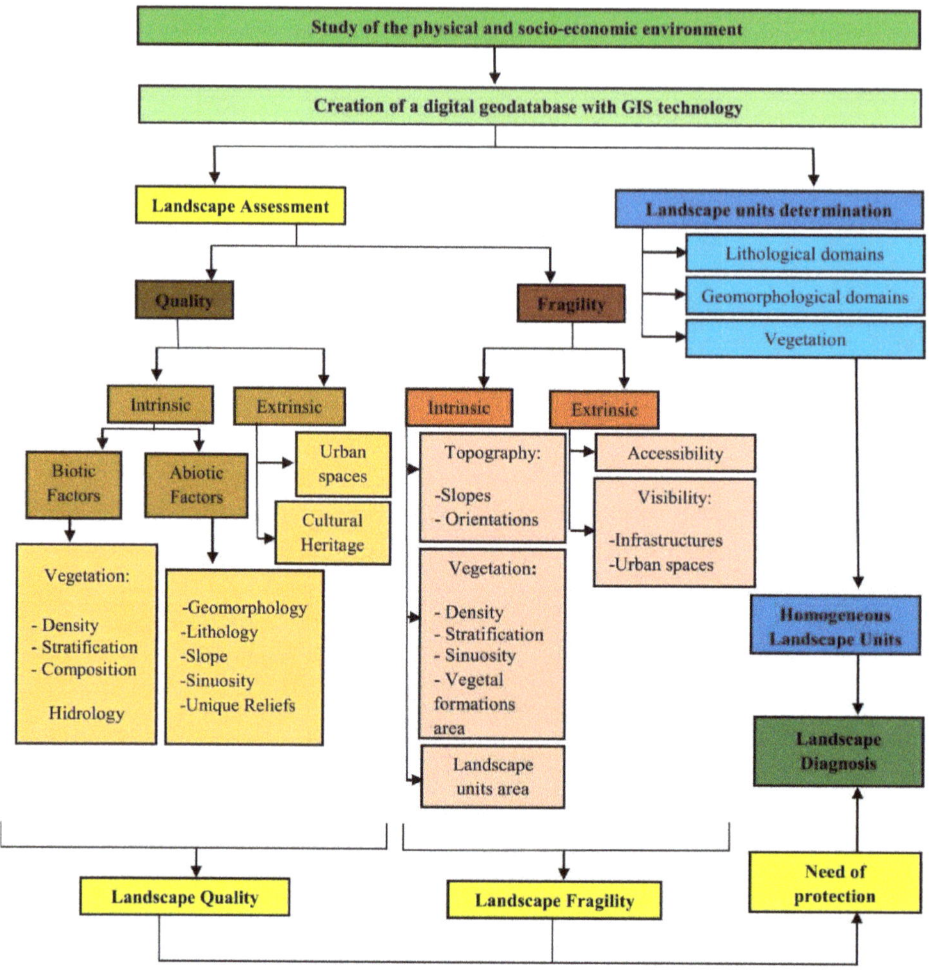

Figure 2. General methodology of the manuscript.

In order to characterize all the parameters analyzed in the landscape assessment, a digital geodatabase was created using GIS, which is based on the study of the physical and socioeconomic environment carried out in the study sector.

2.3. Landscape Units Identification

The study sector was divided into homogeneous landscape units. Three factors were studied, and their combination gives rise to the homogeneous landscape units: geomorphology, lithology and vegetation/use of the soil. In the geomorphological analysis, 13 units were identified that were grouped according to their representativeness and impact on the landscape in six groups: ridges and hills, hillsides, surfaces, escarpments, terraces and valley bottoms. In the geological analysis, the different formations were grouped into lithological groups that have similar landscape effects: slates, quartzites, sandstones, tertiary red sediments, conglomerates and sands, and alluvial deposits. With respect to existing land uses, the following have been defined: wooded plant formations (holm oak and pasture), natural meadows, crops (dry and irrigated), wastelands and urbanized areas [14]. After crossing the cartography, 22 landscape units are obtained. These will be described in Section 3 of Results.

The distribution of these three factors is delimited after field work using GIS techniques and, by superimposing these the homogeneous landscape units are generated. Urban areas are defined along with other land uses, in this case determined by remote sensing in previous studies [14,51].

2.4. Landscape Quality

The intrinsic and extrinsic landscape quality has been studied for its determination. An indirect methodology has been followed based on the manipulation of the geographical information elaborated and collected in the field campaign and data collection. The different parameters studied, both in the case of intrinsic and extrinsic quality, are treated with GIS techniques and by means of map algebra; by making a summation of all parameters, the intrinsic and extrinsic quality is determined. The final quality is obtained from the sum of the intrinsic and extrinsic quality values. For all cases, the resulting values are segmented into five intervals of equal range, which correspond to the quality values: very high, high, moderate, low and very low [38]. The size of the pixel was 20 m in all studied cartography.

2.4.1. Intrinsic Quality (IQ)

The IQ is the visual attraction for the observer that derives from the natural characteristics of each point in the territory [2]. Each of the abiotic and biotic parameters studied in this analysis are weighted for their determination.

The main abiotic factors are geology and geomorphology, which are key elements in the study of the landscape since they strongly condition the arrangement of the reliefs and the forms of the terrain and, therefore, they have a clear influence on the perception of the landscape in addition to conditioning the establishment and development of the biotic elements [52]. The parameters analysed to determine the impact of the abiotic factor on the intrinsic quality of the landscape are the following: geomorphological domains, slope of the terrain, sinuosity of the terrain, lithology and unique reliefs. The scores assigned to each component of these parameters are shown in Table 1. These scores are based on the field work carried out and on studies preferably analyzed in the bibliography [20–23,53–56]. The geomorphological domains are the different forms that the terrain takes as a result of the action of the forming and modeling agents of the relief. The slope evaluates how abrupt the terrain is. The sinuosity refers to the disposition that the lines of the terrain take, determining its curved character. To do this, GIS techniques are used to determine the resulting polygons between the contours and calculate their perimeter and area. Using an index that relates both parameters (sinuosity = perimeter^2/area), the values of the curvature of the terrain lines are obtained, which are reclassified in equal intervals. Rough terrains, with a diversity of shapes and sinuousness, are visually very attractive. The effect that the lithological factor confers on the landscape, in addition to conditioning the forms of the terrain, is determined by the chromatism that each lithology presents (light and striking colors being more highly valued than dark and dull ones). Finally, the presence of elements with a singular imprint on the relief leads to an increase in its visual perception (lithostructural reliefs related to geological structures, such as folds or faults).

On the other hand, in relation to biotic factors, vegetation and hydrology are the main elements in the intrinsic evaluation of landscape quality [57,58], since evaluating the effect of other elements such as fauna, is an added difficulty due to their mobility. The effect of vegetation varies according to its structure; that is, how the different plant components are distributed within the formation to which they belong (it depends on the density or quantity of elements per unit of surface area and the number of strata present in each plant formation) and the composition or types of existing plant formations. On the other hand, the bodies and courses of water provide added visual and sound values when contemplating the landscape in the spaces near them.

Table 1. Weights assigned to the components of each analyzed parameter (linked to abiotic factors) for the determination of intrinsic landscape quality.

P	Components of Each Parameter	Quality
Geomorpholo-gical domains	Ridges and hills	Very high
	Escarpments	High
	Surfaces	Moderate
	Hillside	Low
	Terraces	Very low
	Valley bottoms	Very low
Slope	>30%	Very high
	20–30%	High
	15–20%	High
	10–15%	Moderate
	5–10%	Low
	<5%	Very low
Sinuosity	<5000	Very high
	5000–10,000	High
	10,000–15,000	Moderate
	15,000–25,000	Low
	>25,000	Very low
Lithology	Red materials: red sediments and white limestone crusts	Very high
	Sandstone: ochre, beige, yellowish and greenish colors	High
	Quartzite: white-grey and reddish colors	Moderate
	Slates: brown and greyish shades	Low
	Conglomerates and sand: dark brown colors	Very low
	Sand, silt and clay: brown or grey shades	Very low
	Unique reliefs: ridges and escarpments	Very high

The biotic parameters studied in the analysis of intrinsic quality are the following: density, stratification and composition in relation to the vegetation and proximity and visibility of water bodies with respect to hydrology. The plant density of a site is determined by the fractional vegetation cover (FVC) where, depending on its value, we find different plant formations, which are weighted according to their characteristics, with dense formations being more highly valued than scattered ones. Stratification refers to the variety of growths or plant strata present in each of the existing formations (herbaceous, shrub and tree) so that the more growths there are in a formation, the more diverse the forms and, therefore, the greater the quality of the landscape. The composition of the vegetation is related to the grouping of a set of related plant elements to give rise to different formations, with woodlands being more valuable than shrubs and herbaceous ones. With regard to hydrology, the visual and sound impression of water masses is limited to an environment close to them so, depending on their entity, these areas of influence must be determined by means of the GIS, being 20 m for streams, while in the case of the Tormes, due to its greater importance, an analysis of its visibility is carried out to determine from which areas it is observable and influences the perception of the landscape. The scores of each component of the parameters analyzed are located in Table 2.

2.4.2. Extrinsic Quality (EQ)

The EQ analyzes the impact that external anthropic components, such as infraestructures, have on the landscape, especially those for housing (urban centers), transport and the different elements that make up the historical, artistic and cultural heritage. First, the impact of the landscape on human settlements is assessed through a visibility analysis, which allows the identification of natural landscapes visible from urban centers. These areas are weighted with a value of 2. Secondly, the analysis of the historical, artistic and cultural heritage is carried out [59], where a total of 50 cultural heritage sites, five livestock trails and 60 archaeological sites were identified. In this analysis, livestock routes and archaeological sites (whose areas of influence, of 100 and 300 m, respectively, have been evaluated with a value of 2) have been taken into account as have the properties of cultural interest, around which an area of influence of 200 m was estimated, with these zones being assigned a value of 6.

Table 2. Weights assigned to the components of each analyzed parameter (linked to biotic factors) for the determination of intrinsic landscape quality.

P	Components of Each Parameter	Quality
Fractional vegetation cover (FVC)	>40%	Very high
	10–40%	High
	5–10%:	Moderate
	<5%	Low
	0: urban areas and croplands	Very low
Stratification	3	High
	2	Moderate
	1	Low
	0	Null
Composition	Woodland	Very high
	Scrubland	High
	Mixed crop-tree formations	Moderate
	Croplands and grasslands	Low
	Urban spaces and areas without vegetation	Null
Hydrology	Visibility and proximity to Tormes river	High
	Proximity to minor streams	Low

2.5. Landscape Fragility

A procedure analogous to quality analysis has been followed for its determination, studying both intrinsic and extrinsic fragility. Using GIS techniques, a series of parameters are analyzed (related to topographical, plant and surface factors of the landscape units), whose components are weighted according to their contribution to fragility, with the final fragility being the sum of the intrinsic and extrinsic ones.

2.5.1. Intrinsic Fragility (IF)

The intrinsic fragility of the landscape has been analyzed according to the study of the parameters: slope of the land, orientation of the land, area of the landscape units, plant stratification, plant density, extension of the plant units and sinuosity of the vegetation [38]. Table 3 includes the weights of the parameters. In relation to the slope, the areas with high values are more fragile, since an alteration of the existing conditions and forms will have a great impact on the valuation of the landscape. On the other hand, the perception of the impacts related to the orientation or spatial arrangement of the elements is due to the way in which the light falls on it, so that the most fragile areas are the most illuminated (sunny), while the shady areas present shadows and are protected from direct solar exposure on a large number of occasions. The average situations are the east and west exposures, although these considerations must take into account the variations in sunshine that occur throughout the year due to the movement of the Earth. With regard to vegetation, stratification can be related to plant diversity, so that the most vulnerable areas will be those with little diversity (fewer layers). The plant density allows for better cushioning of impacts as there is a greater number of specimens, so the areas with lower density will be the most fragile. The surface area of a plant mass also has an impact on the fragility values, since the smaller the surface area, the greater the emphasis on impact and, therefore, the greater the fragility. With regard to the sinuosity of the vegetation, the aim is to evaluate its forms and, especially, to evaluate the edge effect, since it is in these external areas of the plant mass that the impacts are most visible, so that the greater the sinuosity, the greater the fragility. Finally, the extension of each landscape unit is important in order to evaluate its fragility, since units with a small surface area will be more sensitive to impacts than those with a larger surface area. The extension (ha) of each unit has been calculated and, depending on its contribution to the total surface area studied, it is weighted with values ranging from 0 for the largest units to 8 for those with a smaller surface area.

Table 3. Weights assigned to the components of each parameter analyzed for the determination of intrinsic landscape fragility.

P	Components of Each Parameter	Fragility
Slope	>30%	Very high
	20–30%	High
	10–20%	Moderate
	5–10%	Low
	<5%	Very low
Orientation	South	Very high
	East	High
	West	Moderate
	North	Low
	Plane	Very low
Stratification	3	Low
	2	Moderate
	1	High
	0 (urban areas)	Null
Fractional vegetation cover (FVC)	>40%	Low
	10–40%	Moderate
	5–10%:	High
	<5%	Very high
	0: urban areas and croplands	Null
Vegetal sinuosity (S)	Shrub areas (S = 309)	Very low
	Mixed (cropland and woodland) (S = 324)	Low
	Cropland, grassland and fallows (S = 1062)	Moderate
	Woodland (S = 3513)	High
Vegetation units surface (VS)	Shrub areas (VS = 21.6 ha)	High
	Mixed (cropland and woodland) (VS = 911.3 ha)	Moderate
	Woodland (VS = 1923.2 ha)	Low
	Cropland, grassland and fallows (VS = 22,770 ha)	Very low
Area of each landscape unit in relation to the average extension		

2.5.2. Extrinsic Fragility (EF)

The determination of the extrinsic fragility of the landscape is based on the study of two parameters: accessibility and visibility. Accessibility to each point of the sector is analyzed on the basis of its proximity to urban centers and linear infrastructures, as these areas have better access and, therefore, greater fragility to new impacts. The GIS has calculated an area of influence of 500 m around the urban centers and infrastructures and has assigned them a value of fragility of 4, while the sectors of worse accessibility receive a value of 0. On the other hand, the visibility of the environment is calculated from the places with the greatest transit of people (municipalities and main infrastructures) and, therefore, where the greatest observations are made. The GIS has identified, by calculating visual basins, those places in the territory where the impacts or alterations would be most visible as they are highly exposed to places with high traffic, with the most visible areas showing high values of fragility (value 4) compared to the low fragility represented by the less exposed areas (value 0).

2.6. Landscape Need of Protection

Each sector studied has been assigned a type of protection class, from one to five, depending on the quality and fragility values it presents, confronted in a double-entry matrix (Table 4), so that class one represents the areas with the greatest absorption capacity and, therefore, the least need for protection, while class five represents the areas most susceptible to alteration and therefore requiring conservation efforts.

Table 4. Matrix for allocating landscape need-of-protection classes [36].

		Very Low	Low	Quality Moderate	High	Very High
	Very Low	I	I	III	IV	IV
	Low	I	I	III	IV	IV
Fragility	**Moderate**	I	II	III	IV	V
	High	II	II	III	V	V
	Very High	II	II	III	V	V

2.7. Landscape Diagnosis

The landscape diagnosis consists of an evaluation of the landscape situation in each of the identified landscape units. Using GIS techniques, the quality, fragility and need for protection cartography is superimposed on the mapping of landscape units and the extension (ha) and percentage (%) of each degree (very high, high, moderate, low and very low) of quality, fragility or need for protection in each unit is determined.

3. Results

3.1. Landscape Units

After crossing the cartography of geomorphology, lithology, land use and urban areas (Figure 3), 24 landscape units are obtained (Table 5) along with their distribution, which is presented next to that of the urban nucleuses (Figure 4). Due to these common characteristics of the spaces belonging to the same unit, which in turn are distinguished from the rest of the units, they allow the landscape units to be used as land management units in the territorial planning processes.

Table 5. Landscape units present in the study area.

N°	Name of the Landscape Unit
1	Ridges on quartzite with wooded vegetation
2	Degraded surface on slates with crops
3	Structural surface on sandstone with grassland
4	Polygenic surface on sandstone with crops
5	Hillsides over slope deposits with crops
6	Hillsides on slates with grasslands
7	Slopes on sandstone with crops
8	Hillsides over red materials with crops
9	Hillsides on sandstone with mixed crop/tree formations
10	Hillsides on red materials with mixed crop/tree formations
11	Hillsides over sandstone with wooded vegetation
12	Hillsides over red materials with wooded vegetation
13	Escarpments on sandstone with grasslands
14	Escarpments on sandstone with wooded vegetation
15	Fluvial incision on slate with grassland

Table 5. *Cont.*

N°	Name of the Landscape Unit
16	Terraces with conglomerates and sands with irrigated crops
17	Terraces with conglomerates and sands with crops
18	Terraces with conglomerates and sands with mixed crop and tree formations
19	Terraces with conglomerates and sands with tree formations
20	Valley bottoms of sands, silts and clays with grasslands
21	Valley bottoms of sand, silt and clay with wooded vegetation
22	Semiendorheic areas of silt and clay with grassland
23	Tormes river
24	Urban areas

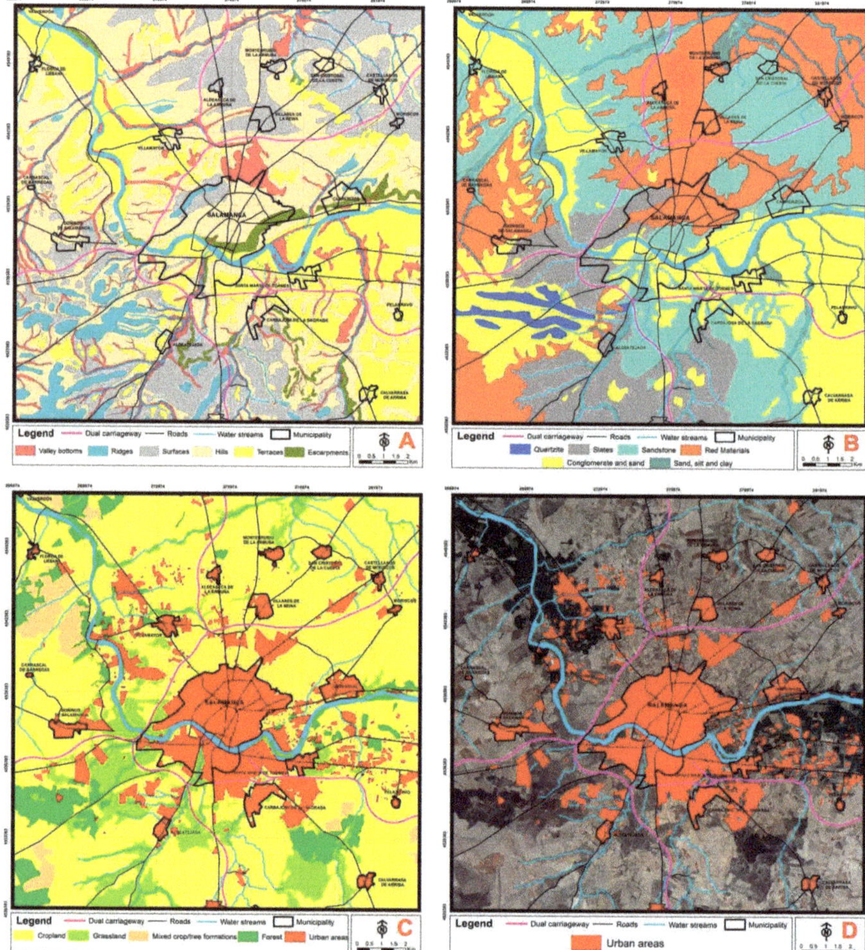

Figure 3. (**A**) Geomorphological domains, (**B**) lithological domains, (**C**) vegetation-land cover and (**D**) urban areas.

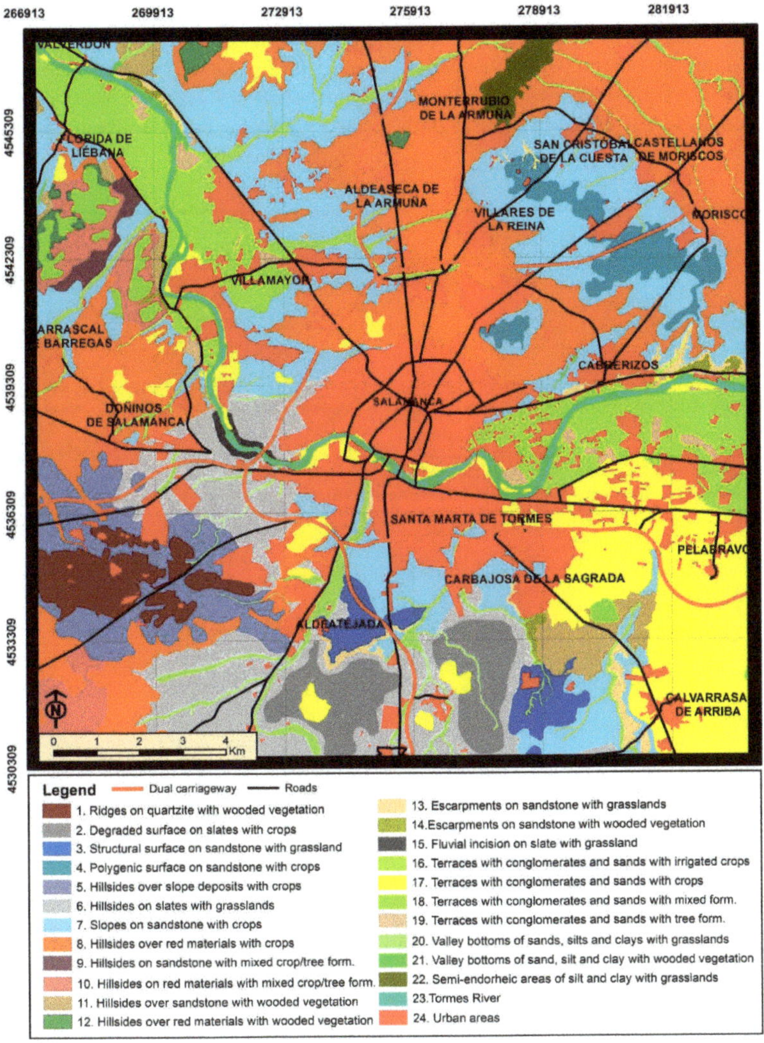

Figure 4. Landscape units map.

3.2. Landscape Quality

After mapping the weighting of the parameters studied for the determination of intrinsic and extrinsic quality (Figure 5), they are added up to obtain the intrinsic and extrinsic quality, from which the final landscape quality is obtained (Figure 6). The areas of very high quality correspond to the most outstanding geological and geomorphological elements as well as to the most diverse and striking reliefs of the environment, such as in the area of Los Montalvos (southwestern sector of the study area), escarpments along the Tormes and forest areas irregularly distributed throughout the sector. The high-quality areas present a diversity of forms and are usually spaces surrounding the higher quality areas. Therefore, they appear in the southwestern sector, the central-southern sector and the northwestern quadrant, linked to large slopes and spaces with morphological highlights, where forest ecosystems generally appear. The areas of moderate quality cover a large part of the sector studied, mainly occupying the northeast quadrant, although they also appear in the rest of the sector although to a lesser extent the further south. The low-quality areas appear to be mainly linked to the dynamics of the Tormes (the whole of the river plain and the terrace sectors of the south-eastern quadrant) and are also represented in areas of the southwestern sector in undulating slate and red materials. The very low-quality areas are mostly concentrated in the eastern sector of the area studied, coinciding with very monotonous terrace spaces that have undergone a high level of urban development in recent years, their presence being anecdotal in the rest of the sector.

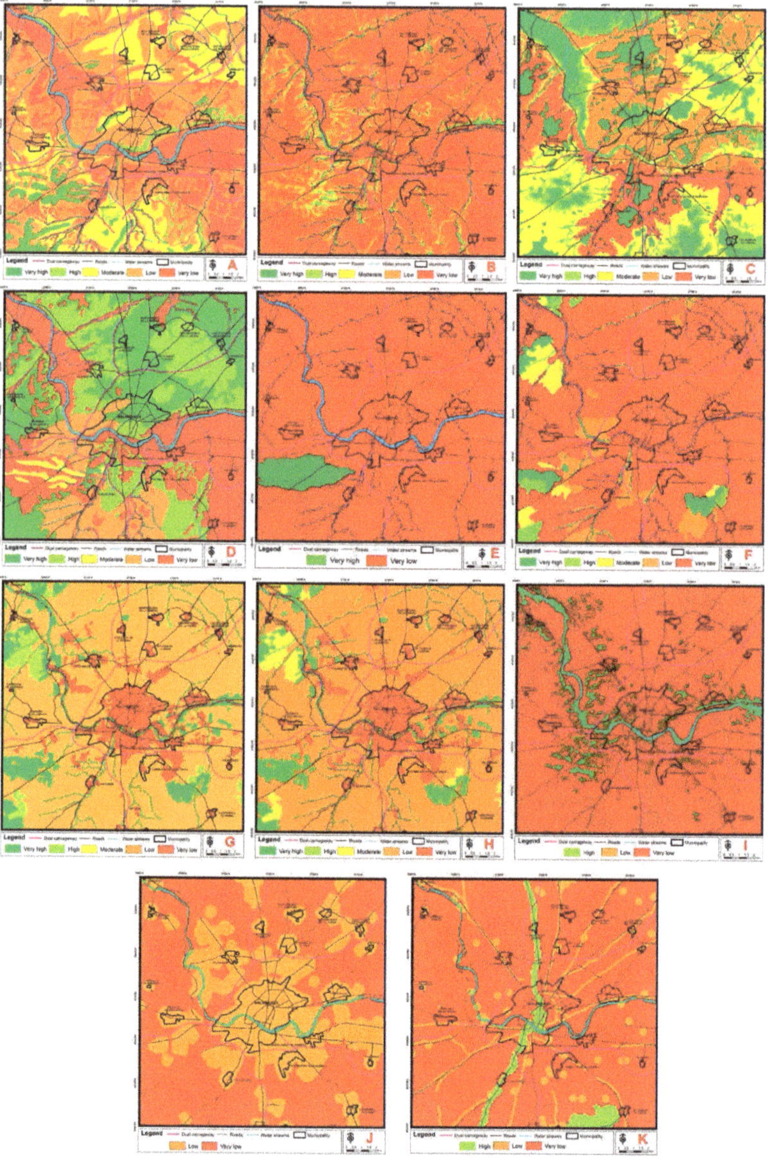

Figure 5. Studied parameters in the determination of quality Intrinsic: (**A**) geomorphology, (**B**) slope, (**C**) sinuosity, (**D**) lithology, (**E**) unique reliefs, (**F**) fractional vegetation cover (FVC), (**G**) stratification, (**H**) composition and (**I**) hydrology; and Extrinsic: (**J**) accessibility and (**K**) visibility.

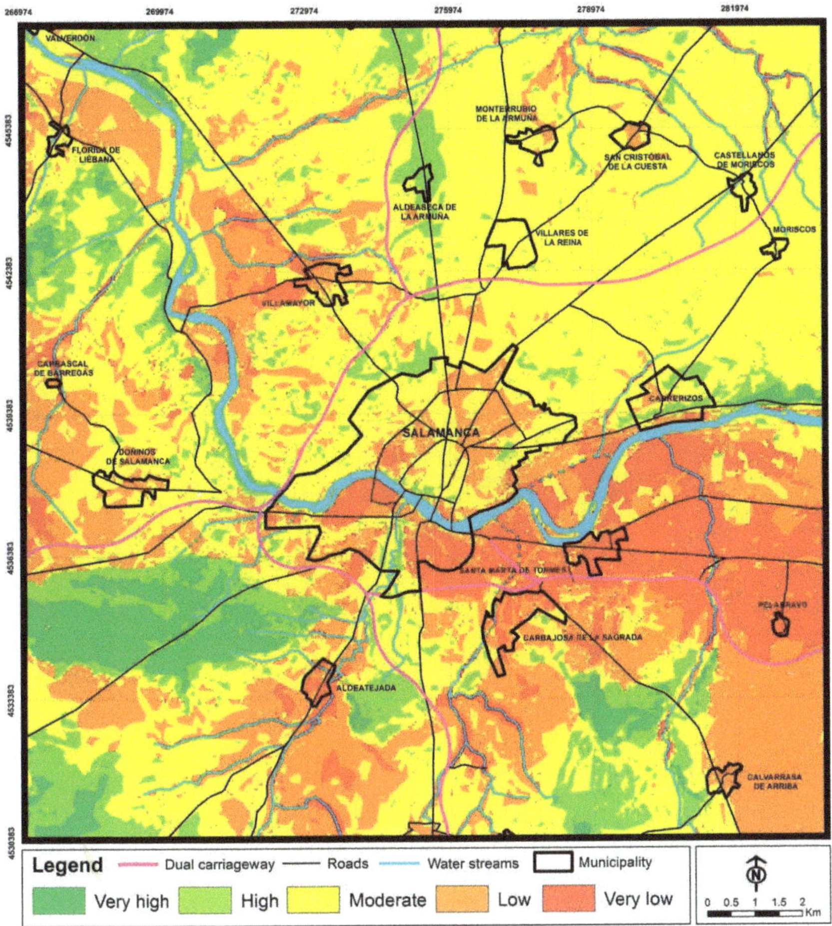

Figure 6. Landscape quality map.

3.3. Landscape Fragility

After carrying out the mapping related to the weighting of the parameters studied for the determination of intrinsic and extrinsic fragility (Figure 7), the sum of these parameters is elaborated to obtain the intrinsic and extrinsic fragility, from which the final landscape fragility is obtained (Figure 8). The areas of very high fragility generally extend over the most abrupt areas and those that have wooded vegetation, usually corresponding, in turn, to landscape units of scarce extension. The areas of high fragility stand out in the eastern sector, characterized by high anthropization, with a large number of buildings and infrastructures as well as a reduced plant diversity. The areas of moderate fragility cover large areas of the sector studied, especially around Salamanca, and seem to be related to accessibility and visibility from the linear infrastructures. The areas of low and very low fragility correspond to the most inaccessible areas and generally have an important diversity and plant cover, being present in the sectors furthest from Salamanca.

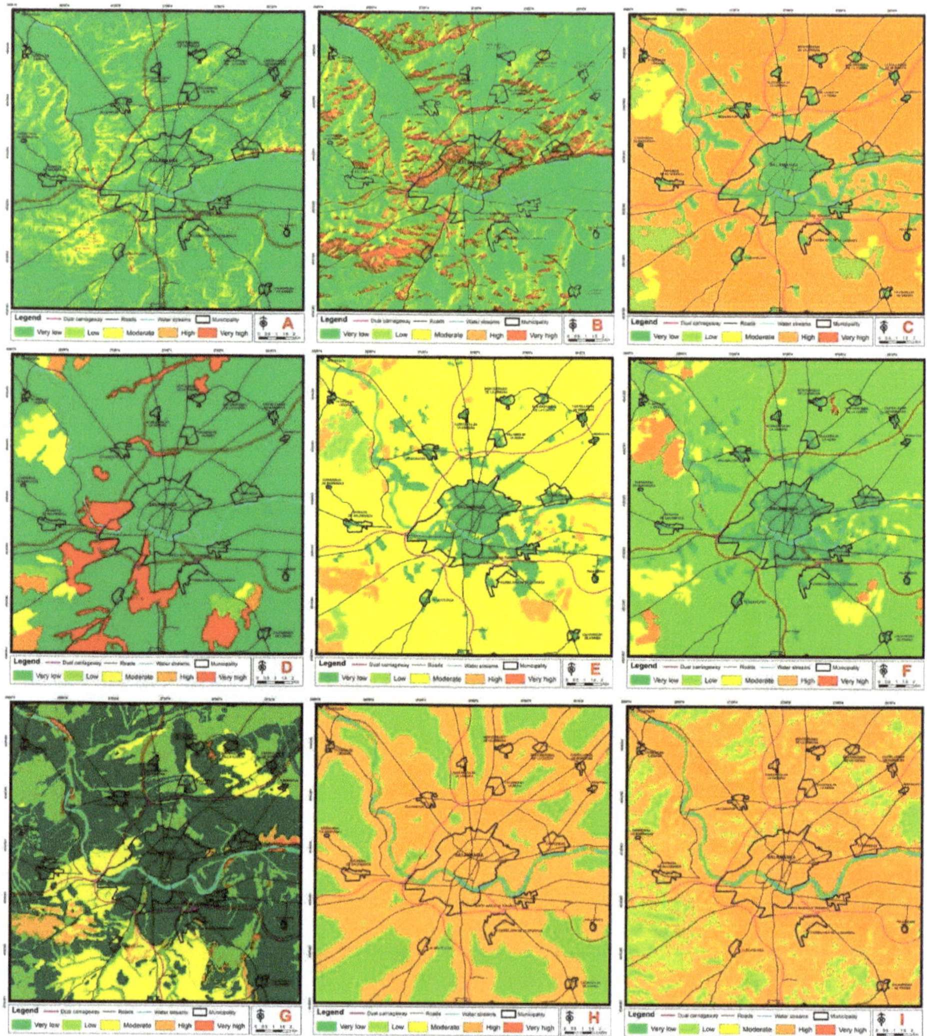

Figure 7. Studied factors for the analysis of fragility Intrinsic: (**A**) slope, (**B**) orientation, (**C**) fractional vegetation cover (FVC), (**D**) stratification, (**E**) vegetation units surface (VS), (**F**) vegetal sinuosity and (**G**) area of landscape units; and Extrinsic: (**H**) accessibility and (**I**) visibility.

3.4. Landscape Need of Protection

From the fragility and quality of the landscape, the need for landscape protection in each sector is obtained (Figure 9). Classes I and II areas are representative of the least valued landscapes, in which the implementation of different activities would not produce a significant impact. They are identified as being linked to the dynamics of the Tormes, which is very frequent throughout the floodplain, as well as in the terrace systems, mainly in the southeastern sector of the area, with its appearance being more dispersed throughout other parts of the sector. On the other hand, the areas most vulnerable to impacts related to the alteration of the landscape are those belonging to Classes IV and V which need protection measures to conserve the most valuable landscapes of the environment. This includes landscapes linked to river escarpments, to the reliefs of Los Montalvos (southwestern sector) or to areas with well-conserved tree vegetation, preserved in areas far from the city of Salamanca. Finally, a large part of the sector is classified as Class III, indicating that no urgent or priority conservation measures are required. It is mainly distributed in the northern half and around the capital, although there are also areas irregularly distributed throughout the rest of the study area.

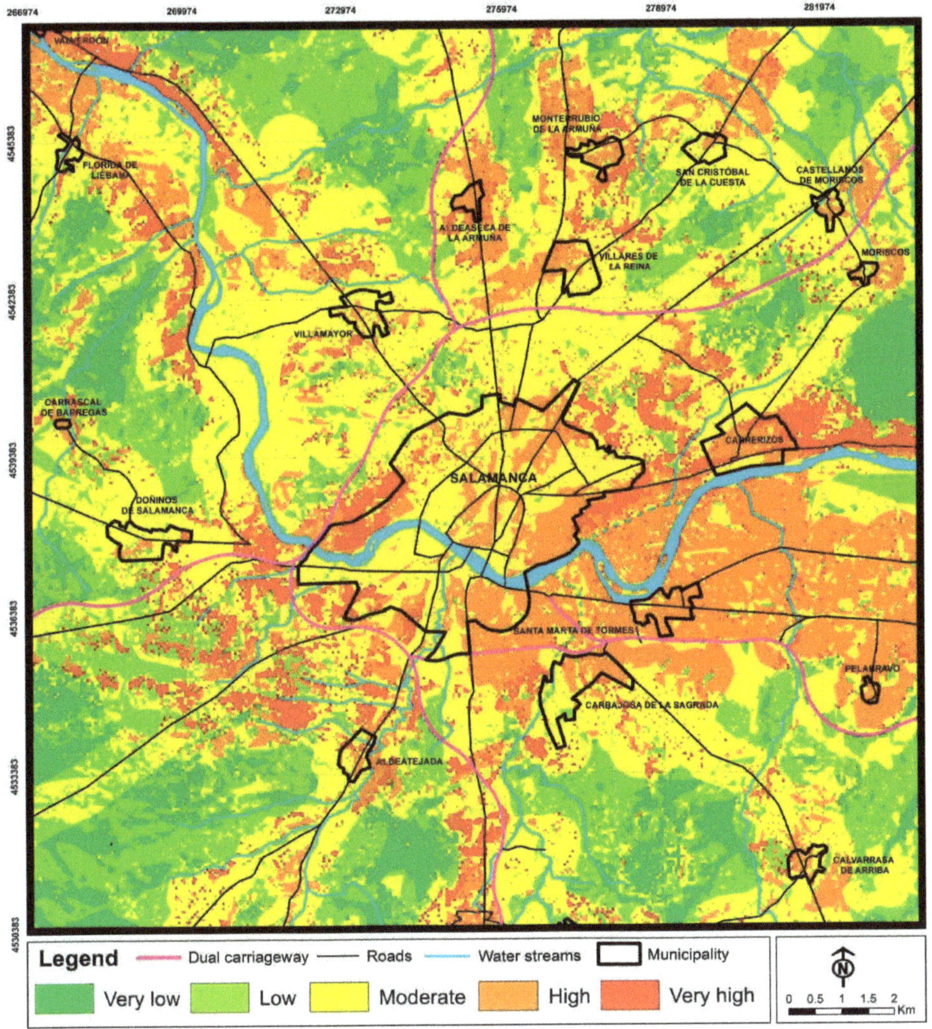

Figure 8. Landscape fragility map.

3.5. Landscape Diagnosis

Table 6 summarizes the percentage (%) for each landscape unit that is related to each of the degrees of quality, fragility and need for protection. In relation to landscape quality, the units with the highest quality values are 1 and 14, which correspond to the greatest slopes and tree-lined plant formations. Units with high quality are those with diverse orography and generally tree-lined vegetation, as is the case with units 3, 9, 10, 11, 12 and 13. The moderate quality level is characteristic of most landscape units. With regard to fragility, units 13, 14 and 15 show the highest values due to the fact that they group together the most abrupt spaces in the sector, although high and very high values are not common in general. Moderate fragility is predominant in the majority of landscape units, with lower values appearing in units with a greater diversity of forms and especially important tree formations (units 1, 2, 3, 4, 10, 11 and 12). As regards protection needs, they are very high mainly in units 1, 11 and 12, which group together the main forests in the area. With high needs are other units of scarce entity but of varied forms and rich in vegetation (units 3 and 10). Units 2, 16, 17 and 20 have low and very low protection needs and correspond to monotonous landscapes characterized by soft or flat carved forms. The remaining units clearly present intermediate protection requirements, in line with the results obtained from the analysis of quality and fragility.

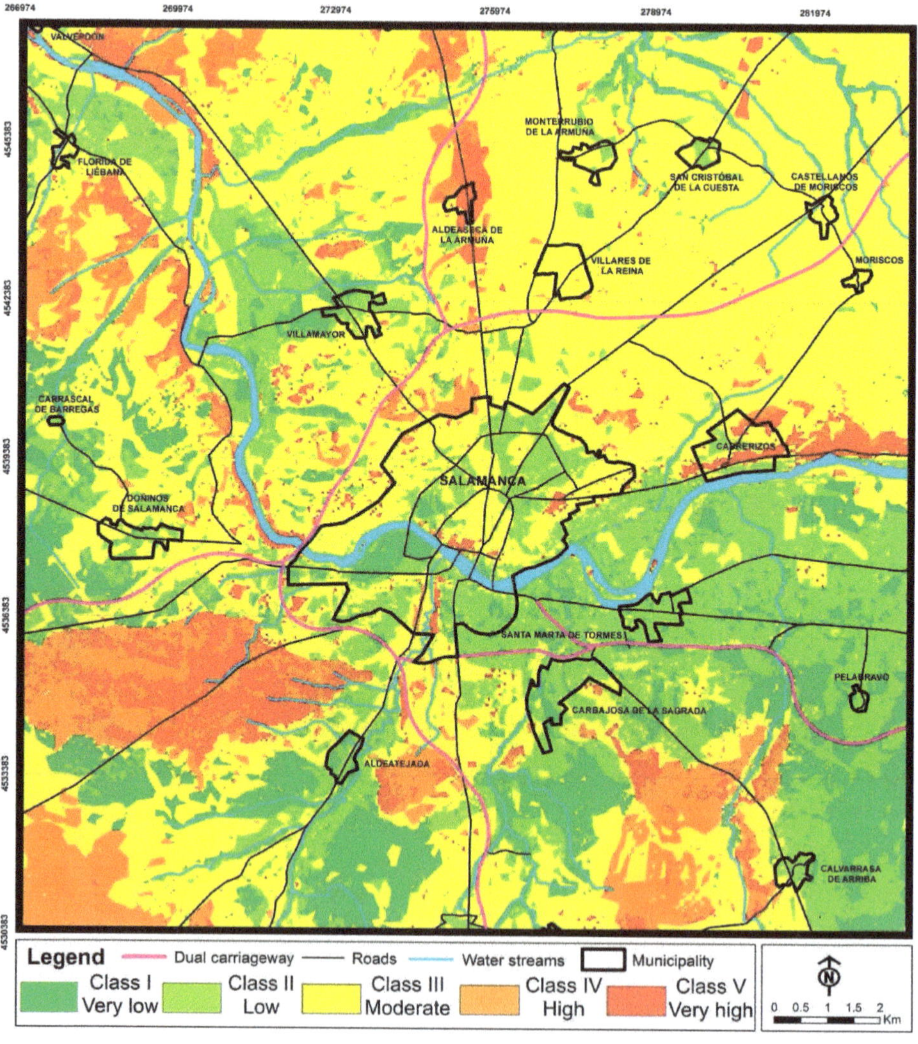

Figure 9. Landscape need-of-protection map.

Table 6. Area (%) of landscape units (according to their total surface area) included in each of the ranges of quality, fragility and need for landscape protection.

LU [1] (n°)	A [2] (ha)	Landscape Quality					Landscape Fragility					Need of Protection				
		VL [3]	L	M	H	VH	VL	L	M	H	VH	VL	L	M	H	VH
1	550	0	0	0	11	89	20	35	22	8	15	0	0	1	57	42
2	759	5	63	30	2	0	23	48	23	2	4	53	15	30	2	0
3	373	0	0	38	57	5	22	42	24	7	5	0	0	38	53	9
4	500	0	1	96	3	0	21	41	31	4	3	1	0	96	2	1
5	1007	4	40	16	24	16	11	43	32	9	5	31	14	16	24	15
6	2320	1	30	62	6	1	6	23	48	11	12	13	18	62	4	3
7	4606	1	13	76	10	0	10	24	51	8	7	5	9	76	7	3
8	6060	1	8	79	12	0	10	28	45	14	3	5	4	79	9	3
9	101	0	0	48	51	1	21	27	23	12	17	0	0	47	33	20
10	632	0	2	25	57	16	31	41	20	5	3	2	0	25	67	6
11	455	0	1	7	62	30	20	33	34	7	6	1	0	8	72	19
12	129	0	0	2	35	63	46	26	16	9	3	0	0	2	78	20
13	244	0	1	31	53	15	0	11	32	23	34	1	1	32	23	43
14	87	0	0	5	21	74	2	9	20	12	57	0	0	5	16	79
15	50	0	2	34	50	14	0	8	32	48	12	0	2	35	20	43
16	1967	19	49	30	2	0	0	8	43	44	5	9	59	30	1	1
17	2490	29	55	14	2	0	5	26	41	27	1	34	50	14	2	0
18	254	0	4	66	25	5	17	37	35	7	4	2	2	66	29	1
19	189	6	12	52	26	4	0	10	40	45	5	3	15	51	21	10
20	894	16	42	38	4	0	16	27	47	9	1	32	25	39	3	1
21	214	9	24	29	33	5	0	11	56	26	7	8	26	29	24	13
22	188	5	33	59	3	0	7	27	55	10	1	16	22	59	2	1
Tot	24069	5	24	54	12	5	10	27	42	15	6	12	17	54	12	5

[1] LU: landscape unit (see Table 5); [2] A: area; [3] VL: very low, L: low; M: moderate; H: high; VH: very high.

4. Discussion

The high diversity of environmental elements, and their synergies, make it very difficult to analyze the landscape, especially in such dynamic environments as periurban ones. Therefore, it is necessary to develop methods to manage this high amount of information. On the other hand, the numerous approaches that have been given to the landscape resource caused the appearance of multiple interpretations of it. It is therefore necessary that current methodologies address common criteria in the approach to analysis, with the European Landscape Convention serving as the basic starting framework. The search for quantitative interpretations of the landscape must also be a requirement when developing new methodologies, so that these results can be incorporated into territorial planning.

The proposed model allows for an objective assessment of the landscape through the implementation of rapid, low-cost parametric mapping using GIS, a widely used tool for landscape analysis [34,35,38,44]. The development of a digital geodatabase with the analyzed data allows their weighting and processing with GIS techniques, allowing the automation of the process and its implementation in other places. In addition, the objectives of the European Landscape Convention are achieved as it allows the identification of landscape units and the analysis of landscape quality, as well as identifying pressures for change. Also, complementary analyses of vulnerability and landscape protection needs are incorporated, which are very valuable in the planning and decision-making processes, and which finally allow a diagnosis of the situation to be made. Using direct methods, the divisions made for each landscape unit can be corroborated and validated in the field. This can also serve to validate the analysis of landscape quality, although the diversity and conditions of observers may appear as limitations. On the other hand, the limitations of the methodology derive from its lesser capacity to value urban landscapes to the detriment of taking into consideration the natural and seminatural landscapes of periurban areas. Likewise, the methodology aims to obtain information of a markedly environmental nature, which allows it to be used in the decision-making stages, which perhaps partly underestimates the perceptive component of the landscape. Furthermore, for the methodological development, part of the subjective information derived from the authors' assessments and scores is also assumed. In relation to this, the preferences of the population established from landscape studies were also incorporated, although this information is considered intrinsically objective as it is representative of large groups surveyed. In addition, the methodology does not promote direct citizen participation, an aspect included in other papers [60].The determination and evaluation of landscape units allowed the identification of areas of landscape interest. Through the quality evaluation, the areas of greatest landscape value were identified, which coincided with the spaces of greatest natural interest in the area, in line with what has been observed by other authors in Spain [61], which may correspond to the general high degree of anthropization of the sector, dominated by agriculture. In addition, these areas of higher quality were identified mainly on the edges of the area studied, far from the city of Salamanca and areas of greater influence, which can be linked to rural spaces where urban pressures have been lower and allowed for landscape conservation.

On the other hand, the vulnerability of landscapes must be taken into account when deciding whether or not to install a specific activity in a given landscape. This aspect is particularly relevant in urban and periurban areas due to the greater speed and diversity of pressures in these areas, with urban expansion being the main pressure on the landscape in and around Salamanca, as shown by the increase of almost 600% in the built area in the period 1956–2018 [14]. In addition, the installation of new industries and the construction of new infrastructures were also identified as important alterations to the landscape in Salamanca and its surroundings. Agricultural practices (crop and fallow land rotation, transformation of agricultural patches, etc.) and the management of natural ecosystems in the surroundings (forest treatments and agro-livestock practices) also have importance in the dynamics of the landscape. Due to the magnitude and speed of transformations in the urban environment [62], it is necessary to establish the protection needs of each type of landscape, which varies according to the capacity of the landscape to cushion the impact that an activity in question would produce. These conservation recommendations are necessary tools that can be incorporated into planning processes [35]. Furthermore, the cross-checking of this mapping of recommendations with that of landscape units allows a diagnosis of the landscape situation to be made. The main contribution of this diagnosis is that it makes it possible to quantify the landscape characteristics of each unit, so that comparisons and discriminations can be made between units that could be qualitatively identical. Therefore, it provides more precise and concrete results that would contribute a greater amount of information to the planning process, facilitating decision making, in line with other multicriteria landscape research [34,35,38]. Furthermore, the use of these homogeneous landscape units as territorial planning units is proposed since, due to their unique and singular characteristics they present clearly identifiable vocational uses of the territory [39].

5. Conclusions

The various landscape units determined by this method can be interpreted as territorial sectors with similar characteristics due to the many constraints they share. At present, environmental assessment procedures require analyses of the quality and fragility of the different entities or sectors of the territory. Therefore, the analysis of landscape units can be incorporated into these procedures not only to strictly determine the attributes of the landscape, but also to characterize the broad outlines of the territorial entities.

The proposed methodology objectively assesses the landscape components in accordance with the provisions of the European Landscape Convention, so its results are compatible with their incorporation into the decision-making processes.

The superimposition of landscape units on landscape assessment cartographies makes it possible to carry out a landscape diagnosis and to better specify the characterization of each landscape unit, facilitating decision-making and landscape management measures in future planning actions. It is therefore an instrument for sustainable management of the space capable of limiting the implementation of development initiatives according to the carrying capacity of the physical environment.

By means of the proposed landscape diagnosis it is possible to quantify the general state of the landscape of a sector and of the corresponding units that make it up, enabling situation, monitoring and comparative reports to be made over time. In Salamanca and the surrounding area, intermediate assessments dominate. The quality of the landscape is greater in the areas furthest from the city, coinciding with the areas of diverse orography that best conserve the natural values of the sector. Almost 20% of the study sector presented high and very high landscape quality, with a majority, however, of low (29%) and, especially, moderate quality areas (54%). In turn, the fragility shows a more irregular distribution, tending to show higher values in areas of unevenness and those close to urban areas and roads. The spaces of low and moderate fragility are predominant (37% and 42% respectively) over the spaces of high landscape fragility (21%). Finally, high protection needs are restricted to the most environmentally and scenically valuable units, although they occupy scarce extensions (17% of the territory) in comparison to those that do not require conservation measures (12% and 17% for very low and low requirements, respectively), with those in need of intermediate protection being the majority (54% of the territory).

Author Contributions: Conceptualization, M.C. and A.M.-G.; methodology, M.C. and A.M.-G.; software, M.C.; validation, M.C. and F.S.-F.; formal analysis, M.C.; investigation, M.C. and A.M.-G.; resources, M.C. and L.M.; data curation, M.C.; writing—original draft preparation, M.C. and A.M.-G.; writing—review and editing, M.C. and F.S.-F.; visualization, M.C.; supervision, A.M.-G and F.S.-F.; project administration, A.M.-G. All authors have read and agreed to the published version of the manuscript.

Funding: This research was funded by projects Junta Castilla y León SA044G18 and the GEAPAGE research group has participated.

Conflicts of Interest: The authors declare no conflict of interest.

References

1. European Landscape Convention. "Council of Europe, Florence". Available online: https://rm.coe.int/1680080621 (accessed on 10 April 2020).

2. Bolós, M. *Manual de Ciencia del Paisaje. Teoría, Métodos y Aplicaciones*; Masson: Barcelona, Spain, 1992; p. 273.

3. Zhang, F.; Ayinuer, Y.; Dongfang, W. Ecological risk assessment due to land use/cover changes (LUCC) in Jinghe County, Xinjiang, China from 1990 to 2014 based on landscape patterns and spatial statistics. *Environ. Earth Sci.* **2018**, *77*, 491–507. [CrossRef]

4. Fairclough, G.; Herlin, I.S.; Swanwick, C. (Eds.) *Routledge Handbook of Landscape Character Assessment: Current Approaches to Characterisation and Assessment*; Routledge: Abingdon, UK, 2018; p. 294.

5. Thompson, C.W. Linking landscape and health: The recurring theme. *Landsc. Urban Plan.* **2011**, *99*, 187–195. [CrossRef]

6. Liu, B.X. Study on the effects of different landscapes on elderly people's body-mind health. *Landsc. Archit.* **2016**, *7*, 113–120.

7. Domon, G. Landscape as resource: Consequences, challenges and opportunities for rural development. *Landsc. Urban Plan.* **2011**, *100*, 338–340. [CrossRef]

8. Smith, M.; Ram, Y. Tourism, landscapes and cultural ecosystem services: A new research tool. *Tour. Recreat. Res.* **2017**, *42*, 113–119. [CrossRef]

9. Willemen, L.; Hein, L.; Verburg, P.H. Evaluating the impact of regional development policies on future landscape services. *Ecol. Econ.* **2010**, *69*, 2244–2254. [CrossRef]

10. Nassauer, J.I. Landscape as medium and method for synthesis in urban ecological design. *Landsc. Urban Plan.* **2012**, *106*, 221–229. [CrossRef]

11. Antrop, M. Landscape change and the urbanization process in Europe. *Landsc. Urban Plan.* **2004**, *67*, 9–26. [CrossRef]

12. Angel, S.; Parent, J.; Civco, D.L.; Blei, A.; Potere, D. The dimensions of global urban expansion: Estimates and projections for all countries, 2000–2050. *Prog. Plan.* **2011**, *75*, 53–107. [CrossRef]

13. World Urbanization Prospects: The 2018 Revision. Available online: https://population.un.org/wup/Publications/Files/WUP2018S-$KeyFacts.pdf (accessed on 25 May 2020).

14. Criado, M.; Santos-Francés, F.; Martínez-Graña, A.; Sánchez, Y.; Merchán, L. Multitemporal Analysis of Soil Sealing and Land Use Changes Linked to Urban Expansion of Salamanca (Spain) Using Landsat Images and Soil Carbon Management as a Mitigating Tool for Climate Change. *Remote Sens.* **2020**, *12*, 1131. [CrossRef]

15. Antrop, M. Background concepts for integrated landscape analysis. *Agric. Ecosyst. Environ.* **2000**, *77*, 17–28. [CrossRef]

16. Termorshuizen, J.W.; Opdam, P. Landscape services as a bridge between landscape ecology and sustainable development. *Landsc. Ecol.* **2009**, *24*, 1037–1052. [CrossRef]

17. Bateman, I.J.; Harwood, A.R.; Mace, G.M.; Watson, R.T.; Abson, D.J.; Andrews, B.; Binner, A. Bringing Ecosystem Services into Economic Decision-Making: Land Use in the United Kingdom. *Science* **2013**, *341*, 45–50. [CrossRef] [PubMed]

18. Panagopoulos, T. Linking forestry, sustainability and aesthetics. *Ecol. Econ.* **2009**, *68*, 2485–2489. [CrossRef]

19. Mayoh, J.; Onwuegbuzie, A.J. Toward a conceptualization of mixed methods phenomenological research. *J. Mixed Methods Res.* **2015**, *9*, 91–107. [CrossRef]

20. Loures, L.; Loures, A.; Nunes, J.; Panagopoulos, T. Landscape Valuation of Environmental Amenities throughout the Application of Direct and Indirect Methods. *Sustainability* **2015**, *7*, 794–810. [CrossRef]

21. Arthur, L.; Daniel, T.; Boster, R. Scenic assessment: An overview. *Landsc. Plan.* **1977**, *4*, 109–129. [CrossRef]

22. Stamps, A.E. Demographic effects in environmental aesthetics: A meta-analysis. *J. Plan. Literature* **1999**, *14*, 155–175. [CrossRef]

23. Ulrich, R.S. Human responses to vegetation and landscapes. *Landsc Urban Plan* **1986**, *13*, 29–44. [CrossRef]

24. Lothian, A. Landscape and the philosophy of aesthetics: Is landscape quality inherent in the landscape or in the eye of the beholder? *Landsc. Urban Plan.* **1999**, *44*, 177–198. [CrossRef]

25. Daniel, T.C. Whither scenic beauty? Visual landscape quality assessment in the 21st century. *Landsc. Urban Plan.* **2001**, *54*, 267–281. [CrossRef]

26. Cañas, I.; Ayuga, E.; Ayuga, F. A contribution to the assessment of scenic quality of landscapes based on preferences expressed by the public. *Land Use Policy* **2009**, *26*, 1173–1181. [CrossRef]

27. Martínez-Graña, A.M.; Goy, J.L.; Cimarra, C. 2D to 3D geologic map transformation using virtual globes and flight simulators and their applications in the analysis of geodiversity in natural areas. *Environ. Earth Sci.* **2015**, *73*, 8023–8034. [CrossRef]

28. Shi, J.; Honjo, T.; Zhang, K.; Furuya, K. Using virtual reality to assess landscape: A comparative study between on-site survey and virtual reality of aesthetic preference and landscape cognition. *Sustainability* **2020**, *12*, 2875. [CrossRef]

29. Tveit, M.; Ode, A.; Fry, G. Key concepts in a framework for analysing visual landscape character. *J. Landsc. Res.* **2006**, *31*, 229–255. [CrossRef]

30. Ode, A.; Tveit, M.; Fry, G. Capturing landscape visual character using indicators: Touching base with landscape aesthetic theory. *Landsc. Res.* **2008**, *33*, 89–117. [CrossRef]

31. Sevenant, M.; Antrop, M. Cognitive attributes and aesthetic preferences in assessment and differentiation of landscapes. *J. Environ. Manage.* **2009**, *8*, 2889–2899. [CrossRef]

32. Martín, B.; Otero, I. Mapping the visual landscape quality in Europe using physical attributes. *J. Maps* **2012**, *8*, 56–61.

33. Van Der Perk, M.; De Jong, S.M.; McDonnell, R.A. Advances in the spatiotemporal modeling of environment and landscape. *Int. J. Geogr. Inf. Sci.* **2007**, *21*, 477–481. [CrossRef]

34. Jeong, J.S.; García-Moruno, L.; Hernández-Blanco, J. A site planning approach for rural buildings into a landscape using a spatial multi-criteria decision analysis methodology. *Land Use Policy* **2013**, *32*, 108–118. [CrossRef]

35. Jeong, J.S.; Montero-Parejo, M.J.; García-Moruno, L.; Hernández-Blanco, J. The visual evaluation of rural areas: A methodological approach for the spatial planning and color design of scattered second homes with an example in Hervás, Western Spain. *Land Use Policy* **2015**, *46*, 330–340. [CrossRef]

36. Veronesi, F.; Hurni, L. A GIS tool to increase the visual quality of relief shading by automatically changing the light direction. *Comput. Geosci.* **2015**, *74*, 121–127. [CrossRef]

37. Criado, M.; Martínez-Graña, A.; Santos-Francés, F.; Veleda, S.; Zazo, C. Multi-Criteria Analyses of Urban Planning for City Expansion: A Case Study of Zamora, Spain. *Sustainability* **2017**, *9*, 1850. [CrossRef]

38. Martínez-Graña, A.M.; Silva, P.G.; Goy, J.L.; Elez, J.; Valdés, V.; Zazo, C. Geomorphology applied to landscape analysis for planning and management of natural spaces. Case study: Las Batuecas-S. de Francia and Quilamas natural parks,(Salamanca, Spain). *Sci. Total Environ.* **2017**, *584*, 175–188. [CrossRef]

39. Soto, S.; Pintó, J. Delineation of natural landscape units for Puerto Rico. *Appl. Geogr.* **2010**, *30*, 720–730. [CrossRef]

40. Bulut, Z.; Yilmaz, H. Determination of landscape beauties through visual quality assessment method: A case study for Kemaliye (Erzincan/Turkey). *Environ. Monit. Assess.* **2008**, *141*, 121–129. [CrossRef]

41. Vizzari, M. Spatial modelling of potential landscape quality. *Appl. Geogr.* **2011**, *31*, 108–118. [CrossRef]

42. Burley, J. *Environmental Design for Reclaiming Surface Mines*; The Edwin Mellen Press: New York, NY, USA, 2001; p. 504.

43. Ayala, R.; Ramirez, J.; Camargo, S. *Valoración de La Calidad y Fragilidad Visual del Paisage en el Valle de Zapotitlán de Las Salinas, Puebla (México)*; Faculdad de Geografia e Historia da Universidad de Madrid: Madrid, Spain, 2003.

44. Vargues, P.; Loures, L. Using Geographic Information Systems in visual and aesthetic analysis: The case study of a golf course in Algarve. *WSEAS Trans. Environ. Dev.* **2008**, *4*, 774–783.

45. von Haaren, C.; Lovett, A.A.; Albert, C. *Landscape Planning with Ecosystem Services*, 1st ed.; Springer: Dordrecht, The Netherlands, 2019; p. 540.

46. With, K.A. The application of neutral landscape models in conservation biology. *Conserv. Biol.* **1997**, *11*, 1069–1080. [CrossRef]

47. Galpern, P.; Manseau, M.; Fall, A. Patch-based graphs of landscape connectivity: A guide to construction, analysis and application for conservation. *Biol. Conserv.* **2011**, *144*, 44–55. [CrossRef]

48. Qi, K.; Fan, Z.; Ng, C.N.; Wang, X.; Xie, Y. Functional analysis of landscape connectivity at the landscape, component, and patch levels: A case study of Minqing County, Fuzhou City, China. *Appl. Geogr.* **2017**, *80*, 64–77. [CrossRef]

49. Instituto Nacional de Estadística (National Institute of Statistics of Spain). Available online: http://www.ine.es/dyngs/INEbase/es/operacion.htm?c=Estadistica_C&cid=1254736176951&menu=ultiDatos&idp=1254735572981 (accessed on 28 June 2020).

50. UNESCO. Available online: https://whc.unesco.org/en/list/381/ (accessed on 28 June 2020).

51. CORINE-Land Cover. Available online: https://land.copernicus.eu/pan-european/corine-land-cover (accessed on 7 April 2020).

52. DiPietro, J.A. *Geology and Landscape Evolution: General Principles Applied to the United States*, 2nd ed.; Elsevier: Amsterdam, The Netherlands, 2018; p. 638.

53. Meinig, D.W. The Beholding Eye. Ten versions of the same scene. *Landsc. Archit.* **1976**, *66*, 47–54.

54. Dearden, P. *Societal Landscape Preferences: A Pyramid of Influences. Landscape Evaluation: Approaches and Applications*; University of Victoria: Victoria, BC, Canada, 1989; pp. 41–64.

55. Kaltenborn, B.P.; Bjerke, T. Associations between environmental value orientations and landscape preferences. *Landsc. Urban Plan.* **2002**, *59*, 1–11. [CrossRef]

56. Dramstad, W.E.; Tveit, M.S.; Fjellstad, W.J.; Fry, G.L. Relationships between visual landscape preferences and map-based indicators of landscape structure. *Landsc. Urban Plan.* **2006**, *78*, 465–474. [CrossRef]

57. Istanbulluoglu, E.; Bras, R.L. Vegetation-modulated landscape evolution: Effects of vegetation on landscape processes, drainage density, and topography. *J. Geophys. Res. Earth Surf.* **2005**, *110*. [CrossRef]

58. Roy, N.G.; Sinha, R. Linking hydrology and sediment dynamics of large alluvial rivers to landscape diversity in the Ganga dispersal system, India. *Earth Surf. Proc. Land.* **2017**, *42*, 1078–1091. [CrossRef]

59. Tengberg, A.; Fredholm, S.; Eliasson, I.; Knez, I.; Saltzman, K.; Wetterberg, O. Cultural ecosystem services provided by landscapes: Assessment of heritage values and identity. *Ecosyst. Serv.* **2012**, *2*, 14–26. [CrossRef]

60. Santé, I.; Fernández-Ríos, A.; Tubío, J.M.; García-Fernández, F.; Farkova, E.; Miranda, D. The Landscape Inventory of Galicia (NW Spain): GIS-web and public participation for landscape planning. *Landsc. Res.* **2019**, *44*, 212–240. [CrossRef]

61. Holgado, P.M.; Rieth, L.J.; Menárguez, A.B.B.; Álvarez, F.A. The Analysis of Urban Fluvial Landscapes in the Centre of Spain, Their Characterization, Values and Interventions. *Sustainability* **2020**, *12*, 4661. [CrossRef]
62. González-Ávila, S.; López-Leiva, C.; Bunce, R.G.; Elena-Rosselló, R. Changes and drivers in Spanish landscapes at the Rural-Urban Interface between 1956 and 2018. *Sci. Total Environ.* **2020**, *714*, 136858. [CrossRef]

MDPI

St. Alban-Anlage 66

4052 Basel

Switzerland

Tel. +41 61 683 77 34

Fax +41 61 302 89 18

www.mdpi.com

Sustainability Editorial Office

E-mail: sustainability@mdpi.com

www.mdpi.com/journal/sustainability